Introduction to Biotech Entrepreneurship: From Idea to Business

Florentina Matei • Daniela Zirra
Editors

Introduction to Biotech Entrepreneurship: From Idea to Business

A European Perspective

Springer

Editors
Florentina Matei
Faculty of Biotechnology
University of Agronomic Sciences
and Veterinary Medicine
Bucharest, Romania

Daniela Zirra
Department of Commerce, Economic
Integration and Business Administration
Romanian-American University
Bucharest, Romania

ISBN 978-3-030-22140-9 ISBN 978-3-030-22141-6 (eBook)
https://doi.org/10.1007/978-3-030-22141-6

This Springer imprint is published by the registered company Springer Nature Switzerland AG.
The registered company address is: Gewerbestrasse 11, 6330 Cham, Switzerland

Preface

We often ask why in some areas the entrepreneurial process is ampler and faster than in others, why business success is more famous or why entrepreneurial initiatives are more common. Please note that there is no simple answer for any of the matters outlined above. However, there are some factors that help us better understand why these differences exist, for example, the size of the profit that can be achieved and the time required to obtain a consistent profit, the frequency of emergence of opportunities and the pace of change in consumer needs and wishes, the degree of novelty of the products and services that are to be created, the degree of risk specific to the field, diversity and accessibility of funding sources, the legislative and institutional framework that exists in the chosen field and the level of availability of necessary resources.

Considering the importance of biotechnologies in the current context of global economic development, it is more than obvious that an effective entrepreneurial process in this area is vital. We must not forget that no matter how innovative, numerous or important are the discoveries and results of biotechnology research, if they are not capitalized on the market and do not reach the end beneficiaries, in this case consumers (regardless of whether we are talking about individuals or firms), all the effort will be in vain and all the resources used to achieve those performances can be considered as wasted.

In response to the biotech business environment, the biotech entrepreneurship study programs are booming in Europe, completed with the highly claimed need for industrial PhD in the domain of life sciences. These are viewed as very important tools in filling the gaps between Europe and Northern-American biotech business. This book addresses mainly the students and teachers involved in such educational programs, and actually, the idea for such a handbook arose during the setup of a new master's program in "Biotechnology and Entrepreneurship". Teachers and students will find in this book valuable teaching and/or learning material, adapted to the biotech business reality in Europe. Meanwhile, the provided information targets academic and industrial researchers in the field, as well as business professionals and potential investors.

Some other books have been written on the same topic, but the most cited are describing biotech business in the USA and Canada or biotech venture in Europe, with a focus on the suboptimal state of the European biotechnology industry especially in the UK, just before the global financial crunch of 2008.

Our handbook is supposed to provide an overview on the opportunities and drawbacks of biotech entrepreneurship in the European context, including a comparative analysis among different countries' perspectives and European regions (Eastern and Western Europe), as well as to identify possible gaps or even advances, comparing with other world economic regions. The book provides information on technical and economical solutions useful for the development of a biotech start-up, capable of generating value and employment, supported with clear examples from the European environment. They approach different fields of biotechnology, from plant and food biotechnology, to food industry and environmental areas, passing through human and animal health biotech opportunities, as well as bioinformatics as source of income. As a must, the book includes key elements of the biotech markets and financial/investment sources, taking also into account the huge importance of the innovation and intellectual property issues.

The readers will find valuable information on how to deal with a start-up in the biotech field, will be able to choose from a long list of potential profitable biotech start-ups and will be in a position to access a collection of successful stories of biotech start-ups developed in the European context. The potential investors and business consultants will be punctually informed on benefits and potentials risks in supporting biotech business.

Bucharest, Romania
8 April 2019

Florentina Matei
Daniela Zirra

Acknowledgements

We would like to take the opportunity to thank all the people who have been involved in the development of this handbook. Special thanks go to all our academic and non-academic partners from Italy, Spain and Belgium, as well as to our Romanian colleagues who answered fast and enthusiastically to our proposal.

We would like to thank all the contributors to the biotech case studies presented in this work and to people accepting to provide written and video testimonials about their experience in biotech start-up.

Part of this work has been funded with support from the European Commission, project no. 2017-1-RO01-KA203-037304, coordinated by UASMV Bucharest, and this work reflects the views only of the authors, and the Commission cannot be held responsible for any use which may be made of the information contained therein.

Last but not the least, we send our deep and sincere gratitude to our families for their continuous and unparalleled love, help and support.

Contents

1 The Potential of Biotechnology to Generate Prosperity

Florentina Israel-Roming and Mihaela Ghiduruş

Abstract

Biotechnology is considered now as a key knowledge-based business that generates products and processes for the growing global needs. Biotechnology offers modern solutions for almost every aspect of human life: economic, social, health and environment. It promotes sustainable economic growth, increasing productivity and diversity and lowering by-products and wastes generation. Modern diagnostic approaches, therapeutic solutions, vaccines and other pharmaceutical products are generated by biotechnology. These achievements are intended to increase the survival rate and to lower the resources and pain associated with a non-suitable treatment. Biotechnology also offers solutions for producing food enriched with specific nutrients, with significant contribution to a proper human health condition and even to malnutrition. Microbial processes are successfully used for improving the environmental quality by biodegradation and bioremediation. Economic prosperity is expected in rural areas or in developing countries based on agriculture, as well as in developed economies where biotechnology engenders "high-tech" solutions.

Keywords

Prosperity · Agricultural biotechnology · Biopesticides · Biofertilisers · Herbicide resistance · Industrial biotechnology · Biofuels · Environmental biotechnology · Health biotechnology

The economy of the twenty-first century has to face important challenges regarding the growth of the global population, the climate changes and the threats related to environmental protection. According to United Nations Report (2017), the current world population of 7.6 billion is estimated to increase every year with about 83 million people. The needs of this growing population can no longer be supported only by the traditional primary and secondary business sectors based on finite or temporally limited resources of food and fuel (Timmis et al. 2017). On the other hand, a greater number of people are associated with pronounced negative environmental effects. Population growth and reduction of available resources, problems arising from the use of traditional physical and chemical technologies, more complicated health problems and accelerated environmental degradation have caused the economic media and the policymakers to focus their attention on biotechnology as a hopeful and promising solution for sustainable development (Caggiano et al. 2005).

F. Israel-Roming (✉) · M. Ghiduruş
University of Agronomic Sciences and Veterinary Medicine of Bucharest, Bucureşti, Romania
e-mail: florentina.israel@biotehnologii.usamv.ro

F. Matei, D. Zirra (eds.), *Introduction to Biotech Entrepreneurship: From Idea to Business*,
https://doi.org/10.1007/978-3-030-22141-6_1

Agricultural biotechnology aims at finding sustainable solutions for better use of limited resources, in terms of quantity, quality and diversity, with less environmental and social impact.

Genetic engineering is based on transferring a specific gene into an organism, from another organism, in order to improve the host performances and resistance to pests, diseases and poor environmental conditions. A highly successful genetic achievement is insect resistance to lepidopteran pests obtained with genes from soil bacterium Bt (*Bacillus thuringiensis*) inserted in corn and cotton genome. Another important step for crop protection consisted in the obtaining of genetically modified herbicide-resistant corn, cotton and soybean. Glyphosate-resistant crops became the most rapidly adopted technology in the history of agriculture due to the positive impact on farmers (Green 2012). According to the United States Department of Agriculture, in 2018 genetically modified varieties with insect resistance cultivated in the USA represented 85% of the total planted acres with corn and 95% of the total planted acres with cotton, while for herbicide-tolerant varieties the percentage was 90 for corn, 91 for cotton and 94 for soybean. In the European Union, there are stronger regulations regarding the use of genetically modified organisms and the environmental risk assessment.

Another biotechnological perspective is the identification of molecular markers associated with a particular gene in a performant organism and their usage for selecting plants and animals that possess the desirable gene, making breeding process more efficient. Comparing with other traditional markers (morphological, cytological and biochemical), they have more information and higher polymorphism. Molecular markers are a useful tool for plant cultivar identification, seed purity test, preservation and evaluation of germplasm resources and molecular-assisted breeding (Fang et al. 2016).

Tissue culture technology permits obtaining of healthy plantlets from high yielding varieties, with positive effects on farmers' budget. Besides the advantage of producing common disease-free plants, this technique is very important especially when speaking about rare plant genotypes or high producers of important value-added compounds (Espinosa-Leal et al. 2018). Comparing with conventional horticulture, micropropagation has significant advantages: production of healthy plants all year round, independent of geographical position, constant level of secondary metabolite conferring uniform quality and extraction yield and not using plant protection treatments. There are a lot of small and medium companies focused on processing plants, especially medicinal and aromatic, for extracting different phytochemicals with potential applications in food or pharmaceutical industry. Because of the low level of plant bioactive molecules, biochemical and biotechnological approaches are currently used for enhancing the yield of the compound of interest or for producing novel secondary metabolites (Gandhi et al. 2015).

Biopesticides represent the biological or biochemical solution to crop pests with high efficiency and less effect on humans and environment. Initially used only in organic farming, biopesticides are more and more present in the conventional one too, mainly because they are efficient and not so vulnerable to developing pest resistance due to the combined action on digestive system, reproduction and repelling. Toxic effects of synthetic pesticides have been claimed by handlers and even by consumers, and the persistence and accumulation in the environment was more than 99.7% (Pino-Otín et al. 2019). Growing demands on these bioproducts triggered diversification and higher production, thus becoming more and more attractive for small companies wanting to start a business in biotechnology as well as for the well-known big companies with activity in plant protection field. Analysing the total costs for developing a plant protection product and the necessary period till approval, Olson (2015) showed that a synthetic one requires USD250 million and 9 years, while for a biopesticide USD10 million and 4 years are needed. Today, the total global market of biopesticides is USD3 billion in value worldwide, representing approximately 5% of the total crop protection market, with an upward trend expected to reach 7% in 2023 (Olson 2015).

Biofertilisers are an alternative to chemical fertilisers and act on the biological characteristics of soil, promoting plant growth. They are living or latent microorganisms that can enhance the availability of inorganic compounds (nitrogen, phosphorus, potassium, sulphur) to the plant by interactions with biotic and abiotic components of the rhizosphere (Igiehon and Babalola 2017). At the same time, these environmental friendly products have a positive effect on the soil organic composition. The market of biofertilisers is increasing significantly due to the orientation of the consumers towards organic products, but also because chemical fertilisers are more expensive. In this way, biotechnology is offering solutions for achieving food security in an organic manner, environmentally safer and, occasionally, cheaper than the conventional ones.

Industrial biotechnology uses microorganisms or enzymes to obtain a large variety of products ranging from primary metabolites like organic acids, alcohols, amino acids, nucleotides and vitamins to complex products such as biopolymers, detergents and biofuels with diverse applications in food sector, chemical and pharmaceutical industry, environment and bioenergy. Developing such new technologies based on biological systems, with respect to low resource-consuming and environmental protection, stimulates scientific research and innovation.

The direct economic effect of the industrial biotechnology sector is defined by its in-house activities, i.e. the people it employs and the turnover and added value it creates as a sector. In 2013, the industrial biotechnology sector employed about 94.000 full-time equivalents (FTEs) for its core activities in the EU28 member states. This employment is quite evenly spread over its different IB product segments. The largest share of employment is generated in the market of bio-based chemicals, followed by bioplastics and biofuels. Also a number of pharmaceutical applications, notably antibiotics, account for a substantial share of industrial biotechnology employment (Debergh et al. 2016).

Enzymes are biochemical molecules showing high catalytic power. These active proteins can be obtained from different sources like microorganisms, plants and animals, but the most attractive are the microbial ones, mostly because they are less expensive and may be subjected to genetic improvement for higher production yields. Modern genetic engineering techniques are also applied for obtaining enzymes with improved characteristics and that are able to act in extreme conditions in terms of pH, temperature and saline concentration, extending their possible applications. In the last 50 years, the use of biocatalysts has increased because of their great potential for producing food, animal feed, detergents and cleaners, biofuels, textiles and leather, cellulose and paper, cosmetics and pharmaceuticals. Enzyme inhibition studies are currently used for developing new and specific therapies, more targeted and with fewer side effects, with important contribution to human health. The continuous expansion of enzyme applications may be considered as an opportunity for developing new business in this field. Enzymatic technologies are more environmental friendly, generating higher quality and safer products with minimum wastes.

Biofuel production has expanded rapidly, encouraged by the regulations concerning renewable energy. According to European Strategy (2010), the target for energy obtained from renewable sources is 20% of total energy till 2020. If the first generation of biofuels was based on starchy (corn, wheat, sugar cane, sugar beet) and oily (soybean, sunflower, rapeseed) raw materials competing with food, the second generation is oriented to renewable lignocellulosic materials and waste oil feedstocks with low economic value. The third-generation biofuel is produced using microalgae fermented with microorganisms, converting CO_2 and producing O_2 (Slade and Bauen 2012). Applying biotechnology on agricultural and food processing wastes and by-products for biofuels has also an important environmental contribution towards mitigating the impact of climate changes. Economic impact of biofuels depends a lot on the availability and price of the used feedstock. That is why bioethanol price in the European Union is about USD70/barrel, in the USA is USD50–60/ barrel and in Brazil up to USD25/barrel (Mollahoseini et al. 2015). Regarding the social aspect, obtaining alternative solutions for fuels and energy requires highly qualified human resources, encouraging academia and industry collaboration and not being dependent on limited fossil carbon resources.

It is estimated that every direct job in the bio-based product industry creates 1.76 jobs in other industries. In a study specifically conducted for the biotech industry, TEConomy calculates that every direct bioscience industry job creates 5.5 additional jobs elsewhere in the economy. For workers in the agricultural feedstock and chemicals sector of the biotech industry, which includes biofuel producers, the multiplier is as high as 18.4 additional jobs for every direct job (TEConomy/BIO 2016; Bio 2017).

Environmental biotechnology applied in solid waste and wastewater treatment or remediation of contaminated soils has proven to be useful and efficient solutions for improving quality of living. Increased waste generation from anthropic activities requires adequate technologies based on microorganisms or enzymes for their biodegradation. Accumulation of toxic compounds in soil, due to industrial or agricultural activities, may be subjected to two main actions. The first one consists of bioremediation and phytoremediation, cleaning techniques based on microorganisms and plants. The second one is prevention of environmental degradation by replacing a conventional technology with biotechnology. Biosensors are a biotechnological solution for monitoring the quality of air, water and soil. Environmental biotechnology can be considered an emerging and growing challenge, stimulating enhanced biodegradation using new microbial strains, enzyme engineering, biosensor development, process engineering, waste assessment and recycling (Gavrilescu 2010). In situ environmental biotechnology may bring prosperity because it creates employment opportunities with increased incomes for highly qualified specialists, gives products with added value by waste valorisation and creates a healthier environment.

Healthcare biotechnology has to identify people's needs and to meet them with specific and efficient solutions like vaccines, antibiotics and molecular treatments. Biopharmaceuticals provide targeted and personalised treatment for serious illnesses like cancer, diabetes and cardiovascular and immune diseases. A dedicated chapter of this book contains detailed information about the most important categories of drugs manufactured using biotechnological methods and about the potential of stem cell/gene therapies. The production of biopharmaceuticals is a growing sector of economy, representing now about one-third of total drugs business. Emerging countries can be attractive for companies to invest in, a large market is available to supply and economic growth is expected to rise rapidly (Moorkens et al. 2017). Finding proper answers to health concerns has a social dimension but an economic one too. Correct diagnostics and suitable and efficient treatment, with lower side effects, are meant to improve the quality of life and to reduce health spending.

Biotechnology contributes to economic welfare in high-income countries, as well as the in low- and middle-income ones. For countries with agriculture-based economy, it is important to improve crop yield by cultivation of new varieties with better performances and pest and disease resistance and to obtain more nutritive food and feed. In developed countries, research activity is more significant, stimulating education by strong collaboration with academia and creating employment opportunities. Biotechnology is now a fast-moving sector with new achievements that have the chance to grow a business.

References

Bio (2017) The biobased economy: measuring growth and impacts. Biotechnology Innovation Organization. https://www.bio.org/sites/default/files/Biobased_Economy _Measuring _Impact.pdf. Accessed 22 Feb 2019

Caggiano R, D'Emilio M, Macchiato M, Ragosta M (2005) Heavy metals in ryegrass species versus metal concentrations in atmospheric particulate measured in an industrial area southern Italy. Environ Bio Monit Assess 102:67–84

Debergh P, Bilsen V, Van de Velde E (2016) Jobs and growth generated by industrial biotechnology in Europe, prepared for EuropaBio-the European Association for Bioindustries. http://edepot.wur.nl/392243. Accessed 22 Feb 2019

Espinosa-Leal CA, Puente-Garza CA, Garcia-Lara S (2018) In vitro plant tissue culture: means for production of biological active compounds. Planta 248 (1):1–18

Europe Strategy 2020 (2010). https://www.ec.europa.eu/info/business-economy-euro /economic-and-fiscal-policy-coordination/eu-economic-governance-monitoring-prevention-correction/european-semester/framework/europe-2020-strategy_en. Accessed 22 Feb 2019

Fang J, Zhu X, Wang C, Shangguan L (2016) Applications of DNA technologies in agriculture. Curr Genomics 17 (4):379–386

Gandhi SG, Mahajan V, Bedi YS (2015) Changing trends in biotechnology of secondary metabolism in medicinal and aromatic plants. Planta 241(2):303–317

Gavrilescu M (2010) Environmental biotechnology: achievements, opportunities and challenges. Dyn Biochem Process Biotechnol Mol Biol 4(1):1–36

Green JM (2012) The benefits of herbicide-resistant crops. Pest Manag Sci 68(10):1323–1331

Igiehon NO, Babalola OO (2017) Biofertilizers and sustainable agriculture: exploring arbuscular mycorrhizal fungi. Appl Microbiol Biotechnol 101(12):4871–4881

Mollahoseini A, Tabatabaei M, Najafpour GD (2015) Biofuel production. In: Najafpour G (ed) Biochemical engineering and biotechnology, 2nd edn. Elsevier, Amsterdam, pp 597–630

Moorkens E, Meuwissen N, Huys I, Declerk P, Vulto AG, Simoens S (2017) The market of biopharmaceuticals medicines: a snapshot of a diverse industrial landscape. Front Pharmacol 8:314. https://doi.org/10.3389/fphar.2017.00314

Olson S (2015) An analysis of the biopesticide market now and where it is going. Outlooks Pest Manag 26 (5):203–206

Pino-Otín MR, Ballestero NE, González-Coloma A, Val J, Mainar AM (2019) Ecotoxicity of a novel biopesticide from Artemisia absinthium on non-target aquatic organisms. Chemosphere 216:131–146

Slade R, Bauen A (2012) Micro-algae cultivation for biofuels: cost, energy balance, environmental impacts and future prospects. Biomass Bioenergy 53:29–38

TEConomy/BIO (2016) The value of bioscience innovation in growing jobs and improving quality of life. Biotechnology Innovation Organization, Washington, DC. https://www.bio.org/sites/default/files/BIO%202016_Report_FINAL_DIGITAL.pdf. Accessed 22 Feb 2019

Timmis K, de Lorenzo V, Verstraete W, Ramos JL, Danchin A, Brüssow H, Singh BK, Timmis JK (2017) The contribution of microbial biotechnology to economic growth and employment creation. Microb Biotechnol 10(5):1137–1144

United Nations, Department of Economic and Social Affairs, Population Division (2017) World population prospects: the 2017 revision, key findings and advance tables. ESA/P/WP/248. https://population.un.org/wpp/Publications/Files/WPP2017_Key Findings.pdf. Accessed 22 Feb 2019

USDA, Economic Research Service (2018) Recent trends in GE adoption. https://www.ers.usda.gov/data-products/adoption-of-genetically-engineered-crops-in-the-us/recent-trends-in-ge-adoption.aspx. Accessed 22 Feb 2019

2 Doing Business in Biotech: European Paradigm Versus American Success

Daniela Zirra

Abstract

Entrepreneurship is the central element of growth and economic development. In the absence of it, however abundant and available the economic resources may be, progress cannot be achieved in any field of activity. The entrepreneurial process presents both common and specific elements, depending on the level of economic development, the pace of technical progress, the social and cultural elements, the applied development strategies and policies, the level of education, the conjectural aspects, etc. For this reason, a comparative analysis of successful factors in entrepreneurship can make a valuable contribution to identifying aspects that can decisively influence the stimulation and sustainable development of entrepreneurship. We refer here to national, regional, or microeconomic level effects and to the level of various fields of activity such as biotechnology. The American entrepreneurial process is characterized as the most advanced in the world, and this is intended to encourage other countries to make sustained efforts to reduce development gaps in entrepreneurship. As far as US entrepreneurship in biotech is concerned, there is an important difference between it and the degree of development of entrepreneurship at European level for several reasons. Thus, identifying the similarities and differences between the European and American perspectives on entrepreneurship in biotech can reveal several directions for action to increase the speed of development of this field in the future.

Keywords

Entrepreneurship · Entrepreneurial process · Biotechnology industry · European approach · American perspective · Comparative analyses · Reduce the gap

2.1 Introduction

Various analyses in the field of investment process show how important are the investments in high-tech industries. For example, since 2000 (Gompers and Lerner 2001: 190), the share of venture-type investments in high-tech industries, in total US investment, is over 70%. This shows once more the importance accorded to stimulating entrepreneurship in the biotechnology industry, as part of economic development strategies, because this industry is founded on the entrepreneurial phenomenon.

At the same time, considering the importance of entrepreneurship for growth and economic development, it is interesting to think about the relationship between entrepreneurial activities and the level of economic development of a

D. Zirra (✉)
Department of Commerce, Economic Integration and Business Administration, Romanian-American University, Bucharest, Romania

F. Matei, D. Zirra (eds.), *Introduction to Biotech Entrepreneurship: From Idea to Business*,
https://doi.org/10.1007/978-3-030-22141-6_2

country. Many analysts show that there is a U-shaped relationship between the two. Moreover, the conclusion presented in various studies (Acs and Szerb 2006: 109) is that, in highly developed countries, entrepreneurship has a positive influence on economic growth, while in developing countries the relationship is inverse.

Entrepreneurship analysis is easier to achieve when the degree of homogeneity of the entrepreneurial process as a whole is high (see USA, Canada, etc.). In the context of pronounced heterogeneity, such as Europe, it is very difficult to make unitary assessments. However, some models have been noted in European economies (Aernoudt 2004: 130–132), such as the profit-oriented Anglo-Saxon model, through the creation of technology parks in countries such as the UK or Finland; the German, non-profit model, focused on stimulating and supporting regional development, opening new businesses by people who have become unemployed, and stimulating and promoting technological transfer; and the model promoted in Latin-Mediterranean countries, dedicated to supporting regional development.

In other respects, a comparative analysis of the entrepreneurial process in Europe versus the USA is at the same time important and useful for several reasons. First of all, the US biotech industry's entrepreneurship is the best in the world, so it can be considered a model to follow. Being developed in innovative performance clusters, the US biotech industry achieves enviable performance. Secondly, Europe has not yet managed to achieve the desired performance level in this area, which is a major drawback to the pace of economic development as a whole. Moreover, many experts appreciate that, in fact, the differences between the two major regions of the globe are increasing. Thirdly, it would be interesting to find some ideas on how Europe could reduce the development gap in this area.

Despite the fact that the importance of entrepreneurship for growth and economic development has been taken into account since the mid-1980s, it was only in 2002 at the Barcelona European Council that it was decided that Europe must become more entrepreneurial and more innovative. Until then, entrepreneurship continued to be approached with caution (Aernoudt 2004: 132). In the beginning, business incubators have experienced rapid development, but most of them were focused on regional development. Therefore, they have benefited from European funds, but they have been used to stimulate high-tech entrepreneurship to an almost insignificant extent.

In the USA, it is very easy to open a new company in biotech, but it is extremely costly to make this company to survive in an aggressive competition and, moreover, develop sustainably, given that the financial market dedicated to the field is heterogeneous and the risk associated with this type of business is significant. In order to have a clearer picture, we have to outline some of the features characteristic of the industry (Ahn and Meeks 2008), besides being an industry of the future with enormous potential: the industry firms are facing global competition, "struggling"—for the same amount of knowledge and the use of the same category of highly qualified employees; in many of the situations, the innovative products achieved exceed the trade boundaries of the countries of origin; there is a synergic relationship between biotechnology, IT & C, and engineering.

The importance of biotech industry and its central place in the overall US development policy is based on the following (Bagchi-Sen et al. 2016: 200): it is a vector of sustainable development in a changing environment, generated by technological progress and globalization; is concentrated in highly performing clusters; and needs a large volume of venture capital (high uncertainty and risk) and highly qualified staff, which are the most difficult barriers to entry if they are not accessible; the main factors affecting its performance are location, access to public programs, trading capacity, and barriers to access to innovation.

Finally, what we need to focus our attention on is that the need to fund a business in biotech industry as a whole is significant and can come from a variety of sources (Shimasaki 2014: 449), such as own funds or obtained from family members and/or from friends; funds attracted from angel investors or from venture investors; government grants; local programs for start-up financing; contributions from foundations and associations interested in research and achievement of expected results; financing obtained through the conclusion of advantageous partnerships; national programs to stimulate and support entrepreneurship in high-tech areas, etc. From this point of view, Europe still has significant difficulties due to relatively cumbersome regulations and the lack of a dedicated capital market for the consistent financing of entrepreneurial activity in this area.

2.2 Why Entrepreneurship in Biotech?

It is no longer a secret for anyone that entrepreneurship in high-tech domains has a positive impact on multiple levels. Firstly, there will be more innovative products at individuals' level, with a higher capacity to meet the needs of consumers. Secondly, it will increase the level of performance and competitiveness of the industry. Third, it can contribute to increasing the level of development of the national economy. Promoting innovation in elite areas, such as the biotech industry, stimulates increased prosperity and economic well-being. The higher the level of development in a country, the higher the innovative and highly creative industries (Acs 2010: 165) and the stimulation of entrepreneurship in these areas is particularly important.

Furthermore, biotechnology industry is considered a domain of the future, which results in more and more voices appreciating that in the future this industry will have an extremely favorable dynamic on a global level (Ahn and Meeks 2008: 20). In this context, we can very well understand why in the biotech industry there is such a large number of stakeholders (from the public sector, the private sector, investors, scientists, managers, institutions operating within the market, etc.), and entrepreneurship is supported and stimulated by various levers.

One of the topics that most often appear in the analysis of entrepreneurship is the causes that make a person want to become an entrepreneur in a certain area. In making such a decision, motivational factors play a very important role alongside the characteristics of the industry and the relevant market. Moreover, since entrepreneurial activity is both challenging and difficult, the decision to enter the entrepreneurial rush is also influenced by a series of personal and contextual factors. Simplistically, for many people, entrepreneurship is the best opportunity to affirm or acquire a certain social status or a better position within the firm, by delivering outstanding performance.

In our case, the frequently asked question is why would anyone want to become an entrepreneur in the biotechnology industry? In this case, the answer can also be simple or complicated. In short, because you have the chance to do something special. Elaborating some more, biotech entrepreneurship is the best opportunity to create innovative and distinctive products and technologies that are sometimes considered revolutionary and bring global recognition, positively affecting a wide range of consumers.

The biotechnology entrepreneur is a special category of entrepreneurs (Shimasaki 2009: 9). As we know, not every new business has an entrepreneurial approach. An entrepreneurial business must have a high degree of innovation and the capacity to generate beneficial effects for both consumers and industry. The biotech entrepreneur has two categories of traits. Some are common to all entrepreneurs (the will to succeed, ability to assume and manage risks, intuition, leadership, lifelong learning, etc.), while other features strongly distinguish them personally in relation to others. In the biotech industry, the entrepreneur is often a renowned researcher with recognized activity and international visibility who makes the decision to become an entrepreneur because he wishes to do so. A fundamental aspect is that, as it is about entrepreneurial-research activities, professional experience is vital to success. The entrepreneur in this industry must be a tenacious researcher with a lot of patience, who, with passion and earnestness, can lead and see through research and achieve tangible high-quality results that will be materialized in revolutionary products that can be used on a large-scale market because it addresses the needs of a large number of consumers.

The biotech entrepreneur is a person able to cope with the unexpected in his field, with a sense of business, great skill in negotiating, the ability to quickly learn new things even if they are not strictly related to the research they are doing, and a high sense of responsibility because its products can affect very many people. Another notable aspect is that the biotech researcher is an "international citizen." He has to interact with a wide range of people in different corners of the world to increase his chances of success and his performance.

Returning to the reasons that drive people to do business in the biotech industry, the answer can be synthesized as such: because you have the chance to do something special, you have the opportunity to gain national and international recognition, to do good things for the benefit of a wide spectrum of users and consumers, and, in a word, to bring enormous satisfaction. Why is this industry so attractive? Because it is a top area, a high-tech field which employs high-quality and undeniable human resources, it offers many opportunities, it represents the future, and it has an enormous growth potential.

In addition, we mention that the biotechnology industry has some particular features (Hine and Kapeleris 2006: 19–20), which differentiates it strongly from other industries: the time needed to develop new products is relatively long (the more complex the research is, and the more implications it has, the longer this period increases); it is highly regulated (particularly in the area of intellectual property rights); it benefits from important allocations and financial resources, from several sources; the knowledge that is used has a high scientific level and a high degree of innovation; it is knowledge and science based, being developed through intensive research activities; and it takes place in clusters, through complex networks of collaboration/cooperation that are extended intra-cluster and internationally; added value results from the circuit created from R&D activities to the market, in three stages (scientific research—technical development—commercialization).

Biotechnology industry is an excellent field for entrepreneurial activity because it has been established and developed, becoming today one of the most incontestable stars of the economy. There is a lot to say about this industry, and many researchers of entrepreneurial science have been intrigued by the rapid growth and the notoriety that it has gained.

For example, the most interesting aspects identified (Shimasaki 2014: 445–446) regarding the biotech industry are:

- It presents a very high level of diversity.
- It offers many challenges.
- It is focused on delivering innovative products to the market.
- Most of the products have the results of research carried out in universities, research institutes, and laboratories as their source.
- It has a rising growth rate, it requires intensive capital, the chances of success and sustainable development of a business in this area are high, and the possibilities of affirmation in an elite field are enormous.

2.3 European Framework of Doing Business in the Biotech Area

There are numerous studies and analyses that have as research objectives the differences that exist between doing business in the USA versus Europe and the factors that generate these differences. In one of the works, Acs et al. (2009: 20–22), we can distinguish some key steps in the recent history of the European transition from management-based to entrepreneurial-based way of doing business:

The first phase is considered to be between the early 1980s and 1990s when all the energy of European firms and decision-makers was focused on the efforts that had to be made to cope with the competitive threat posed by large corporations, neglecting the power that SMEs could have from an economical point of view, especially due to their small size, which gives them great flexibility. It refers to American small businesses, which operate in areas considered "exotic" as IC&T (especially the software sector) and biotechnology industries were and still are considered.

In the second phase, between the early 1990s and the middle of the decade, European decision-makers, whether we are talking about companies or public institutions, began to realize that US entrepreneurial activity is a force that is worth considering due to their ability for sustainable development, based on outstanding economic performance.

The third phase, which lasted from the mid to late 1990s, was the period when the US business-based economy was already well consolidated and continues to pose a growing threat to the heavy economic system in the European countries still centered on the classical way of doing business.

It was only in the fourth phase of the late 1990s that Europe agreed to take urgent and drastic measures as the development gap in the high-tech domains, especially biotechnology industry and IT&C, increased rather than declined. Thus, various initiatives aimed at recalibrating European economic systems on entrepreneurial backgrounds were promoted.

The fifth phase, launched in the early 2000s, brought important changes to the European business paradigm, because entrepreneurship becomes the basis for growth and economic development policies. In 2003, she drafted the Green Paper of Entrepreneurship to empower decision-makers to give entrepreneurship the role it deserves and to mobilize them to take the necessary steps to stimulate entrepreneurial activity, expanded and developed at an ever-faster pace across all segments, especially in knowledge-based or science-intensive and high-tech-based domains.

In the next phase, which begins in the second half of the 2000s, it is becoming increasingly clearer at a European level that entrepreneurship is the safest way to cope with the wave of change generated by technological progress and globalization.

Another characterization of the stages of entrepreneurship in Europe is centered on the same five periods, but the findings are otherwise nuanced (Audretsch et al. 2002: 4–6), from skepticism, recognition and envy to consensus and attainment.

On the European continent, there is no fully accepted general framework used to stimulate biotechnology entrepreneurship, as it exists in the US economy, and the gaps between the two regions are still significant. In 2000, the Lisbon Summit decided that Europe would become the world's most competitive knowledge-based economy by 2010. As we all have seen, this goal is still very far away. In 2007, the European Commission established that biotechnology is extremely important, promoting the concept of knowledge-based bioeconomy (KBBE), introducing into the program to stimulate and develop research activities, the Framework Program 6 (FP6), a special chapter on biotechnology (Theme 2: food, agriculture, fisheries and biotechnology). Further progress did not have the desired pace, although the KBBE Agenda wanted to support biotechnology industry's ability to meet the needs of society (Birch et al. 2012: 7), in the agriculture, food, medical, pharmaceutical, and energy sectors.

In addition, the KBBE Agenda created a common EU-28 identity for bioresearch institutions, because the concepts have been defined and problems identified, threats were recognized and opportunities identified, and the institutional framework and policies to boost this important business area were drawn.

At the end of 2010, in the Knowledge Based Bio-Economy in Europe: Achievements and Challenges, Full Report (14 September 2010: 13, elaborated by the European Commission, European Research Area, Clever Consult BVBA, Belgium), it is stated that in 10 years the allocated funds amounted to 2 trillion Euros and 21.5 million employed people are in fields such as agriculture, water, climate change, fossil fuel, and health.

The most representative challenges (EC 2010: 14–17) for biotech industry in the global economy are the impact of social and demographic developments on agriculture, water problem, effects of human activities on the natural environment, climate change, reduction of fossil resources and energy security, the needs of sustainable development, the rapid pace of technological progress, changes in lifestyle and eating habits, the need for healthy and quality food, and the need to control and prevent epizootic and zoonotic diseases.

The greatest achievements in KBBE support and development (EC 2010: 18–20), between 2005 and 2010, were the establishment of the European Technology Platforms (ETP) and the promotion of research in the field by the Framework Programmes (FP6 and FP7), ERA-NETs (Plant Genomics, System Biology, Industrial Biotechnology, European Network of Transnational Collaborative R&D for SME projects in the field of biotech, etc.), ETPs (Plants for the Future, Food for Life, Biofuels, Agricultural Engineering, etc.), and EU experts (Advisory Group on Food, Agriculture and Biotechnologies, KBBE-Net, KBBE National Contact Point, EU Standing Committee for Agriculture Research).

On 10 June 2018, the tenth anniversary of the Small Business Act was celebrated. On this occasion, the SME Annual Report (mainly aimed at assessing the dynamics and quality of small and medium-sized European businesses) shows that the number of SMEs increased by 13.8% over a 10-year period (2008–2017). In 2017, SMEs in the EU-28 accounted for 22.7% of the non-financial business, and the employment rate for intensive high-tech activities (biotechnologies, for example) was around 4.8%.

The objectives of the Small Business Act for stimulating entrepreneurship were cutting red tape and administrative simplification in opening a new business, encouraging entrepreneurial initiatives and implementing policies that contribute to growth and development, increasing access to the necessary financial resources and markets to improve the marketing of new products, and stimulating research and development activities.

As a result of the implementation of the Small Business Act at EU-28 level, more than 3300 measures were developed and implemented during the reporting period to achieve a more intense and performant entrepreneurial activity. Some of the most important achievements are as follows: productivity increased by 11.6%, the value added by SMEs has increased to over 50% of total value added, the number of new jobs created was over 60% of the total, etc.

2.4 The American Perspective on Biotech Entrepreneurial Process

In the early 1980s, US decision-makers took very seriously the importance of stimulating and developing entrepreneurship in the biotech industry. At first it was intended to increase the entrepreneurial process. Over time, the complexity of the relationships between stakeholders (universities, research institutes and laboratories, firms, mass media, government institutions and agencies, associations and foundations, etc.) has become more and more (Aldrich and Martinez 2010: 415). The networks created between the biotechnology industry players have expanded, refined, and quickly transcended national boundaries. This development has made possible the completion of many win-win partnerships and the development of collaborations that have generated increased performance for all parties involved, have favored the creation of true technological communities, and have ensured the establishment of very strong relationships among stakeholders; the created communities have gained over time a force that is large enough to influence to a certain extent the set of standards and regulations applied.

Below we will make some succinct and concrete references to the biotechnology entrepreneurial process in the USA (Acs et al. 2009: 34–36). As mentioned above, entrepreneurs and their firms are the focus of policy makers' concerns. Here, we have two aspects: stimulating the establishment and development of knowledge-based firms and supporting the development of a performing activity, both in the process of producing new knowledge and/or innovative products and in the process of valorization on the market. In addition, the US economy took the start at the right time.

Once there was awareness regarding the importance of the biotech industry (since 1980), the SBIR (Small Business Innovation Research) program was started, in response to the decline in the US economy's global competitiveness. At that time, 4% of the federal government's annual budgets were or could be used to set up small-scale, knowledge-based, or scientific-based firms.

Moreover, all US companies or R & D agencies were free to allocate some of their budgets for the same purpose, based on a three-step succession. The first step is to identify and fund research projects with a high potential to achieve quality results, with the possibility of their future development. The second step addresses the development of research results so that they can be marketed (results that could respond to consumer needs). The third step refers to the actual marketing of the results, or their valorization on the market, which in our opinion is the most important aspect of any economic activity. However good, innovative, or performing a product is, if its creator does not have the ability to cross the boundaries of the firm with the product he has made, then all his efforts can be considered useless and the resources wasted.

Another aspect worth mentioning is that through the SBIR program, biotechnology industry companies have enjoyed and continue to receive due attention from the National Institutes of Health and the US Department of Defense, being considered of strategic importance for growth and economic development of the country.

The impact achieved by the decision-makers is notable, because the implementation of the program has succeeded: to increase the survival rate of the new firms (the failure rate is quite high due to the specificity of the activities), to improve the rate of growth and business development, and to develop the entrepreneurial spirit of researchers and scientists as many have changed their perspective on how to do business by making the transition from research itself as a process—fundamental research or basic research toward entrepreneurial research; many of the researchers have opened their own businesses and the foundation of the research activity is to obtain innovative products that address an identified need in the market; many of the established firms have had great commercial success, bringing prosperity to their creators, which has created a driving effect for other researchers and scientists in the field.

In order to have insight into the evolution of US biotech industry, it is interesting to consider the stages of cluster development (Feldman et al. 2005: 134–137), specific to this particular industry, as they appear in various studies and analyses. If initially, in the first phase of the entrepreneurial activities, the firms in the field were simple suppliers for different industries, the subsequent policy decisions to support and promote the technological transfer determined the increase of the entrepreneurship in the second stage without the companies being involved in research activities. The third step was decisive, as through the legislative changes regarding the licensing and patenting of the scientific results of the universities, institutes, and research laboratories, it was possible to achieve the transfer of the intellectual property rights to the industrial sector. This transformation has given rise to the establishment and development of small firms that have had the opportunity to carry out their own research activity (initially small research contracts, supported by the government).

The next step is characterized by the increase of the amplitude and importance of the exploratory research activity of the companies, which attracted the attention and captured the interests of the venture capitalists. In the current stage, biotech entrepreneurship is very efficient, many companies have created their own venture capital, the pace of development is sustained, and biotech industry has become "self-sustaining," which is the dream of any industry and any kind of economic activity.

Another secret of US performance in high-tech entrepreneurship is the management and financing of activities through a legislative, institutional, financial, fiscal, etc. framework, characterized by being unitary, developed, and improved over time and coherently. The US Small Business Administration (SBA—created in 1953 by the American Congress when the Small Business Act was launched) is the national institution that addresses all aspects that can contribute to the efficient and rapid development of entrepreneurship in the cutting-edge domains of the national American economy. The SBA was initially intended to stimulate and support high-quality entrepreneurship (knowledge-based, innovative-based, and high-tech).

The first entity created to provide quality financial support to new firms was the Small Business Innovation Research (SBIR—founded in 1980). In addition to SBIR, Small Business Technology Transfer (STTR), the Federal and State Technology Partnership (FAST), and the Tibbetts Award (SME Prize for Excellence in Business) were created in time. SBA manages SBIR and STTR through the Technology Program Office.

SBIR grants three-step funding for new and innovative high-tech businesses. In the first step, up to $ 150,000 is granted for a period of 6 months (at the start, the maximum amount was $ 70,000 for the same period). The purpose of these funds is to evaluate the qualitative level of scientific research and the feasibility of the new business idea, or to develop a prototype of the new product. The second step involves a maximum $ 1 million funding for a 24-month period to deepen research and get the new product in the form in which it may be intended for commercialization. The third step is dedicated to the marketing process. SBIR grants non-financial support and funding is provided either from private funds or from other sources.

FAST provides grants to strengthen the technological competitiveness of newly established small businesses or their activity to give them the chance of sustainable development. There are other notable elements that we want to remember and that have a major positive impact on stimulating entrepreneurship in leading industries. First, there is no need for the researcher to have a company set up in order to benefit from the first step of SBIR funds. Secondly, emphasis is placed on performance in R&D activities and not on equipment or on the efforts to market a technology that has already been discovered. Last but not least, the biotechnology industry is focused on developing new products that respond to market needs.

Why is the US economy so efficient in biotech industry entrepreneurship? The answer is very simple. First, the USA can be considered the cradle of entrepreneurship, which is encouraged, supported, and respected almost since always. Secondly, high-tech sectors are at the heart of decision-makers' concerns in local and national governments and are considered to be the main driver for growth and economic development in a global, aggressive, and dynamic competitive environment. Last but not least, the "salt and pepper" of US advantage over Europe and other areas of the world in the field of biotechnology industry are the associated entrepreneurial activities, which enjoy a great deal of attention and support, on multiple levels. The results of research activities need to "go out into the world" and enjoy a wider range of consumers in order to generate prosperity for both individuals and society. This was very well understood and promoted at the level of all decision-makers in the American economy.

2.5 Key Elements of Developing Entrepreneurship in Biotech: A Brief Comparative Analysis

Based on the expansion of globalization, fueled by the wider use of new IT & C technologies, all economic activity has seen significant changes in the conduct of business. The more developed and powerful the economy is, the easier it adapts to change, and this is an exciting factor for accelerating and diversifying change. Therefore, we must not be surprised that the US economy has adapted very quickly to the new business environment (Acs et al. 2009: 19), making it relatively easy to transition from the classical managerial economy to the new way of doing business, based on stimulating and developing entrepreneurship.

Europe also makes sustained efforts to develop and strengthen the entrepreneurship-based economy, but the process is more difficult for at least two reasons. On the one hand, there are significant differences in economic development both between European countries and at a regional level within each national economy. On the other hand, the way of doing business in Europe

Table 2.1 Evolution of entrepreneurship in 2009–2016

		Increasing rate 2016/2009 (%)		
No	Region/Country	Number of SMEs	Gross value added	Employment
1	EU-28	12.4	22.3	2.3
2	USA	8.9	54.2	9.2
3	Japan	−24.4	No data	−11.6

Source of data: Eurostat, National Statistical Offices, DIWEcon

is a tribute to the classical tradition, in which the management-based economy still has an important share.

SME Annual Report 2017/2018 presents some relevant statistical data on SME performance in 2009–2016 in the USA and EU-28 (Table 2.1).

The same report shows the share of start-ups in high-tech activities in 2017 (EU-28) was the following: 19.1% IT/software development; 18.5% software as a service; 6.5% biotechnology, nanotechnology, and medical technology; and 4% green technology (source of data: EU Start-up Monitor).

SME Annual Report (for EU-28 area) took into account the capital structure analysis used to finance business. The conclusion reached was that in the period 2015–2017 the share of venture capital was 54% in the USA and 40% in the EU-28 (source of data: Preqin), and the DIWEcon forecast for SME activity (in EU-28) in 2019 is 1.3% growth in employment and 4.3% in added value.

As a general description, biotech industry is one of the areas of activity that has been developed based on the concept of the Triple Helix (Etzkowitz 1998: 829), because it was formed as a result of collaboration between universities, government, and businesses. Performance that was obtained in clusters specific to this industry has had a major positive impact on the regions where they have been formed. This confirms once again that the Triple Helix is a sine qua non condition for both the sustainable development of biotech industry and regional development as a whole.

American universities have close ties of collaboration with economic activities, contributing fully to increasing the performance of the national economy. They are characterized by increased multidisciplinary characteristics and can access funds from multiple sources, including the business area, to fund their research activities. Faculty members have an increased mobility because the labor market is favorably regulated, and they enjoy a high level of independence in doing business. In order to provide strong support for biotechnology entrepreneurship, the patent system is heterogeneous, providing the best ways to protect intellectual property rights tailored to each individual case. To all this information, we add that the US Government is highly decentralized, which is a great advantage in adopting good policies for growth and economic development in all the states.

In Europe, the relationship between universities and the business world is lagging behind in the USA for several reasons, such as legislative barriers or cultural traits, which have a relatively unfavorable view of "marketing" the results of scientific research. In addition, European researchers have low mobility and the possibilities to get funding for research projects are relatively limited. Another negative aspect for high-tech entrepreneurship is that the division of innovative and research work is concentrated to a significant extent at the level of universities and other research entities (institutes and laboratories).

An important disadvantage in terms of enhancing biotech entrepreneurship is that, unlike the US public structure, governments are centralized, exerting strong hierarchical control.

2.6 Reducing the Gap Between the European and American Entrepreneurial Process in Biotech

Europe has to recover a serious gap in comparison to the USA in the field of high-tech entrepreneurship, especially in terms of biotechnology industry. Experts and researchers of science and entrepreneurship have some recommendations (Acs and Szerb 2006: 121) in this respect, for example:

- Emphasis on the formation and development of human capital, especially in countries where the level of income for R&D personnel is low and is not motivating to achieve high performance
- Accelerating the high-quality technological transfer rate in countries where researchers' income is satisfactory but the R&D infrastructure is obsolete
- Developing simplified policies, rules, and procedures that are designed for innovative start-ups (high-tech start-ups)
- Reforming the labor market so as to stimulate the high-tech workforce (universities, institutes, and research laboratories) in order to become more involved in the entrepreneurial process and to generate change in the R&D activities (adopting and disseminating a new vision in the field of R&D, so that the products resulting from these activities have a higher degree of addressability to current and prospective consumer needs—market-based R & D
- Improving the institutional framework of funding so as to become more attractive to those looking for venture capital

In the Expert Group Report: Review of the EU Bioeconomy Strategy and its Action Plan (2017), a relevant analysis is included on the strategy developed to stimulate and develop bioeconomy at a European level. It aimed at investing in research, stimulating innovation and capacity building, and strengthening cooperation between stakeholders in the field.

Although the objectives were not clear and well-grounded enough, some positive developments have been achieved, as the growth potential of bioeconomics has grown, based on a coherent policy framework. Also, for the field of bioconomics, a formal framework for monitoring progress is not used, but the increase in the visibility of bioeconomics is noticeable (Stakeholders' Manifesto for the Bioeconomy in Europe, Utrecht 2016), just like the strengthening of public–private partnerships in R&D activities and increasing the importance of the biotech industry for economic development.

Finally, in order to increase the pace of development and the level of performance of European entrepreneurship in the biotech industry, a plan of measures must be developed and implemented which takes into account the leveraged start-up feature (Taylor 2011) of new companies, which has the following meaning:

- Access to the research-development infrastructure of universities, research institutes, or laboratories, which facilitates the setting up of companies at minimal cost and allows the reduction of design and production periods for new products.
- Access to highly qualified workforce due to their proximity, so well-known scientists are involved in companies' research activities, and cooperation and collaboration networks between participants to the process are extended both inside and outside the clusters.
- Access to the resources needed to market the results of research, because in support of the companies, a number of entities are actively contributing to the technological transfer, to the implementation and enforcement of the intellectual property rights, and even to the stimulation and development of new businesses (start-ups).
- Access to a wide range of public and private (which is very important) funding sources, bearing in mind that the funding of research activities, particularly in basic research, at the initial stage, is very expensive and the risk that research does not lead to a new and innovative product (which is suitable for commercialization) is very high.

2.7 Conclusions

The problems faced by mankind are extremely diverse and increasingly complex and the biotech industry can offer the right solutions to most of them (food, safety, environmental protection, energy, etc.). For this reason, there are a number of questions (Ahn and Meeks 2008: 21) related to how the biotech industry can be supported:

- Which are the factors that influence the development of biotechnology?
- Why it is so important to locate activities?
- What are the factors that stimulate competitiveness in the field?
- What effects can different actions have on biotech entrepreneurship and productivity?
- How important are its innovation and management in the biotech industry?

The entrepreneur in biotechnology is of a particular breed (Shimasaki 2014: 446) because he must be both a high-class researcher and a high-skill businessman. The basics for an entrepreneur to be successful in the biotech industry are as follows: a great passion for his work, because it can inspire those with whom he collaborates to be more committed and more involved in the activity, in order to increase performance; a very clear perspective on the research objectives and the way forward to achieve the desired results; not afraid of risks or failure and he leans on what he has to do with perseverance and responsibility; and a particular capacity to attract the necessary funds for growth, diversification, and business development.

There are some aspects that we can mention to emphasize once again why the biotech industry is so important in the contemporary economy. First of all, the potential for sustainable growth and development in the future is enormous. Then, for example, many sectors of the US biotech industry have become "self-sustainable," because the multiplier effect of new business has as its main source the fact that the consolidated companies in the industry have the capacity to finance other entrepreneurial approaches. Third, the degree of innovation is extremely high, so the industry has all the prerequisites to provide prompt answers to complex issues in various areas. Finally, being a science-based industry, it is a basic area for the growth and development of a knowledge-based economy.

References

Acs ZJ (2010) High impact entrepreneurship. In: Acs ZJ, Audretsch DB (eds) Handbook of entrepreneurship research. An interdisciplinary survey and introduction, 2nd edn. Springer, New York, NY, pp 165–182. https://www.researchgate.net/profile/Sharon_Gifford/publication/226441377_Risk_and_Uncertainty/links/53e28e910cf275a5fdd9ea2e/Risk-and-Uncertainty.pdf#page=412

Acs ZJ, Szerb L (2006) Entrepreneurship, economic growth and public policy. Small Bus Econ (2007) 28:109–122. Springer. https://doi.org/10.1007/s11187-006-9012-3. https://search.proquest.com/docview/220909353?pq-origsite=gscholar

Acs J, Aldrich HE, Audretsch DB, Baumol WJ, Boko S, Gatewood EJ, Johannisson B, Reynolds PD, Sabel C, Thurik RA, Lundström A (2009) The role of SMEs and entrepreneurship in a globalised economy. Expert report no. 34 to Sweden's Globalisation Council. https://www.government.se/49b731/contentassets/8efd3c3a4c844f88883513fa451760bd/the-role-of-smes-and-entrepreneurship-in-a-globalised-economy

Aernoudt R (2004) Incubators: tool for entrepreneurship? Small Bus Econ 23(2):127–135

Ahn MJ, Meeks M (2008) Building a conducive environment for life science-based entrepreneurship and industry clusters. J Commer Biotechnol 14:20–30. https://doi.org/10.1057/palgrave.jcb.3050076. Published online 27 Nov 2007. https://link.springer.com/content/pdf/10.1057%2Fpalgrave.jcb.3050076.pdf

Aldrich HE, Martinez MA (2010) Entrepreneurship as social construction: a multilevel evolutionary approach. In: Acs ZJ, Audretsch DB (eds) Handbook of entrepreneurship research. An interdisciplinary survey and introduction, 2nd edn. Springer, New York, p 387–430. https://www.researchgate.net/profile/Sharon_Gifford/publication/226441377_Risk_and_Uncertainty/links/53e28e910cf275a5fdd9ea2e/Risk-and-Uncertainty.pdf#page=412

Annual Report on European SMEs (2017/2018). In: Hope K (ed) The 10th anniversary of the Small Business Act, SME performance review 2017/2018, Contract number: EASME/COSME/2017/031, November 2018, European Union

Audretsch DB, Thurik R, Verheul I, Wennekers S (eds) (2002) Understanding entrepreneurship across countries and over time, in entrepreneurship: determinants and policy in a european-us comparison, Economics of science, technology and innovation, vol 27. Kluwer Academic Publishers, Dordrecht, pp 1–10. https://books.google.ro/books?hl=en&lr=&id=zjQ1MGthzesC&oi=fnd&pg=PA2&dq=Business,+Biotechnology,+Entrepreneurship,+European,+paradigm,+American,+success,+Perspective,+Key+elements,+Gap&ots=n5VO9YQ1cw&sig=RL3NYj9CWly7wDZJMN97ntX2qV8&redir_esc=y#v=onepage&q&f=false

Bagchi-Sen S, Smith HL, Hall L (2016) The US biotechnology industry: industry dynamics and policy. Environ Plan C Gov Policy 22(2):199–216

Birch K, Levidow L, Papaioannou T (2012). Self-fulfilling prophecies of the European knowledge-based bio-economy: the discursive shaping of institutional and policy frameworks in the biopharmaceuticals sector. J Knowl Econ. https://s3.amazonaws.com/academia.edu.documents/30668917/KBLLTP_Self-fulfilling_KBBE_JKE_12.pdf?AWSAccessKeyId=AKIAIWOWYYGZ2Y53UL3A&Expires=1549647509&Signature=7iqPcxbP9XLTFGgqrm8w5dxXNFg%3D&response-content-disposition=inline%3B%20filename%3DSelf-Fulfilling_Prophecies_of_the_Europe.pdf

Etzkowitz H (1998) The norms of entrepreneurial science: cognitive effects of the new university-industry linkages. Res Policy 27:823–833. Elsevier. http://www.oni.uerj.br/media/downloads/1-s2.0-S0048733398000936-main.pdf

Expert Group Report: Review of the EU Bioeconomy Strategy and Its Action Plan (2017) European Commission, Directorate General for Research and Innovation, Directorate F – Bioeconomy, Brussels. Publications Office of the European Union. https://gbs2018.com/fileadmin/gbs2018/Downloads/Expert_group_report.PDF

Feldman M, Francis J, Bercovitz J (2005) Creating a cluster while building a firm: entrepreneurs and the formation of industrial clusters. Reg Stud 39(1):129–141

Gompers P, Lerner J (2001) The venture capital revolution. J Econ Perspect 15(2):145–168

Hine D, Kapeleris J (2006) Innovation and entrepreneurship in biotechnology, an international perspective. Concepts, theories and cases. Edward Elgar, Cheltenham. http://citeseerx.ist.psu.edu/viewdoc/download?doi=10.1.1.463.4354&rep=rep1&type=pdf

Shimasaki CD (2009) The business of bioscience: what goes into making a biotechnology product. Springer, New York. https://doi.org/10.1007/978-1-4419-0064-7_2. https://www.springer.com/cda/content/document

Shimasaki CD (ed) (2014) Biotechnology entrepreneurship. Starting, managing, and leading biotech companies. Elsevier, Waltham, MA. https://www.ott.nih.gov/sites/default/files/documents/pdfs/Ferguson-Chapter14-BiotechEntrepreneur-2014.pdf

Taylor J (2011) The 'leverage' start-up model, trade secrets, a blog from nature biotechnology. http://blogs.nature.com/tradesecrets/2011/06/09/the-leverage-start-up-model

The Knowledge Base Bio-Economy (KBBE) in Europe: achievements and challenges, full report (14 September 2010), European Commission, European Research Area, coordinated by Clever Consult BVBA (Belgium). http://www.mercadosbiotecnologicos.com/documents/the_knowledge_based_bioeconomy_kbbe_in_europe.pdf

Creating Products and Services in Plant Biotechnology

3

Monica Boscaiu, Miquel Estruch, Ana Fita, Mariola Plazas, Jaime Prohens, Adrián Rodríguez-Burruezo, Mercedes Verdeguer, and Oscar Vicente

Abstract

This chapter presents a brief description of the most relevant applications of plant biotechnology, with examples of specific techniques that can be used to provide commercial products and services to customers and represent, therefore, business opportunities to entrepreneurs. Start-up companies can base their activities on the development of *molecular markers*, their use in MAS (marker-assisted selection) for molecular breeding programmes or in genetic fingerprinting applications; on the use of *mutagenesis* for generation of mutant collections in species of interest, the development of novel mutant detection methods and the phenotypic or molecular analysis of the mutants; or on applications of in vitro *culture* techniques. These are all services that can be offered to small breeding companies and research laboratories that do not have in-house facilities to perform these activities. Small start-ups cannot afford to bring a novel *transgenic* crop variety to the market; they can, however, develop and characterise the initial transgenic lines with traits of interest and transfer them (for a price) to big transnational companies, which can undertake all further field tests and marketing procedures. It can also be profitable to use GM plants as *biofactories* for the production of high added-value recombinant proteins or other biomolecules for different industries. Other topics included in this chapter are the applications of *'omics' technologies* in plant biotechnology, the new business opportunities opened by the customers' increasing interest in the products of *organic agriculture* and the possible commercial exploitation of some of the thousands of compounds—*phytochemicals*—synthesised by plants.

Keywords

Molecular markers · Mutagenesis · In vitro culture · Transgenic plants · 'Omics' technologies · Organic agriculture · Phytochemistry

M. Boscaiu · M. Verdeguer
Mediterranean Agroforestry Institute (IAM, UPV), Universitat Politècnica de València, Valencia, Spain

M. Estruch
Viromii Innovation S.L., Valencia, Spain

A. Fita · M. Plazas · J. Prohens · A. Rodríguez-Burruezo
Institute for Conservation and Improvement of Valencian Agrodiversity (COMAV, UPV), Universitat Politècnica de València, Valencia, Spain

O. Vicente (✉)
Institute for Plant Molecular and Cell Biology (IBMCP, UPV-CSIC), Universitat Politècnica de València, Valencia, Spain
e-mail: ovicente@upvnet.upv.es

3.1 Introduction to Plant Biotechnology

Plant biotechnology represents a broad sector that offers many opportunities for entrepreneurs. Plants have always been essential for humankind, not

F. Matei, D. Zirra (eds.), *Introduction to Biotech Entrepreneurship: From Idea to Business*,
https://doi.org/10.1007/978-3-030-22141-6_3

only because they generate most of the oxygen we breathe and are our primary sources of energy—fossil fuels have been mostly formed by the decaying of plants—and food. Plants also provide us with raw materials for different industries, to produce, for example, paper, textiles, adhesives, dyes or lubricants, and contain secondary metabolites that we use in cosmetics, perfumery, as medicines or in the chemical industry.

Plant biotechnology directly impacts all the fields and industries listed above. However, as some of these applications are included elsewhere in this book, this chapter focuses on how the different plant biotechnology tools and techniques can be used to create products and services with commercial potential. Research and development activities in plant biotechnology (unfortunately, not some of its applications) have enjoyed a certain degree of freedom, if compared to the restrictions on human and, to a lesser extent, animal biotechnology; this has led to the development of a wide range of techniques and tools that can be applied to generate innovations, which may ultimately become the basis for a commercial company.

The plant biotechnology market is enormous. In 2015, the market for agricultural biotechnology, which is only one sector within the whole plant biotechnology business, accounted for more than 20×10^9 US $, and is growing by around 10% each year; it is expected that it will reach 39.5×10^9 US $ by 2022 (Wise Guy Reports 2017). The industry is attracting much attention, and the number and amount of investments in plant biotechnology companies are rapidly growing. Therefore, this is an appropriate moment for entrepreneurs with background knowledge on plant biochemistry, molecular biology, plant genetics and genomics, or related scientific areas, to start their own business by creating products and services that deliver value to companies or consumers.

This chapter aims to inspire future entrepreneurs, by presenting different examples and real-life applications of plant biotechnology tools and techniques. The chapter is organised into different sections, describing specific technologies that can be used to provide commercial products and services. These methodologies include molecular markers, mutagenesis, in vitro culture techniques, transgenesis, 'omics' technologies and also organic agriculture and phytochemistry; at the end of the chapter, there is a section on perspectives and how the future of plant biotechnology is being shaped.

3.2 Applications of Molecular Markers in Plant Biotechnology

3.2.1 What Are Molecular Markers?

As its name implies, a molecular marker is a characteristic or difference (polymorphism) in a biomolecule that can be used to reveal one or more features of a cell or individual carrying this marker. Molecular markers are a useful tool with many applications in plant biotechnology because they can be associated with a particular phenotype (marker-assisted selection) and/or used to establish relationships and evaluate differences among individuals, populations and species (genetic fingerprinting). Although the term molecular markers can refer to different types of biomolecules (e.g. proteins, nucleic acids), in recent times, it is mostly referred to DNA markers, for which there have been many technological developments in the last few decades. Consequently, the cost of development and utilisation of molecular markers has decreased dramatically in the last few years, thanks to the advances in genomics and bioinformatics.

3.2.2 Advantages of Molecular Markers in Plant Biotechnology

One significant advantage of the application of molecular markers in plants is that they allow increasing the efficiency of many processes in plant science and breeding without relevant ethical issues. Molecular markers do not increase genetic variability, as other techniques such as mutation, genetic modification or gene editing can do; instead, they allow selecting and using the variation already available. This issue is important, as the application of molecular markers to provide services and products in plant biotechnology is not constrained by special regulations that apply to the techniques for increasing the variation mentioned above. This facilitates that entrepreneurs establish companies that do not require special and cumbersome regulations, other than those applicable to a company working with chemical reagents and live plant materials (and, where appropriate, plant pests and diseases).

Another clear benefit of DNA molecular markers is that, once we have a molecular marker linked to a trait, the selection for the trait can be performed without the need of observing the corresponding phenotype. This has many advantages because the selection of plants can be done at nurseries, using seedlings or small plantlets, for traits that are expressed in adult plants, and also allows selection for tolerance or resistance to diseases without the need to manage pathogenic agents. For example, selection of tomato plants resistant to quarantine diseases such as *Tomato spotted wilt virus* (TSWV) can be performed by using molecular markers linked to resistance gene *Sw5*, without the need of inoculating plants with the virus. This possibility considerably reduces the costs involved in growing, phenotyping and selecting the plants and eliminates those related to management of pathogenic agents. By increasing the efficiency of selection, molecular markers allow reducing the costs and speeding breeding programmes.

The increasingly reduced cost of screening with DNA markers is also a clear advantage that has facilitated and promoted their use. Nowadays, most laboratories can perform molecular marker genotyping in-house, even when only basic equipment is available. However, high-throughput genotyping of plant samples with molecular markers is frequently done by specialised companies at very low costs. For example, some recent technologies, like Single Primer Enrichment Technology (SPET) genotyping, allow obtaining over 10,000 single nucleotide polymorphism (SNP) markers for less than 20 €/sample (i.e. 0.002 €/SNP marker). The use of emerging molecular marker technologies such as SPET has allowed the expansion of some initially small companies such as IGA Technology Services (Udine, Italy) (www.igatechnology.com).

3.2.3 Types of Molecular Markers

Many different DNA molecular markers exist, based on the type of polymorphism which is targeted. For example, in restriction length polymorphisms (RFLPs), random amplified polymorphisms (RAPDs) and amplified fragment length polymorphisms (AFLPs), the polymorphism targeted is on the length of fragments generated after shearing the DNA with restriction enzymes, whereas in simple sequence repeats (SSRs) the target is the different number of repeats of short sequences of one or a few nucleotides, and in SNPs the target is a difference in a single nucleotide. Also, one important characteristic that influences the possible uses of molecular markers is related to their dominant or co-dominant nature. Thus, co-dominant markers are more informative, as they distinguish between homozygote and heterozygote genotypes, while dominant markers only distinguish between the homozygote recessive genotype on the one side and the homozygote dominant and the heterozygote on the other. The technology used for detecting the molecular marker is also relevant, as it greatly influences the cost. For example, RFLPs were among the first developed DNA molecular markers, and their detection was based on hybridisation with labelled (quite often radioactively) probes, which is much more expensive than PCR-based detection methods. The reproducibility of molecular markers within and between laboratories is also an important characteristic that affects their applications in plant biotechnology. The more reproducible a molecular marker is, the better, in particular for applications based on obtaining highly reliable specific genetic fingerprints. Many other subsequent markers are based on the use of the polymerase chain reaction (PCR) technique, which produces multiple (often millions) copies of specific target fragments of DNA. The abundance in the genome is also an essential factor to take into account when choosing a molecular marker. In this respect, thanks to transcriptome and genome sequencing and re-sequencing projects, it is frequent to have millions of SNP markers available. However, for other markers, such as SSRs, their abundance in

Table 3.1 Main characteristics of some of the most common DNA-based molecular markers

Characteristic	RFLPs	RAPDs	AFLPs	SSRs	SNPs
Target	Fragment length	Fragment length	Fragment length	Fragment length in tandem repeats	Nucleotide
Mode of action	Co-dominant	Dominant	Dominant	Co-dominant	Co-dominant
Reproducibility	High	Medium	Medium	Very high	High
Detection technology	Hybridisation	PCR	PCR	PCR	PCR
Genome abundance	High	High	High	Medium	Very high

the genome is generally much lower; nevertheless, even in this case, it is not uncommon to have thousands of SSR markers in species for which genome or transcriptome sequences are available. A summary of relevant characteristics of some of the most common DNA molecular markers is provided in Table 3.1.

3.2.4 Technical Requirements for Applying Molecular Markers in Plant Biotechnology

Different types of facilities, depending on the type of marker and the throughput level (low, medium or high), are needed for the use of molecular markers. For PCR-based markers, the facilities needed are relatively simple, and modest laboratories can have all the equipment required (centrifuge, thermocycler and other basic laboratory instruments) to perform low- to medium-throughput genotyping using markers such as SSRs. The use of high-throughput systems based on genotyping-by-sequencing (GBS) techniques usually requires expensive equipment, including sequencers which may cost over 200,000 € and that require specialised personnel and a high intensity of use to redeem the initial investment. In addition, the data analysis of the vast amount of sequencing data (frequently ranging from GB to TB) requires bioinformatics servers with enough capacity and capabilities to manage such amount of data. Because of these high investments costs and the availability of many DNA sequencing providers—from large companies such as Macrogen (www.macrogen.com) or Beijing Genomics Institute (www.bgi.com) with thousands of employees to smaller enterprises such as Stab Vida (www.stabvida.com) or SecuGen (www.secugen.com)—many laboratories and private companies are outsourcing the services of sequencing. These sequencing companies frequently offer other services, such as the extraction of DNA and the evaluation of its concentration, quantity and quality, as well as the bioinformatics analysis of data. In consequence, today it is even possible for a small start-up company to offer services related to the application of molecular markers without the need of a wet lab, as most of the work related to obtaining the markers can be outsourced to other DNA sequencing provider companies. In fact, as the price per marker offered by these providers is so low, many labs which have the required facilities and equipment are increasingly using the services of sequencing companies, as they are cheaper than obtaining the markers in-house.

3.2.5 Entrepreneurship Opportunities: Development of Molecular Markers

Development of new technologies for obtaining molecular markers is a clear entrepreneurship opportunity that can be economically very rewarding. For example, Cetus, the company that developed the widely used PCR using the heat-resistant polymerase enzyme from *Thermus aquaticus* (*Taq*), sold the PCR and Taq patents to Roche (www.roche.com) in 1991 for 300 × 10^6 US $. More recently, the highly valued patent on GBS—a high-throughput approach used for genotyping by many laboratories and companies all around the world—that has been in dispute between several companies has been finally assigned to KeyGene (www.keygene.com). The exploitation of the patent on GBS has increased the cost per sample, which has sparked the interest for developing alternative genotyping solutions not covered by this patent, such as SPET by NuGEN (www.nugen.com) or Capture-seq by RAPiD Genomics (www.rapid-genomics.com). The development of microarrays, in which thousands of markers, typically SNPs, could be scored for each sample, has been largely overcome by the technologies of genotyping based on sequencing.

Development of new markers in non-model species or neglected crops is also a potential field for entrepreneurs, as the availability of molecular markers is essential for genetic mapping, marker-assisted selection, genetic fingerprinting and other applications. Although some types of molecular markers do not need previous genomic information, in other cases this is required or of great utility for developing specific types of markers. For example, this is the case with SSRs, where the flanking sequences are needed in order to design the primers for PCR amplification. Older techniques for developing SSR markers relied on the use of libraries enriched in repeated sequences, but with the advent of next-generation sequencing (NGS) technologies, the sequencing of transcriptomes or genomes typically allows detecting thousands of SSR markers and hundreds of thousands or millions of SNPs. Given that the current cost of a high-quality transcriptome can be around 600 €, the development of new markers for species in which little genomic information is available is a field of increasing interest and which requires little investment.

3.2.6 Entrepreneurship Opportunities: Marker-Assisted Selection and Molecular Breeding

Marker-assisted selection and molecular breeding are one of the most evident entrepreneurship opportunities for the application of molecular markers. Given that molecular markers are highly heritable and not affected by environmental conditions, they are very reliable for performing selection of genes, QTLs, or genomic regions linked to the marker(s). In some cases, markers can even be functional, so that the polymorphism detected in the marker is responsible for the difference in a phenotypic trait. In these cases, linkage of the marker and the trait is complete. The most significant advantage of marker-assisted selection is that selection of a trait controlled by a gene or affected by a QTL does not require scoring the phenotype of the plant.

There are many potential uses of molecular markers in marker-assisted selection and molecular breeding. One of the simplest ones is the selection of plants that carry a certain allele in a breeding programme. For example, in a backcrossing programme for introgressing an allele of a gene conferring resistance to a disease in a variety or pure line, during the backcross process the plants carrying the resistance allele (in heterozygosis) have to be selected. By using marker-assisted selection, this can be made in the nurseries, and only the selected plants are grown to obtain the subsequent generations, and no inoculations and observation of symptoms are required. The advantage is even greater when the allele to be introduced is recessive, as in these cases the phenotype associated with the resistance allele is observed only in homozygous plants. An extension of this application is the selection of multiple markers at the same time, so that several genes, QTLs or markers can be scored simultaneously, increasing the efficiency and speed of the selection and breeding process. Also, when many molecular markers (typically hundreds or thousands) spread throughout the genome and positioned in a genetic map are available, they can be used to further increase the efficiency of the selection by simultaneously selecting for the gene(s) and/or QTLs of interest on the one side and for the genetic background of the recipient variety on the other. This allows decreasing the required numbers of generations in a backcross breeding programme.

When markers associated with a trait of interest are not available, the use of a bulked-segregant analysis (BSA) strategy, or the construction of genetic maps and phenotyping for the trait or traits of interest, allows the discovery of associations of markers and gene(s) or QTLs for traits of interest, i.e. the development of useful molecular markers, which can then be exploited commercially. An extension of the utilisation of molecular markers is the Breeding by Design strategy, which combines genetic mapping, high-resolution chromosome haplotyping and phenotyping to select superior individuals with the desired combination of genes and/or QTLs. An additional service that can be offered by companies is a genomic selection in which markers linked to all target genes or QTLs are available or developed, and the breeding values of individual plants are predicted based on the combinations of molecular markers they present. This strategy, which has not yet been extensively applied in plant breeding, offers the possibility for new business opportunities.

3.2.7 Entrepreneurship Opportunities: Genetic Fingerprinting

Many potential services and business opportunities are related to genetic fingerprinting, which consists of obtaining a genetic profile of individuals with molecular markers. Those molecular markers should be highly repeatable and stable, particularly when used for certification purposes. Among the potential applications of genetic fingerprinting, we can mention the detection of fraud and misuse of plant materials, the certification of authenticity of plant material, the certification of non-GMO material, the evaluation of diversity and relationships among and between samples or varieties and, when combined with the availability of phenotyping data, genome-wide association studies (GWAS).

The detection of fraud (i.e. unauthorised use of a plant variety, or illegal propagation) with morphological methods is problematic and can be ambiguous, as many morphological traits are influenced by the environment. Therefore, the use of specific genetic fingerprints that uniquely identify a variety can be used as proof to demonstrate an illegal use of plant material. Similarly, certain varieties need authentication for particular uses (e.g. local varieties with specific characteristics), and here molecular markers can also provide a solution. An additional application of high interest of molecular markers is the detection of GMOs, whose cultivation and use as food are strictly regulated in many countries, including the European Union. In this case, it is common to use different PCR-based molecular markers for detecting particular genes and/or regulatory elements such as promoter or terminator sequences. A number of companies, such as SGS (https://www.sgs.com/) or Ecogenics (www.ecogenics.ch), offer genetic fingerprinting services for these purposes.

A widespread use of genetic fingerprinting is to evaluate genetic homogeneity or diversity in plant material of a sample, variety or population, which may be of interest to evaluate purity and uniformity. Also, genetic fingerprinting is useful for the establishment of relationships between different varieties or species, either for practical or basic research purposes. When phenotypic data of large collections of germplasm are available, genetic fingerprinting of the collection allows applying a GWAS analysis, which is a potent tool to detect genes and QTLs controlling the trait(s) of interest.

3.3 Mutagenesis and Its Applications

3.3.1 History

Species have had to adapt to new environments and growing conditions to ensure their survival, so it has been essential that they possess the ability to generate different allelic variants in populations, which has generated the current biodiversity. The domestication of species has always been accompanied by the selection of those genetic variants (or mutations) most favourable for cultivation, collection or the quality of fruits, seeds, tubers and so on. Over time, this continued selection in different parts of the world has resulted in different varieties adapted to the soil and climatic conditions and management practices of the region.

Plant breeders attempt to increase and accelerate these events by inducing mutations

(Suprasanna et al. 2015) and selecting rare desirable traits. Genomes have always been exposed to alterations, and errors in replication occur every time a cell copies its information. These errors can be neutral, harmful or even lethal, or they can confer a competitive advantage. The production of mutations involves the exposure of plants or seeds to physical mutagenic agents (e.g. X-rays, UV light, gamma radiation, protons or neutrons) or mutagenic chemicals (e.g. ethyl methanesulfonate (EMS), sodium azide), which induce random changes in DNA sequences throughout the genome. Moreover, environmental factors, such as a viral infection, can also alter the genome. These changes, which can be difficult or impossible to predict, generate variability that may include interesting traits, to be selected by the breeder (Nuffield Bioethics Council 2016).

The mutation may consist of changes in the position of a single nucleotide, or sometimes (more frequently after X-ray radiation) more complex changes, such as major DNA rearrangements (inversion, translocation) or the removal of DNA fragments (deletion). Because mutations are random, very large populations of mutant plants are required to identify rare and useful mutations, and once identified, backcrossing is used to eliminate unwanted traits. However, cultures derived from the reproduction of mutations are likely to continue exhibiting DNA alterations beyond the specific mutation that provided the desired trait.

3.3.2 What Are Mutations and Mutants?

A mutation is any change or alteration in the DNA sequence of an organism that can be inherited by its offspring. The organism carrying the mutation is considered a mutant. For the most part, these changes only affect a single nucleotide or a very small region of the DNA of the millions of nucleotides that make up the genome, but they are responsible for the genetic variability that we can directly observe or can be detected applying different technologies, such as by using molecular markers.

3.3.3 What Is Mutagenesis?

Simply put, mutagenesis is the production of mutations on DNA, cloned or not. However, this definition, although certain, does not fit the biological reality, since mutation is associated with those changes produced spontaneously and mutagenesis involves the artificial generation of mutants using a specific biological technology.

It is, therefore, best to define mutagenesis as one of the most commonly used biotechnological techniques for generating variability in plants and other organisms, with the aim to stably modify the genetic material, so that it is transmissible to daughter cells arising from cell division, and to the offspring of the mutant organism.

3.3.4 Effects of Mutagenesis

Alterations of the genetic material may include small changes such as replacement, insertion or removal of a pair of nucleotides in the DNA sequence (collectively known as point mutations), or more significant changes such as insertions or deletions of large DNA fragments, DNA reversals or translocations (the rearrangements of parts between non-homologous chromosomes). Even the exchange of a single base pair can have substantial, sometimes lethal, effects, depending on the position in which it occurs and whether it can be compensated for (e.g. by the other allele).

The consequence of these natural alterations is that genomes at the species level are dynamic, including genes present in every individual ('core' genome) and genes present only in a subset of individuals (accessory or dispensable genome), which collectively constitute the pan-genome. Dispensable genes may constitute a significant proportion of the pan-genome, about 20% in soybeans (Li et al. 2014).

3.3.5 Types of Mutagenesis

Mutations can be classified according to different criteria (Table 3.2). Considering the origin

Table 3.2 Classifications of the mutations by different criteria

Origin	Spontaneous	Changes in the DNA sequence produced endogenously (by the cellular machinery and metabolism)
	Induced	Produced by exogenous (external) agents, whether natural or artificial
Cause	Adaptative	The mutation is induced to deal with specific conditions, for example with a particular stress factor
	Random	Mutations are random and occur before the stress. Stress serves as a selective condition to show the effect (positive or negative) of the mutation
Location	Autosomal	Occur in autosomal chromosomes
	Linked to sex	Occur in sex chromosomes (XY)
	Somatic	Produced in somatic cells
	Germinal	Produced in germ cells. They are the only mutations that can be inherited by offspring
Molecular nature	Substitutions	No change in the number of base pairs
	Insertions or deletions	Imply changes in the number of base pairs
Phenotypic effect	Loss of function, null or knockout	Disable the function of a protein or gene
	Gain of function	Provide a new activity
	Recessive/dominant	Depending on its phenotypic expression in heterozygosis
	Morphological, biochemical or behavioural	Depending on the phenotypic effect
	Regulatory	Affect gene expression
	Lethal	Cause the death of the individual
	Conditional	Their effects are only seen under certain conditions, e.g. at high temperatures
	Neutral	Have no apparent effect: polymorphisms (very useful for animal and plant breeding)

of the mutation, it can be spontaneous (due to replication errors, tautomeric changes, duplications or deletions, polymerase slippage, depurination or deamination reactions, oxidative damage, etc.) or induced (by base analogues, alkylation agents, acridine dyes, UV light, ionising radiation, such as X-rays or gamma rays, or biological organisms, such as viruses or bacteria).

In addition, depending on the cause of their generation, mutations can be adaptive or random. Furthermore, mutations can appear in different areas of the genome: in coding regions of genes, in untranslated regulatory sequences, in promoters, enhancers or introns. They can be autosomal or sex-linked and can affect somatic or germinal cells. If the mutations are molecular in nature, they consist of base pair substitutions (point mutations, nonsense mutation, sense mutation and silent mutation). Mutations can be recessive or dominant and also regulatory, lethal, conditional and neutral. They may have, or not, a phenotypic effect, such as loss of function or knockout of some gene or gain of function, altering morphological, biochemical or behavioural phenotypes. All these different classifications are summarised in Table 3.2.

Since it is clear that mutagenesis implies an intentional action of mutant production, this section will focus on induced mutations. The standard mutagenesis procedures, for example, chemical mutagenesis, generate mutations in random positions of the (generally unknown) sequence of the target DNA. If the sequence is known, there are now methods allowing the modification of a gene in a specific position, generating the same type of mutations obtained spontaneously or by conventional mutagenesis. The most precise and efficient of these methods is genome editing with the CRISPR/Cas9 system. However, against all scientific arguments, the products obtained by CRISPR technology are (legally) not considered as 'mutants', but as 'transgenics' in the EU, according to the decision of the European Court of Justice of July 25, 2018. Therefore, controlled 'mutagenesis' by genome editing will not be discussed here.

3.3.6 Applications of Mutagenesis

It must be borne in mind that these modifications are essentially random in the genome, and the work of genetic improvement is to identify the traits of interest and transfer them to varieties that can be cultivated; these processes can provide avenues of business for entrepreneurs. As an example, the pink grapefruit was obtained from white grapefruits mutagenised by radiation; the four-leaf clover and even the comestible lemon are other examples of mutagenesis in plants.

Collections of mutants have been long since generated in many crops of interest. The major application of these collections has been their use for the functional annotation of the genes of different species. They are also very important because when problems arise, due for example to abiotic or biotic environmental stresses, these collections are an excellent source of material to evaluate possible resistances or tolerances. Availability of these sources of variability is essential for the genetic improvement of plants and very useful in breeding programmes (Fig. 3.1).

There are currently 3282 varieties of more than 200 species known to have been obtained by induced mutagenesis (IAEA/FAO database https://mvd.iaea.org). For example, we have found three eggplant varieties generated by chemical mutagenesis and one by physical means. Varieties obtained by mutagenesis have generated many business opportunities in the past and are now widely used in agriculture.

3.3.7 Mutant Analysis

When mutants are available, and their quality is known, different screenings can be made for agronomic characters of interest. For this goal, two strategies can be used: (1) first, a mutant phenotype of interest is identified in a population, followed by the isolation of the mutated gene responsible for the phenotype (direct genetics), or (2) after mutagenesis, allelic variants in the gene of interest are identified using, for example, TILLING (Targeting Induced Local Lesions in Genomes), followed by the characterisation of the mutant phenotypes of those variants, in relation to the trait under evaluation (reverse genetics).

Fig. 3.1 Example of the evolution of some wild species into cultivated species (A: *Solanum anguivi*; B: *S. aethiopicum*; C: *S. dasyphyllum*; D: *S. macrocarpon*)

TILLING allows analysing mutations in a specific gene in many individuals (McCallum et al. 2000). As a 'reverse genetics' technique, its most important limitation is that the genomic sequence of the species must be known to find mutations in the particular gene of interest. The method combines a standard and efficient mutagenesis procedure, using a chemical mutagen such as ethyl methanesulfonate (EMS) or acridine orange, with a sensitive DNA screening technique that identifies mutations in a single base pair (point mutations). One version of this technique is the Eco-TILLING that exploits natural variability within a population, searching for spontaneous (as opposed to induced) mutations in the samples, and is widely used for genetic analysis of populations.

3.3.8 Entrepreneurship Opportunities: Development of Collections of Mutants in Species of Interest

Generating mutant collections is a business opportunity offered by mutagenesis. In the current climate change scenario, mutants with enhanced tolerance to abiotic stresses, such as drought, salinity or high temperatures, which could help crop adaptation to the rapid environmental changes taking place, would be especially interesting. In addition, also because of global warming, a large number of pests and pathogens are spreading to areas where they have never been seen before, threatening to reduce even more global agricultural production. Individuals with increased resistance to pests and pathogens could also be identified in the mutant collections. Many other traits, contributing to improving crop yields or the quality of the harvested product, are possible targets of mutagenesis.

In most cases, plants showing specific traits such as enhanced tolerance to biotic or abiotic stress could be generated by genetic transformation or genome editing. The use of these technologies for generating commercial crop varieties is, de facto, prohibited in the EU, and subjected to very long, expensive and cumbersome regulatory procedures everywhere, which only big transnational agrochemical companies can afford. Therefore, one of the main advantages of mutagenesis is that the varieties produced are subjected to a much less restrictive regulation, allowing small start-up companies to undertake commercial projects in this field. While mutagenesis by ionising radiation (X-rays or gamma rays) requires complicated and expensive infrastructure and equipment, chemical mutagenesis, or T-DNA insertional mutagenesis with *Agrobacterium tumefaciens*, can be efficiently performed if basic, relatively simple tissue culture and plant growth facilities are available.

3.3.9 Entrepreneurship Opportunities: Development of Mutant Detection Methods and Mutant Analysis

Once a collection of mutants has been generated, those individual mutants with the expected phenotypes must be selected, for example, based on altered morphological and physiological characteristics such as growth parameters, photosynthetic activity or water relations. In the last few years, advances in robotics, computing, remote sensing and imaging technologies have allowed the development of fully or partially automated, high-throughput plant phenotyping platforms that can be used for mutant detection. This application is offered as a service by private companies, such as LemnaTec (Germany) (https://www.lemnatec.com/plant-phenotyping/), CropDesign, a company of the BASF group based in Belgium (https://www.basf.com/be/en/who-we-are/Group-Companies/cropdesign.html), or KeyGene (the Netherlands) (https://www.keygene.com/technology/2-the-digital-phenotype/), among many others. Similar platforms are also established in universities and research institutes, for example, those associated in the European Plant Phenotyping Network (EPPN^{2020}, https://eppn2020.plant-phenotyping.eu/). Obviously, small start-up companies cannot compete with these large-scale phenotyping services, but can develop targeted methods (e.g. sample analyses to quantify specific metabolites, proteins or enzyme activities) to detect mutants with particular characteristics, which can be offered to small breeding companies or research laboratories which do not perform this kind of analyses in-house. These services can be extended to the molecular analysis of the isolated mutants, depending on the mutagenesis method employed, and outsourcing NGS and other 'omics' analyses to specialised companies, if necessary.

3.4 In Vitro Culture

3.4.1 History

Plant tissue culture methods started to be established as early as 1900 and spread and become popular over the 1960s. Since then, many protocols and applications have been developed. Nowadays, it is impossible to think of large-scale plant propagation, generation of virus-free plant clones, creation of new plant varieties, quick generation of pure lines, plant transformation, conservation of endangered plants species or production of high price plant metabolites, among many other applications, without involving totally or partially some in vitro culture steps (Shahzad et al. 2017).

3.4.2 What Is Plant Tissue Culture?

Plant in vitro culture consists in cultivating under axenic conditions whole plants or, more commonly, parts of a plant (so-called explants). Many plant cells are totipotent so that a differentiated cell can dedifferentiate, reverting to a meristematic stage in which it can divide and differentiate again to form a new plant organ; for example, sections of a leaf blade can produce somatic embryos. Ultimately, it is possible to regenerate whole plants in vitro from those dedifferentiated cells. This process is called morphogenesis and is the base of a wide range of applications. Two main morphogenetic routes are used for in vitro plant regeneration: (1) the embryogenic route, in which the dividing cells organise into an embryo, which may give rise to the whole plant later on, and (2) the organogenic route (Fig. 3.2a), in which the dividing cells organise to form a plant organ (mainly shoots and roots, but can also be flowers, tubers...). In many cases, however, the cells fail to differentiate from the very beginning and grow in a disorganised manner, forming a callus. When the first divisions of a dedifferentiated cell form

Fig. 3.2 (**a**) Organogenesis of tomato shoots growing from a callus, (**b**) microcultured explants, (**c**) tobacco protoplasts, (**d**) tobacco anthers with embryos and plantlets

first a callus and then organs or embryos, it is said that the regeneration was indirect. When in vitro culture methods do not succeed in regenerating the desired organ and/or the whole plant in a particular species or genotype, it is said that the plant is recalcitrant.

3.4.3 Micropropagation: A Quick Method to Reproduce Plants

Micropropagation consists of cloning plants using the morphogenetic properties of the plants so that it is possible to produce thousands of plant clones in a short period and a relatively small space (Fig. 3.2b). Also, some micropropagation and/or micrografting techniques favour the elimination of virus from plants. Therefore, micropropagation allows the production of healthy plantlets at an industrial scale; this planned and controlled way of multiplying plant material is very beneficial for the agriculture sector. Actually, the establishment of micropropagation companies is one of the engines for increasing agricultural production in many regions of the world (Sonnino et al. 2009). Nowadays, micropropagation is especially used for the production of ornamental plants, fruit trees, such as citrus spp., and many other woody plants. This technique is by far the most widespread application of plant tissue culture, and one of the most profitable ones as long as a good market and socio-economic study is done prior to the establishment of the company.

3.4.4 Tissue Culture Techniques for Improving Plant Breeding Processes

A breeder's work is based on selecting individuals with the best combination of traits. To perform this task, the breeder needs genetic diversity. Plant tissue culture offers a wide range of techniques to increase plant diversity. For example, by *protoplast fusion* (Fig. 3.2c) plant protoplasts (i.e. plant cells without their cell wall) are forced to fuse to produce either hybrids (allopolyploids) or higher ploidy products (autopolyploids); it should be remembered that many of our crops are polyploids (e.g. potato, wheat, strawberries). In some cases, the fusion of the two cells is not complete producing cybrids, which consist of cells with the nucleus of one protoplast and the cytoplasm of the other. Generation of cybrid plants has been a milestone for obtaining cytoplasmic male sterile (CMS) lines, which are very useful for hybrid seed production in self-pollinated species, as well as for the production of seedless *citrus* fruits. Another way to create genetic diversity is by generating somaclonal mutants, which can be forced to appear under in vitro conditions. Other useful techniques to generate diversity, such as *plant genetic transformation* and *induced mutations,* are explained elsewhere in this chapter and will not be mentioned here.

In some cases, breeders may find hybridisation barriers when using wild relatives to introduce interesting traits into elite cultivars; in such cases, in vitro *pollination* and *embryo rescue* can be very useful. In addition, embryo rescue can also speed up the breeding process by shortening the time from fruit set to planting the offspring (Manzur et al. 2014). In any case, *double-haploid* production is one of the most effective techniques to accelerate the breeding process, as it allows obtaining homozygotic plants for all their *loci* in a single generation and thus the quick production of pure lines. Generation of double-haploid plants is based on the regeneration of haploid plants from the plant gametes and its diploidisation, either spontaneous or induced by drugs such as colchicine. In most cases, haploids are formed from the immature male gametes, through androgenesis in in vitro anther or isolated microspore cultures (Fig. 3.2d), but can also be obtained from the female gametes by gynogenesis. In vitro *selection* is another technique which reduces breeding costs by allowing the analysis of thousands of plants in a small and very controlled space; it has been instrumental to select genotypes resistant to bacterial or fungal toxins or tolerant to salt stress.

Fig. 3.3 (**a**) Production of microtubers of potato, (**b**) alginate bead containing a shoot node ready to be cryopreserved

3.4.5 Other Plant Tissue Culture Applications

The production of secondary metabolites in plant tissue cultures is a fascinating technology with good business perspectives, and it is covered in Sect. 3.8. In addition to the previous applications, tissue culture can be used to preserve plant material, either by storing the plants in chambers in slow-growing conditions or by cryopreservation at low or very low temperatures (Fig. 3.3). Since this book focuses on practical aspects of plant biotechnology, we do not deal here with the wide array of applications of tissue culture techniques in basic research to investigate different aspects of plant cell physiology, biochemistry and molecular biology.

3.4.6 Entrepreneurship Opportunities: Setting a Micropropagation Company

According to the Financial Tribune (2015), the global market for tissue culture micropropagation is estimated at 10^9 plants per year. Even if there are already many companies doing micropropagation in Europe, it is still a profitable business since there is a wide range of plants for which protocols have been developed. In vitro plants are exported/imported easily all over the world because they grow in axenic conditions and reduced space. Therefore, the target market for the produced plants can be local, national or international. Actually, in many breeding companies, where tissue culture is routine, the in vitro tissue culture work is carried out in their European facilities and then the plants are sent overseas to perform the more labour-intensive activities such as the crosses and field work to produce the marketable seeds.

The main constraint in starting a business related to micropropagation is the initial investment needed, although it is much lower than for other industrial activities. Apart from consumables (glassware/plasticware, substrate, reagents), some infrastructure and basic equipment are required to set up an in vitro culture facility convenient for micropropagation (Table 3.3 and Fig. 3.4).

3.4.7 Entrepreneurship Opportunities: Tissue Culture Services for Breeding or Research

Large breeding companies have their own in vitro culture laboratories, but small companies do not.

Table 3.3 List of minimum infrastructure in a micropropagation culture company

Rooms	Equipment
Sterilisation room	Autoclave
Dissection room	Flow-cabinet
Washing and media preparation room	Glass-bead sterilisers or lighters for flaming
Culture room	Climatic chambers
Weaning room and greenhouse	Fertigation system

Fig. 3.4 (**a**) Industrial autoclave, (**b**) operators working at the culture room, (**c**) climate chamber with micropropagated plants in flasks, (**d**) greenhouse with acclimated plants

Therefore, such companies need to outsource the tissue culture work required in their breeding programmes, thus opening business opportunities to start-ups specialised in this area. Some of the services they can provide are the production of double-haploid plants, embryo rescue, cryopreservation or in vitro conservation of plant material, among others. These services may also be interesting to public research institutions if they do not have the required infrastructures or personnel, or need to handle sporadically high amounts of plant material. Also, a company specialised in plant tissue culture work could also provide complementary services to breeding companies and research institutions, such as cytometry diagnosis, or consulting services regarding the improvement of protocols.

3.5 Transgenic Plants

Plant biotechnology applications discussed in the previous sections—of molecular markers (Sect. 3.2), mutagenesis (Sect. 3.3) and in vitro culture techniques (Sect. 3.4)—are focused mostly on plant breeding; constitute the basis of the activity of many companies, both large and small; and provide interesting business opportunities to entrepreneurs. However, if we refer to the global economic impact of plant biotechnology, it is largely dependent on the commercial cultivation of transgenic crops. In this section, we will briefly describe the methods most commonly used for the generation of transgenic plants and present an overview of updated data on the cultivation of biotech crops and their economic impact, with mention of the specific situation in Europe, as well as future trends and entrepreneurship opportunities in this sector.

3.5.1 What Is a 'Transgenic' Plant?

A simple *technical* definition of a transgenic or 'genetically modified' (GM) plant would refer to a plant which contains one or a few foreign genes, of any origin, stably integrated into its genome, and thus transmitted to the plant progeny as any other endogenous gene. Since the aim of the genetic transformation is to confer a specific (and predesigned) phenotype to the transgenic plant, the introduced genes are accompanied by the appropriate regulatory sequences (promoters and transcription termination signals recognised by the host plant) to allow expression of those genes.

From a *regulatory* point of view, however, the definition of a GM plant focuses on the *method of generation* rather than the *final product*, since genetic engineering techniques must be used in the process of genetic transformation for a plant to be legally transgenic. If classical breeding approaches—natural or forced sexual crosses and selection—are used to introduce foreign genes, or even complete genomes, in a plant variety, the product is not considered as transgenic. Moreover, if the genetic information of the plant is modified through spontaneous or induced mutations, the mutant plant is also not subjected to the regulations applied to GM plants, even though it is evident that it has been 'genetically modified'. There is, however, an exception (in Europe): in case mutations—indistinguishable from the latter—are produced by genome editing techniques (CRISPR/Cas9), then the plants obtained are considered transgenic, according to the recent EU legislation.

3.5.2 How Are GM Plants Generated?

A three-step process is commonly used to obtain a transgenic plant. First, the foreign DNA, cloned in appropriate expression vectors, is introduced into plant cells, generally in vitro (in cell cultures or plant explants), for which several procedures are available (see below). Second, through non-homologous recombination mechanisms, the foreign DNA gets integrated into the plant genome, at a single *locus* or several *loci* of the plant chromosomes, in one or more copies. Third, whole plants are regenerated in vitro from the transformed cells; depending on the starting plant material, different morphogenetic processes can be employed to obtain the transgenic plants: most often by organogenesis, but also through somatic embryogenesis or androgenesis. This last step usually represents the bottleneck of the entire procedure of plant genetic transformation, which has not yet been successful, or is still very difficult, in many 'recalcitrant' species or genotypes, because of the lack of efficient in vitro regeneration protocols (see Sect. 3.4, on in vitro cultures).

The construct used for plant transformation contains a selection marker gene conferring resistance to an antibiotic or herbicide; its expression during the process of in vitro regeneration, which is carried out under the appropriate selective conditions, allows elimination of the cells that have not been transformed. Therefore, the regenerated plant should be, in principle, the expected genetically modified product. Nevertheless, the transgenic nature of the original transformants must be confirmed by molecular analyses, as well as the expression of the transgene and its transmission to the plant progeny, after selfings and backcrosses. Finally, homozygous lines carrying a single copy of the transgene, which should behave as a dominant Mendelian character, are selected. Those plants will be subjected to agronomic analyses, in greenhouses and experimental fields, before they go through all further field tests and procedures required by regulatory agencies to be authorised for commercial production.

3.5.3 Plant Genetic Transformation Methods

Over the years, different procedures have been established for the transfer of DNA to plant cells (see, for example, Potrykus 1991), in some cases just for intellectual property reasons; that is, trying to develop (and patent) novel plant transformation methods. They included, among others, direct gene transfer to protoplasts, electroporation of intact cells, microinjection, vortexing with silicon carbide fibres or uptake of DNA mediated by laser irradiation or ultrasounds. These methods have been shown to be effective for DNA transfer and, sometimes, for the generation of stably transformed plants, but are not routinely used. In fact, all commercially grown transgenic crops have been initially produced by one of two transformation methods: *Agrobacterium*-mediated transformation or biolistics.

3.5.3.1 *Agrobacterium tumefaciens*-Mediated Transformation

Agrobacterium tumefaciens is a soil bacterium that infects many plant species—mostly dicotyledonous—inducing the formation of tumours (crown galls) due to transfer and expression in the plant cells of genetic information from the bacteria, which can thus be considered as a *natural genetic engineer*. The genes responsible for the formation of crown galls are present in the T-DNA, a region of the Ti (for tumour-inducing) plasmid of the bacteria. The Ti plasmid also contains the *vir* (virulence) region, including all genes controlling transfer to the plant cells and integration in the plant genome of the bacterial T-DNA. Upon activation by phenolic compounds segregated by wounded plant cells, proteins encoded by *vir* genes recognise the T-DNA borders—short direct repeats of ~ 25 bp flanking the T-DNA region—and make and release a single-strand copy of the T-DNA, which is protected from nuclease degradation and transferred to the plant cells, where it is localised to the nucleus and integrated in the cell chromosomes.

Two characteristics of the natural DNA transfer system of *A. tumefaciens* allowed using it for the controlled genetic transformation of plants, making possible, for the first time, the generation of transgenic plants (Herrera-Estrella et al. 1983). First, the products of the *vir* region act in *trans*, and there is no need to have the T-DNA and the *vir* region in the same plasmid. Second, only the T-DNA borders, not the T-DNA itself, are recognised by the transfer machinery, so that any DNA substituting the original T-DNA sequences will be transferred as far as it is flanked by the T-DNA borders. Plasmids of relatively small size, the so-called *binary vectors*, were developed, containing all DNA sequences to be transferred—usually, the foreign gene of interest and the selection marker gene, both with the appropriate regulatory sequences, plant promoters and transcription termination signals—cloned in between the T-DNA borders; outside this artificial T-DNA, the plasmid contains all elements for replication and selection in bacteria.

For plant transformation, the binary vector is introduced into an *A. tumefaciens* strain containing a plasmid with the *vir* region, but no T-DNA (a 'disarmed' Ti plasmid). Once the *vir* region is activated, the foreign DNA construct cloned in the binary vector in between the T-DNA borders will be transferred to plant cells by the same mechanism used in nature by *Agrobacterium*. In practice, the simplest procedure for *Agrobacterium*-mediated plant transformation is known as the 'leaf-disc' method (Horsch et al. 1985), in which the *Agrobacterium* strain bearing the binary vector is co-cultivated for a short time, say 30 min, with small pieces of leaves; then the plant material is incubated for 24–48 h in agar plates, to allow T-DNA transfer. After eliminating the bacteria with a suitable antibiotic, the leaf pieces are cultured in a fresh solid medium under selection conditions, for the in vitro regeneration of the transgenic plants.

It is important to highlight that generation of transgenic plants by this method is technically very simple and should be affordable for most laboratories (and small start-ups), as only basic laboratory equipment and standard tissue culture and plant growing facilities are required.

3.5.3.2 Particle Bombardment (Biolistics)

Initially, only natural hosts of the bacterium could be transformed by *A. tumefaciens*. This excluded, for example, the cereals, which are some of the most important crops worldwide, and prompted the search for alternative transformation procedures. Several *direct gene transfer* methods were soon developed for the transfer to plant cells of naked plasmids containing the appropriate constructs; among them, particle bombardment was the most successful and most effective one (Klein et al. 1987).

To generate transgenic plants by *particle bombardment* or *'biolistics'* (short for 'bioballistics'), tungsten or, later on, gold microparticles of ~1 μm diameter are coated with plasmid DNA, accelerated to high speed using a specifically designed device—popularly known as 'DNA gun'—and shot into plant explants or cells cultures. The microprojectiles penetrate the cell wall, and the plasmid is released inside the cells and can get integrated into the plant genome.

Biolistic transformation is widely used in plant biotechnology to generate transgenic plants and was the key for the transformation of cereal crops. Nevertheless, whenever possible, *Agrobacterium*-mediated transformation is recommended, as particle bombardment requires more complicated and expensive equipment and is not as precise and efficient.

3.5.4 Biotech Crops

Transgenic, genetically modified (GM) or genetically engineered (GE) crops—nowadays commonly referred to as 'biotech crops'—were first grown commercially in 1996. Since then, their cultivation area worldwide has increased yearly at a very rapid rate, making biotech crops the fastest agricultural technology ever adopted. In 2017, 17 million farmers in 24 countries—most of them developing countries—grew biotech crops in almost 190 million hectares, which represents a 122-fold increase over the area cultivated in 1996; other 43 countries, not growing themselves GM plants, imported the products of those biotech crops. Referring still to 2017, eleven countries cultivated more than 10^6 ha of GM crops, led by the USA and followed by Brazil, Argentina, Canada and India; these five countries, with a combined area of about 173×10^6 ha, contribute to >90% of the worldwide growth of biotech crops (ISAAA 2017).

The situation in the European Union is, however, different. After several years of a moratorium and the banning of cultivation of biotech crops in many EU countries, at present, there is only a transgenic crop grown in Europe, an old Bt maize variety from Monsanto (MON810), approved in 1998, before the moratorium. It is cultivated only in Spain (124,000 ha in 2017) and Portugal (7300 ha); this means that the total area in the whole EU represents <0.07% of the global biotech cropland. Imports from third countries, such as transgenic soybean or maize for animal feed, are allowed, for the benefit of farmers from the USA, Brazil or Argentina.

Apart from the European anomaly, these data show the enormous success of this technology. Biotech crops have higher yields than the corresponding conventional crops and allow a more stable production and a reduction of energy costs and labour, thus providing higher profits to farmers. It has been estimated that between 1996 and 2016, GM crop cultivation represented an increased productivity of 658×10^6 tonnes—compared to the expected production of conventional crops grown in the same area—with a value at farm level of 186×10^9 US $ (Brookes and Barfoot 2018).

At present, over 99% of biotech crops, in terms of cultivated area, are limited to only four major species: soybean, maize, cotton and rapeseed (canola). Less than 1% of the total area is used to grow minor crops, such as alfalfa, sugar beet, squash, potato, sweet pepper, poplar, papaya or tomato. In addition, all these transgenic plants express almost exclusively only two major traits, independently or combined: herbicide tolerance (HT) and insect resistance (IR). Tolerance to the herbicide glyphosate is conferred by expression of a bacterial gene encoding a form of the target enzyme, 5-enolpyruvylshikimate-3-phosphate synthase (EPSPS), which is not inhibited by the herbicide. Insect resistance is based on the expression of Bt proteins from *Bacillus thuringiensis*, with natural insecticide activity. Other (minor) traits include virus resistance, or modifications of the quality of the harvested products: anti-allergy, increased content of 'healthy' compounds, delayed fruit softening, phytase production, and non-browning phenotype, among others. A paradigmatic example of these 'minor' traits is the *golden rice* (http://www.goldenrice.org/), a transgenic variety synthesising β-carotene (pro-vitamin A) in the seed endosperm, which should help to reduce the large number of cases of blindness, general susceptibility to disease and early death of many small children in developing countries where rice is the staple food and vitamin A deficiency is widespread. Commercial cultivation of golden rice has recently started in Bangladesh.

Although adoption rates of four main GM crops are close to saturation in the major biotech countries, production can still be increased at the global level by introducing biotech crops in countries which have not yet adopted the technology. Most importantly, there is still a considerable potential for increasing agricultural yields by the large-scale cultivation of additional crops and commercialisation of plants with new traits. In the present scenario of climate change, developing crop varieties with enhanced tolerance to drought, soil salinity and high temperatures will be of special interest, as they can contribute significantly to the needed increase of crop yields and food production.

3.5.5 Transgenic Plants as Biofactories

Apart from generating crop varieties with enhanced agronomic characteristics or nutritional value, GM plants can be engineered to be used as *biofactories* for the production of recombinant proteins of interest for different industries. *Molecular farming* with the so-called third generation of transgenic plants includes, among others: (1) the production of pharmaceuticals, such as drugs, therapeutical proteins, vaccines or antibodies ('bio-pharma' or 'molecular *pharming*'); (2) production of industrial recombinant enzymes, for the detergents, leather tanning, paper, adhesives, paints or other chemical industries; (3) proteins commercialised for experimental research, such as avidin, β-glucuronidase, aprotinin, trypsin, lysozyme or bovine serum albumin; (4) specialised food products: food additives, functional food, dietary supplements; and (5) biodegradable plastic, for example, polyesters of 3-hydroxy acids, such as polyhydroxy butyrate (PHB).

In addition, plant metabolism can be engineered to modify the composition of starch, cell walls or fatty acids in lipids, for the improvement of plants as raw material for industrial processes, in the production of adhesives, paper or lubricants. Furthermore, GM plants can be designed for enhanced activity in soil phytoremediation processes for the decontamination of heavy metals or organic solvents.

3.5.6 Entrepreneurship Opportunities: Developing New GM Crop Varieties

The extremely long, expensive and cumbersome procedures necessary to bring a new transgenic crop variety to the market, once the original transformants have been obtained and characterised, have created a de facto oligopoly of large multinational companies, which are the only ones which can afford the commercialisation of such varieties. These companies include, for example, Bayer CropScience, BASF, Dow AgroSciences, DuPont/Pioneer, Limagrain, Monsanto and Syngenta; there is an ongoing process of merging between some of them, for example, Monsanto and Bayer, to create even bigger and more powerful corporations. A small start-up cannot undertake the whole process of generating and marketing a new GM crop variety; in any case, with the present regulatory framework, it would be practically impossible to get authorisation for the commercial cultivation of the obtained variety in Europe. However, that small company can carry out the initial steps of the process, maybe in collaboration with groups from universities or research institutes: design of the transgenic plant to express a trait of commercial interest based on published information or initial experimental results; isolation of the required gene(s), if necessary; genetic transformation and characterisation of the generated transgenic lines; and legal protection of the process and product. In this way, the company can collect a portfolio of products and patents that can be sold or licensed, if one of the big companies considers them of interest for global marketing. This strategy was followed by Plant Genetic Systems, founded in 1982 by Marc Van Montagu and Jeff Schell (who developed the *Agrobacterium*-mediated transformation system), as a spin-off of the University of Ghent. Plant Genetic Systems was bought in 1996 by AgrEvo, later transformed into Aventis Crop Science, and finally acquired by Bayer to form Bayer CropScience. Similarly, CropDesign, also a spin-off of the University of Ghent, in this case together with the Flanders Institute of Biotechnology (VIB), was founded in 1998 and acquired by BASF Plant Science in 2006.

3.5.7 Entrepreneurship Opportunities: GM Plants Used as Biofactories

The possible use of transgenic plants for the production of recombinant proteins or chemical compounds of commercial interest has not encountered such a strong social rejection, not even in Europe, as the cultivation of biotech crops for food and feed use. This is probably due, in part, to the fact that '*molecular farming*' is carried out under containment conditions in greenhouses, not in open fields—thus avoiding the alleged (although untrue) adverse effects of GM crops on the environment. Also, some of the products of the plant biofactories have diagnostic or therapeutic applications in human medicine, so they do not arouse strong opposition (at least publicly) by radical environmental organisations and green parties.

This application of transgenic plants may provide excellent business opportunities for start-ups if the entrepreneur can identify a product with commercial possibilities and a comprehensive market analysis is performed before starting the company's activity. Recombinant proteins with pharmaceutical activity represent the products with highest added value, but the long and expensive procedures required to get a new drug in the market should be taken into account, as they can make the company unviable. That happened, for example, with the Canadian SemBioSysGenetics Inc., which went bankrupt when they already had two products in clinical trials, human insulin and recombinant Apo AI (the lipoprotein associated with HDL, 'good cholesterol') produced in transgenic safflower. Other companies have been more successful, as they chose to produce and market other products, not subjected to such strong regulations. One of them is ORF Genetics, from Iceland (https://orfgenetics.com/), which sells endotoxin-free growth factors and cytokines for stem cell technology and medical research, and human epidermal growth factor for cosmetic applications (anti-ageing skin serum), all synthesised in seeds of transgenic barley plants. Other examples are given in Sect. 3.9, 'Perspectives/future developments' since this is a business area of growing interest.

3.6 'Omics' Technologies

'Omics' technologies represent modern biotechnological tools, with numerous theoretical and practical applications. They are high-performance techniques that allow the analysis of a considerable volume of data. These technologies are essentially based on the (generally) non-targeted identification of all the genetic products (transcripts, proteins and metabolites) present in a biological sample. Numerous scientific advances are due to omics, such as the identification of signalling molecules associated with cell growth, cell metabolism or cell death. In addition to the theoretical advances achieved by these new technological approaches, they triggered a fundamental change in biomedical research, improving the diagnosis and treatment of diseases, contributing to the development of new drugs (Billeo 2005) and starting the new epoch of personalised medicine (Ibrahim et al. 2016). The 'panomic arsenal of omics' is extremely valuable not only in biomedicine but also in other fields

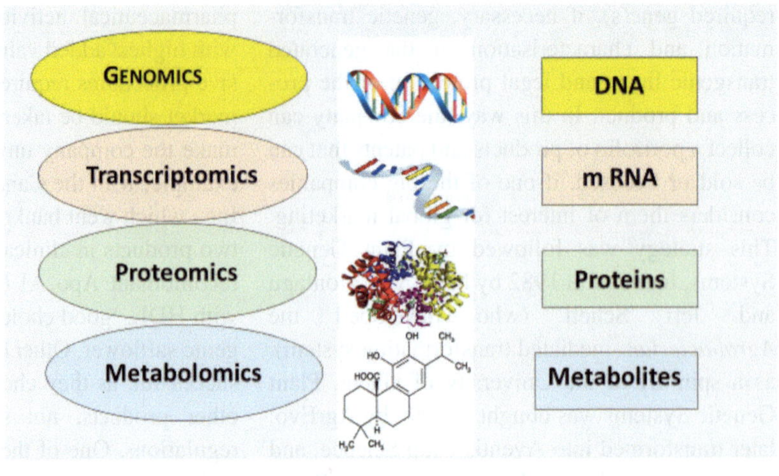

Fig. 3.5 Major 'omics' technologies

such as food sciences and agriculture (Van Emon 2015). The combination of these modern technologies is translated in better quality, taste and nutritional composition of food, an important role in crop protection, better understanding of insect resistance to pesticides, of plant resistance to herbicides, or plant breeding (Van Emon 2015).

Omics represent a constellation of experimental disciplines, such as genomics, proteomics, transcriptomics, metabolomics, regulomics, spliceomics, lipidomics, phenomics, ionomics, microbiomics, metagenomics, phenomics, pharmacogenomics, toxicogenomics and introgressionomics, among others (Fig. 3.5).

3.6.1 Genomics

Genomics is the large-scale study of the genome, the complete set of an organism DNA, including all its genes and non-coding regions (Fig. 3.5). From the first sequenced plant genome, that of *Arabidopsis thaliana* (The Arabidopsis Genome Initiative 2000), followed by rice (International Rice Genome Sequencing Project 2005), up to date the genomes of about 350 land plant species have been completed. The completion of the genome sequence of many crops has profound implications for agriculture, shedding light on the adaptation by natural and artificial selection to environmental constraints. The high-performance whole genome sequencing machines permit fast genomic analysis at low costs. The knowledge of plant genome allows the understanding of the genetic and molecular basis of all biological processes in plants that are relevant to the species. By using genomics, thousands of genes can be quickly analysed in parallel, and complex crop traits, such as yield and yield stability may be unravelled. *Agrigenomics* or agriculture genomics is a new approach in plant biotechnology, with a high utility in molecular breeding and marker-assisted selection. The knowledge of genomics is a valuable tool for crop improvement. It contributes to obtaining higher quality germplasm and also to crop protection. The identification of genes that control economically important traits provides the basis for new progress in the genetic improvement of crop species, complementing traditional methods based on assisted crosses. Genomics is also useful for the biopharmaceutical industry and the conservation of biodiversity by identification of the most relevant genome segments in relation to adaptation and evolution (Wang et al. 2017).

3.6.2 Transcriptomics

Transcriptomics is the study of the transcriptome, the sum of all of the RNA transcripts, including both mRNA and non-coding RNA (ncRNA) expression in a cell (Fig. 3.5). The genetic information contained in the genome is expressed through transcription in the intermediary molecules of mRNA, whereas ncRNAs play additional regulatory roles. Transcriptomics give the possibility to study gene expression in different tissues, conditions or time points, explaining how genes are regulated. The first transcriptomic analyses date back to the late 1990s, using Northern blots and quantitative PCR, but due to the rapid evolution of technical methods such as next-generation sequencing (NGS) the functional elements of the genome are better understood. Transcriptomics represents an ideal tool for analysing the relations between the genotype and the phenotype and determines how the pattern of gene expression changes due to abiotic or biotic stresses, enabling the description of metabolites, transcription factors and stress-inducible proteins in stress-tolerant plants. The study of plants' responses to abiotic and biotic stresses is getting special relevance, as agriculture is already affected by global warming in many parts of the world. Transcriptomics allows the identification of genes and pathways associated with responses to exogenous stresses in plants (Pandit et al. 2018). The major limitation of the 'classical' transcriptomics analysis using DNA arrays is that it requires knowing the genomic sequence of the organism under study; therefore initially it could only be performed in model species. Although at present many plant genomes have been sequenced, the improvement and reduction in the price of NGS technology have led to the substitution of RNA hybridisation with the probes in the DNA arrays, by direct massive RNA sequencing. Thus, transcriptomics analyses can now be carried out in all species, even if no genomic data are available.

3.6.3 Proteomics

Proteomics is the study of the proteome, the total set of proteins in a tissue or organism (Fig. 3.5). Proteins are sequences of amino acids assembled according to templates of DNA and RNAs with structural or functional roles in cells. The proteome varies from cell to cell and in time, the majority of proteins suffering post-translational modifications, which produce different functional types of the same structure. For this reason, proteomics is more informative than genomics when dealing with environmental effects, and it is continuously gaining a protagonist role in environmental monitoring and human health risk assessment. The most popular proteomics methods are based on the combination of two-dimensional gel electrophoresis with mass spectrometry (MS). MS-based proteomics is becoming the standard approach for systematic characterisation of post-translational modifications, such as phosphorylation, glycosylation, acetylation or methylation, among others (Ke et al. 2016). The identification of these modifications is relevant in the study of the phytochemicals produced by plants and of the effect of external factors on their composition. Through protein expression profiling, responses to stimuli such as pathogens or insect attack or abiotic stresses can be analysed, and functioning of particular proteins can be elucidated. The highly complex interactions between plants and microorganism, such as symbiosis, can be better understood. Proteomics also contributes to the unravelling of mechanisms of resistance, mode of action and biodegradation of pesticides, being a tool for the development of more effective and safe pesticides. The hundreds of thousands of different proteins in plants contribute to the texture, yield, flavour and nutritional value of food products. Proteomics is also a useful tool for testing food authenticity, food security and safety.

3.6.4 Metabolomics

Metabolomics is one of the younger omics approaches (Bino et al. 2004) and refers to the study of the 'metabolome', a term coined in 1998 to designate the qualitative and quantitative analysis of all small molecules in an organism, with focus on intermediary metabolites, secondary metabolites and signalling molecules (Fig. 3.5). The metabolome is vast, including a huge number of molecules, which have disparate physical properties and are involved in many metabolic pathways and therefore its analysis is extremely complex; most metabolic profilings are usually performed by mass spectrometry (MS) or nuclear magnetic resonance (NMR). The metabolome is extremely dynamic, fluctuating according to environmental and internal conditions. Metabolomics became an important diagnostic tool in medicine, but it is now used also in many other fields. Metabolite changes can be detected in response to variations in environmental conditions such as light, temperature, humidity, soil type and salinity, but also to pest attacks or the use of fertilisers or pesticides; specifically, monitoring of metabolic changes provides a good picture of how abiotic or biotic stresses affect crops. Metabolomics studies may lead to lesser pesticide usage, optimisation of trait development in agricultural products or increasing nutritional quality of food crops.

Food metabolomics evaluates food quality, in terms of composition and authentication, and allows the assessment of how processing and storage are affecting the bioactive compounds in food. The effect of environmental factors (climatic, pollutants) on plants can be monitored by quantifying the changes in their metabolites. Of the around 200,000 known plant metabolites, new and improved medicinal and nutritional products will be discovered by the modern analytical techniques (Bino et al. 2004; Dunn and Ellis 2005; see an example of a metabolomics analysis in Fig. 3.6).

3.6.5 Plant 'Omics' Technologies on the Global Market

The use of genomics, proteomics, transcriptomics and metabolomics is necessary for feeding an ever-increasing human population, living in a world affected by climate change. Application of the arsenal of omics tools may reduce both the time and expense of food production and enhance the nutritional value of food products and could contribute to sustainability, productivity and clean technologies in agriculture. Opportunities of omics applications in agriculture are huge; in fact, agriculture represents a commercial market where every omics technologic advance has been commercialised in the global market. This is justified by the growing consumer demand, and the necessity to mitigate environmental challenges, but also by a more relaxed regulatory environment in agriculture as compared to healthcare. Due to technical advances, omics toolsets are now getting cheaper and can be used without big investments. Next-gen farms are a growing category of companies that use omics tools to farming methods in locations and settings that cannot support traditional farming. Start-ups in data analysis about plant composition and genetics are improving seed research and plant breeding, and start-ups using big data and predictive analytics contribute to increasing efficiency in farming or optimise herbicide and pesticide application, among many other uses. Omics is essential in the better understanding of the molecular mechanisms of insect resistance to pesticides and the plant tolerance to herbicides for better pest management. Not only research analyses but also bioinformatics tools are extremely important, as a massive amount of information is generated. In this section, we have tried to give a general overview of the many business opportunities of omics technologies in plant biotechnology. Some specific applications in plant breeding have been discussed in Sect. 3.2, on molecular markers, and Sect. 3.3, on mutagenesis.

3.7 Organic Agriculture

At first glance, biotechnology and organic agriculture do not appear to match properly. However, there are many biotechnological strategies and tools that can be applied to improve considerably organic agriculture and are acceptable in the frame of its regulation. In the present section, first, we will review a list of biotechnology and breeding applications and their suitability or not to organic agriculture, including controversial situations from the perspective of the organic sector (Table 3.4). Finally, we will offer a range of currently successful examples applied to organic agriculture.

All kinds of genetically modified organisms (GMOs), which currently include (in the EU)

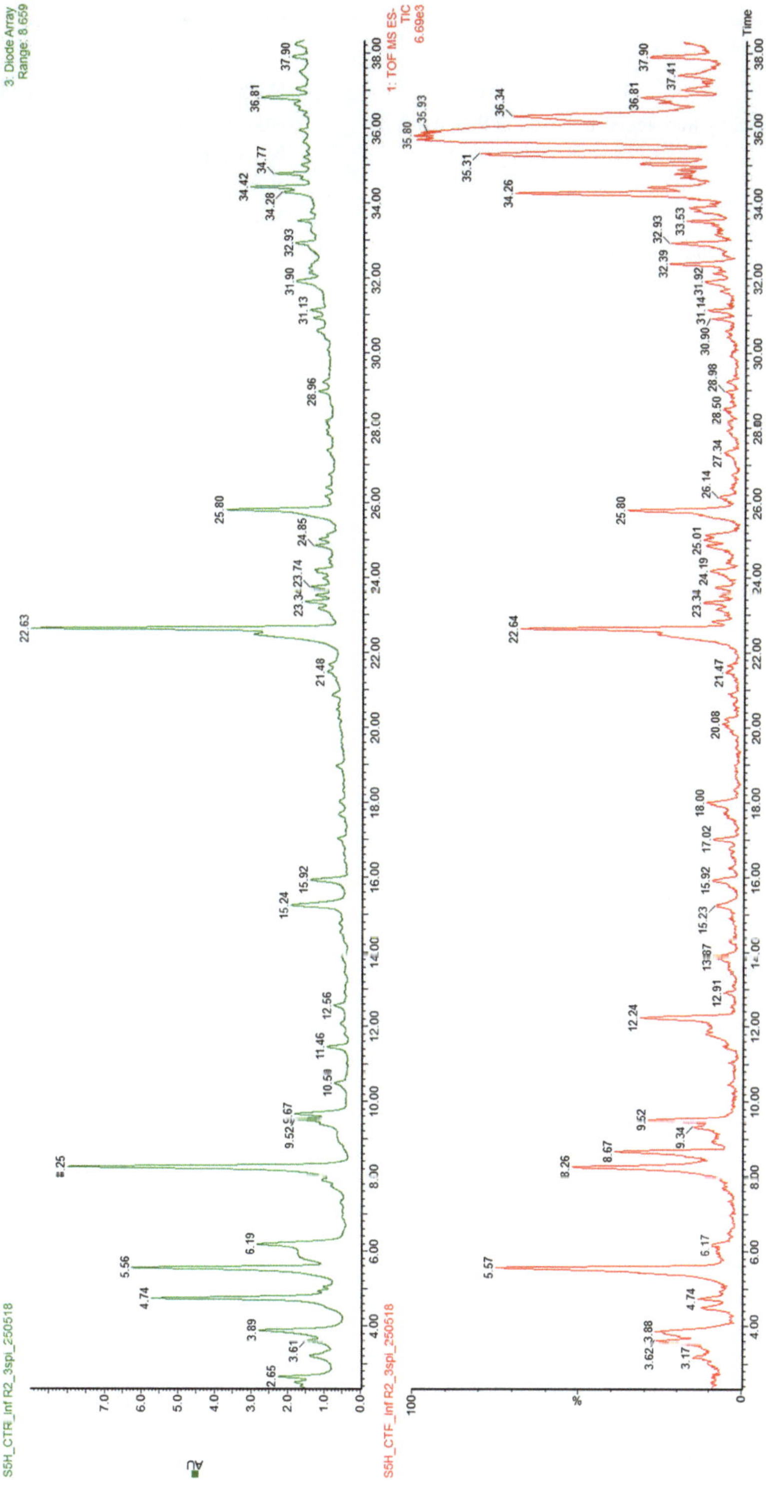

Fig. 3.6 Chromatogram of metabolic extracts in tomato leaves (courtesy of Dr. María Pilar López Gresa, IBMCP; Valencia, Spain)

Table 3.4 Examples of breeding strategies and biotechnological tools and their suitability or acceptance in the organic sector

Breeding technique/strategy	Organic breeding	Organic cultivation
GMOs		
Transgenesis	Not accepted	Not accepted
Pre-breeding boosted by transgenes, removed in final cultivars	Not accepted	Controversial
Gene drives	Not accepted	Not accepted
Cisgenesis	Not accepted	Not accepted
Zinc fingers	Not accepted	Not accepted
Transcription activator-like effect nucleases	Not accepted	Not accepted
Gene silencing by interference RNA	Not accepted	Not accepted
Induced mutations		
Eco-TILLING (study of natural variation)	Accepted	Accepted
Gene edition: CRISPR/Cas	Not accepted	Not accepted
Induced mutation by EMS, radiation, oligonucleotide-directed mutagenesis (ODM)	Not accepted	Controversial
Polyploidisation (chemical mutants)	Controversial	Controversial
In vitro-based techniques		
Protoplast fusion (within species)	Controversial	Controversial
Protoplast fusion (between species)	Controversial	Controversial
Cytoplasm male sterility (CMS) from protoplast fusion	Not accepted	Controversial
Hybrids from protoplast fusion-derived CMS lines	Controversial	Controversial
Double haploids obtained by in vitro culture of anthers, ovaries, microspores, etc. and duplication by colchicine	Controversial	Controversial
Ovary and embryo rescue	Accepted	Accepted
In vitro selection	Under discussion	Accepted
DNA polymorphisms		
Genotyping by sequencing (GBS)	Accepted	Accepted
Marker-assisted selection (MAS)	Accepted	Accepted
Experimental populations (MAGIC, RILs, NILs)	Accepted	Accepted
Metabolomics	Accepted	Accepted
Conventional techniques		
CMS from nature or wide crosses	Accepted	Accepted
Hybrids, pre-breeding and cultivars	Accepted	Accepted
Inbreeding lines, cultivars	Accepted	Accepted
Bridge crossings	Accepted	Accepted
Vegetative (clone) propagation	Accepted	Accepted
Single-seed descent, backcrossings, etc.	Accepted	Accepted
Mixed populations: composite cross pops., dynamic pops	Accepted	Accepted

Adapted from 2017 IFOAM Position Paper on Compatibility of breeding techniques in organic systems

CRISPR/Cas9-derived plants (see Sect. 3.5. on transgenic plants), are not acceptable in organic breeding and organic cultivation. This is mainly due to genome alteration and violation of cell integrity as a functional unit. Only if transgenes were used in pre-breeding to boost the breeding process (e.g. early flowering genes), and then removed from the final cultivar, could these technologies be considered. Nevertheless, due to the bad reputation of GMOs in the organic sector, this should also be discarded.

Regarding induced mutations, as mentioned above CRISPR/Cas9 is not acceptable due to its recent classification as GMO, while other mutations and mutation-derived cultivars are controversial. Thus, Eco-TILLING could be acceptable in research to study natural variation. By contrast, despite chemically or radiation-induced mutations are tolerated in organic cultivation, they affect the genome and cell integrity and are therefore not fully compatible with the principles of organic farming; these mutagenesis methods are currently controversial and under discussion in the organic sector. Also, polyploids derived from synthetic colchicine treatments (as an increasing ploidy agent) are in the same controversial situation, and the use of colchicine from natural plant sources could be advisable as an alternative.

In vitro-based techniques and applications are also controversial and under discussions due to their biotechnological perception in the organic sector. Nevertheless, several techniques are fully compatible with organic principles. Thus, the use of in vitro rescue of ovaries or embryos, applied to prevent potentially abortive interspecific hybrids or seed dormancy, is accepted in organic breeding and organic cultivation. The suitability in organic breeding of cell or tissue cultures exposed to in vitro selection under stress agents is under discussion as these individuals are not exposed to real soil or environment conditions in part of the breeding/selection process, although the populations derived from these in vitro cultures are acceptable for organic cultivation. Finally, protoplast/cell forced fusion techniques and derived materials (including cytoplasm male sterility lines and potential F1 hybrids), as well as double haploids from in vitro regeneration of haploid cells and further duplication with synthetic colchicine, are very controversial. Thus, even though some materials may have been cultivated in organic systems, the organic sector and markets strongly disagree as they are against principles of organic farming.

The techniques related to the analysis of DNA polymorphisms are fully compatible with organic breeding and production systems. In fact, they are used profusely to boost breeding programmes in organic agriculture. For instance, genotyping by sequencing (*GBS*) can be used to establish phylogeny studies in ancient landraces adapted to organic cultivation, DNA fingerprinting, the genetic distance among breeding lines to plan crossings searching heterosis, etc. Also, marker-assisted selection of traits of interest is used commonly to facilitate the breeding process or, on the contrary, to search DNA polymorphisms linked to traits of interest in response to organic conditions in highly segregating populations or introgression populations—e.g. multi-parent advanced generation inter-cross (*MAGIC*), recombinant inbred lines (*RILs*) and near isogenic lines (*NILs*). Finally, metabolomic studies are also applied in populations grown under organic conditions to search for specifically produced metabolites or higher accumulation of metabolites.

Most conventional/traditional breeding techniques are fully compatible with the organic sector and principles. Hybridisation, in both pre-breeding populations and final cultivars, inbreeding lines and cultivars, bridge crossings (i.e. use of a bridge species to facilitate hybridisation between two other distant species), clonal propagation (e.g. fruit trees cultivars, artichoke, potato), mixed populations, single-seed descent strategy, backcrossings, etc., are commonly used in organic breeding and production. Only in the case of CMS lines based on naturally originated mutations, we can find a controversial situation. Thus, despite these natural CMS lines and derived hybrids are on the whole compatible with organic principles, the bad reputation of CMS derived from forced cell fusion makes the organic markets and consumers not very receptive to any product labelled as CMS (Table 3.4).

3.7.1 Entrepreneurship Opportunities: Services and Products in the Organic Sector

Biotechnology tools have several opportunities and gaps to be filled in the organic breeding and production systems. In fact, there are several successful examples: from applications of DNA polymorphisms to in vitro culture, and also exploiting the root and soil interaction of organic systems.

DNA polymorphism analyses can be very helpful in (1) the study of phylogeny and relationships among landraces, heirlooms, ecotypes adapted to organic cultivation; (2) the study of genetic diversity within population varieties, heirlooms or ecotypes; (3) to develop DNA fingerprints in high added value cultivars, protected designations of origin (PDOs) and protected geographical indications (PGIs), among others; and (4) monitoring of genetic diversity evolution in mixed populations as a result of adaptation to different organic environments and/or in response to climatic change/resilience. Also, MAS is extremely helpful for breeders during the breeding process for organic in order to identify which individuals are carrying genes of interest at very early stages, such as resistance to diseases, quality traits (e.g. capsaicinoid content in peppers, fruit colour). Also, next-generation sequencing (NGS) of DNA can be applied in studies of DNA sequences corresponding to the microbiota present in organic soils in interaction with the rhizosphere of plant crops, differences against conventional production systems, etc. Thus, soil metagenomics studies have been revealed to be very useful to explore the complexity of soil microbiology, particularly in organic soils. In addition, DNA analyses can be applied to detect plant material frauds: not accepted as the presence of GMO plants, or controversial cultivars or breeding materials such as of double haploids (0% heterozygosis) or CMS from cell fusion.

In vitro culture also offers opportunities like large-scale multiplication of clone cultivars and breeding populations for organic breeding and nurseries in vegetables, fruit trees and ornamentals (e.g. artichokes, asparagus, allium, tulips, roses, citrus, apples, prunus), in particular if in vitro media are formulated with natural components (e.g. coconut water, agar). Also, in vitro rescue of interspecific hybrids offers the opportunity to facilitate the process of organic breeding by introgressing genes between distant species.

In addition, in comparison to intensive conventional systems, organic farming is a kind of low input agriculture, which involves the cultivation of plants under low or nil fertilisation using synthetic fertilisers. Thus, plants must adapt to low availability of mineral nutrients, with organic matter being the main source of mineralisation (apart from the use of Leguminosae species able to fix atmospheric nitrogen in the roots). For that reason, roots must provide an efficient response in order to mobilise nutrients, facilitate the exchange of nutrients and/or to develop a favourable environment in the rhizosphere. In this regard, the use and/or development of useful mycorrhization in plants in the nursery or in the seeds of vegetables may increase the adaptation and performance of plants under organic farming conditions.

There are many examples of successful companies focused on biotechnologies with applications to organic agriculture. Thus, PROVIVI (https://provivi.com/) is a recently founded company, which develops new environmentally friendly biopesticides, mainly based in natural pheromones. Its biopesticides are very selective, targeting problematic pests through the disruption of their mating cycles while preserving beneficial species. In the same market line, VESTARON (http://vestaron.com/) also develops bioinsecticides from natural peptides, very effective and with no known resistances to date. Particularly successful are their products against thrips in greenhouse crops like ornamentals and vegetables, with similar or higher efficiency to that of conventional, not organic-friendly, pesticides.

In the frame of microbial applications to organic agriculture, several companies like NewLeaf Symbiotics (https://newleafsym.com/), INDIGO (https://www.indigoag.com/) and AgBiome (http://agbiome.com/) have developed successful products. NewLeaf includes products based on symbiotic bacteria, i.e. pink-pigmented facultative methylotrophs (PPFMs aka M-trophs). PPFMs in symbiosis to plants consume methanol, a waste compound from plant metabolism, and in return secrete key nutrients essential for plant growth. INDIGO offers microbial seed treatments for a range of crop species like maize, cotton, rice, soybean and wheat. The company, based in massive algorithm studies from plants and soils worldwide, including a plethora of growing conditions, has identified and predicted which strains of microbes work better in certain conditions. With this information, INDIGO predicts and prepares seed treatments, from only one microbe strain to cocktails of microbes, with the best-expected performance in terms of boosting plant traits, like higher efficiency in the use of water or nutrients and/or lower pesticide uses. AgBiome is using new knowledge of plant-associated microbiome to create new products against pests (insects and nematodes) and diseases. They have created a unique, biological fungicide, a highly effective microbe-based product very efficient against a plethora of fungal diseases and excellent non-refrigerated shelf life.

Finally, WISErg (https://wiserg.com/) is changing the concepts of food waste management by creating organic fertilisers from food waste. Food scraps are harvested from a range of grocery stores, restaurants, bars, cafeterias and other sources and stored in the Harvester. Based on innovative technology, scraps are prevented from putrefying, avoiding associated pests, greenhouse gas emissions and odours. After the nutrient recovery process is completed, the material is transported to facilities where it is processed into liquid fertiliser approved for use on certified organic crops.

3.8 'Phytochemistry': Plants as a Source of Compounds of Commercial Interest

Plants are a rich source of active compounds—phytochemicals—economically important because of their applications in different industries. They are secondary metabolites, with a wide array of biological functions, which the plants synthesise through diverse biochemical pathways. Many have protective roles in the plants, acting as antioxidants, free-radical scavengers or direct UV light screens in the mechanisms of response to abiotic stresses that generate oxidative stress in the plants, or as antiproliferative agents defending the plant against pathogenic microorganisms (bacteria, fungi, viruses and viroids) or nematodes (Kennedy and Wightman 2011). Other roles include feeding deterrence, for which many phytochemicals are bitter and/or toxic to potential herbivores, and this toxicity often extends to direct interactions with the herbivore's central and peripheral nervous systems (Rattan 2010). Secondary metabolites are also involved in the allelopathic interactions between plants (Lotina-Hensen et al. 2006) or the establishment of symbiotic relationships, by attracting with colours and scents pollinators and animals responsible for fruit and seed dispersal; they can as well provide indirect defences for the plants by attracting natural enemies of their herbivorous attackers.

Phytochemical compounds are very diverse chemically, and the distribution of specific types of secondary metabolites is often restricted to taxonomically related species (Gandhi et al. 2015). They can be subdivided into a number of distinct groups by their chemical structure and synthetic pathways. The largest and most prevalent of phytochemical groups are the alkaloids, terpenes and phenolic compounds.

The great diversity of plant secondary metabolites with different chemical structures is very interesting for different industries as they have shown to possess pharmacological activities and are useful as a source of pharmaceuticals, nutraceuticals, flavouring agents, food protectors, fragrances, cosmetics, insecticides, fungicides and other plant protection products, dyes and drugs.

However, when working with secondary metabolites, there are some difficulties to face, as the limited availability of raw material and also that sometimes its exploitation could cause an imbalance in ecosystems. One of the key objectives of plant biotechnology is to develop eco-friendly ways for the large-scale production of pharmacologically active and other high added-value compounds. Moreover, biotechnology could be used to generate novel chemical compounds, with enhanced or new bioactivities, through activation of silent or cryptic metabolic clusters (Gandhi et al. 2015). Powerful molecular tools have been used to exploit microbial biochemistry to produce novel compounds (Prather and Martin 2008). Biotechnological interventions have also played a major role in the improvement of crop yields and quality. Despite such progress in plant molecular biology, only limited applications of biotechnology have been seen in medicinal and aromatic plants (MAPs). In MAPs, generally there is a paucity of available molecular information, and standardised protocols for transgenesis and marker-assisted selection are also not readily available. However, for MAPs, the use of hairy root cultures and bioreactors for the production of secondary metabolites has become popular (Srivastava and Srivastava 2007).

Some examples of biotechnological strategies for obtaining plant secondary metabolites are summarised in Fig. 3.7.

Plant secondary metabolites are usually produced in low quantities; often they get accumulated in specific plant organs, at distinct developmental stages, in a particular agro-geo-climatic zone, or only in response to some external signal, such as exposure to a specific stress condition (Chemler and Koffas 2008). Many commercially interesting metabolites like taxol, artemisinin or forskolin, among many others, are very difficult to synthesise chemically, and the process is economically unviable. Plant tissue culture at an industrial scale presents itself as a commercially viable alternative for production of phytochemicals, considering the increasing demand for metabolites of interest, the long time required for certain slow-growing plants, the continuously reducing land availability for large-scale cultivation of plants and the destruction of wild populations of medicinal plants through exploitation (Gandhi et al. 2015).

Only a few biosynthetic pathways, such as those involved in the production of cinnamic acid derivatives, anthraquinones, berberines, shikonins or anthocynanins, are efficiently expressed in suspension cultures (Chiang and Abdullah 2007). Other compounds, such as morphinan alkaloids, tropane alkaloids (e.g. hyoscyamine and scopolamine), quinoline alkaloids and dimeric monoterpene indole alkaloids (e.g. vinblastine and vincristine), among others, are expressed only in traces in suspension cultures (Berlin 1997). Large-scale efforts to increase their expression through medium engineering and use of elicitors have not yielded results that can lead to commercial exploitation of tissue cultures for production of these compounds (Gandhi et al. 2015).

Organ culture has been explored for the production of those phytochemicals that cannot be obtained at profitable levels in suspension cultures. This is the case of morphinan alkaloids of *Papaver somniferum* (Papaveraceae), dimeric indole alkaloid (anhydrovinblastine, a direct precursor of vinblastine and vincristine) of *Catharanthus roseus* (Apocynaceae) and

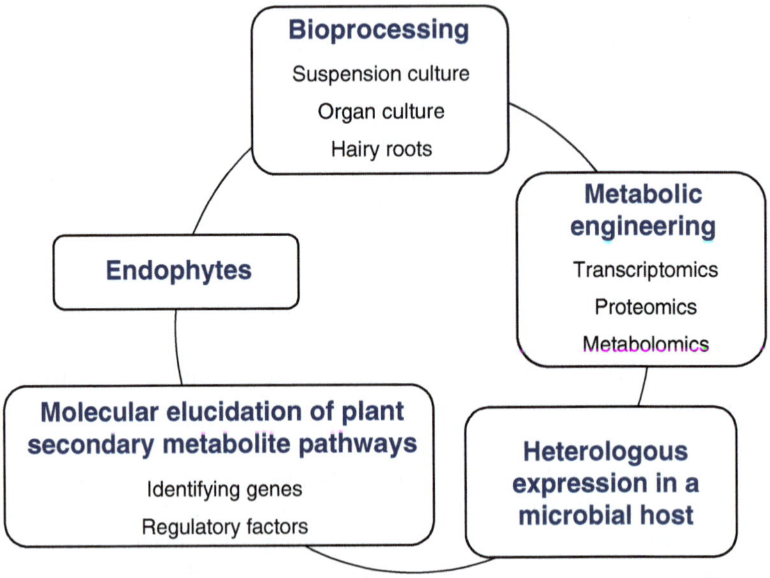

Fig. 3.7 Applications of biotechnology for obtaining plant secondary metabolites (adapted from Gandhi et al. 2015)

sesquiterpene lactone (artemisinin) of *Artemisia annua* (Asteraceae), which are better produced in shoot cultures (Endo et al. 1987; Liu et al. 2003; Tisserat and Berhow 2009). Similarly, root cultures produce higher amounts of tropane alkaloids, such as hyoscyamine and scopolamine, as compared to suspension cultures (Berlin 1997).

Plant cell tissue and organ culture (PCTOC) techniques permit the manipulation of growth and production of the phytoconstituents in the microenvironment of in vitro cultures (Gaosheng and Jingming 2012). This can be achieved through alteration of various growth parameters and factors, to optimise production independently of geographical or seasonal variations. The application in mass propagation has made in vitro culture an effective technique for large-scale production, to overcome the effect of interfering compounds that could affect the productivity of the molecules in field-grown plants, and providing an efficient year-round system without seasonal constraints. The unique opportunity to profile phytochemicals produced by plants in the microenvironment of culture vessels and the simplicity of extraction from in vitro material make it easier for commercial application. However, impediments in the plant cell culture systems, including physiological heterogeneity, slow growth, genetic instability or product secretion, which could lead to low levels of the target molecules, need to be addressed for efficient production. The requirement of aseptic conditions, shear mixing and sensitivity, wall adhesion and light requirements are operational challenges that need to be overcome in scale-up production (Isah et al. 2018).

The most studied class of plant secondary metabolites using PCTOC production systems are alkaloids, including about 12,000 different compounds; their production is, in most of the reported cases, restricted to certain plant families (Newman and Cragg 2016). Only a few of these metabolites have reached worldwide commercial production success, for example, paclitaxel, vinca alkaloids, ginsenoside, saponins, protoberberines, scopolamine, echinacea's polysaccharides, many flavonoids, steroids and shikonin. However, restriction of their production to certain developmental stages and low yield from many of the plant species limit production. Exploiting DNA technology has emerged as an alternative that needs to be harnessed to improve production efficiency by engineering in plant cells the biosynthetic pathways of the molecules of interest (Yue et al. 2016).

Hairy root technology has found application in the scale-up production of many pharmaceutical plant secondary metabolites, such as ginsenoside and some alkaloids, and is increasingly getting considerable attention. Hairy root cultures show in many cases a rapid and plagiotropic growth, with branching on phytohormone-free medium. It has been extensively used to study root nodules formation (Hu and Du 2006) and has found application in different species for the production of pharmaceutical plant secondary metabolites. This can be attributed to their ability to produce the compounds over successive generations without loss of biosynthetic capacity or genetic stability (Giri and Narasu 2000). Other advantages of this technology are the genotypic and biochemical stability, the ability to grow on PGR-free cultivation media, cytodifferentiation, fast growth and low doubling time, the ease at which established hairy root cultures could be maintained and the ability to biosynthesise a range of compounds (Abraham and Thomas 2017). Transformed hairy root cultures can be developed by inoculating plant cell, tissue or organ cultures with *Agrobacterium rhizogenes* strains (Palazon et al. 1997).

Among the metabolites with greatest commercial interest are the taxoids, which are secondary metabolites synthesised by *Taxus* spp. and found in the foliage and bark of these trees. The main pharmacological taxoid is Taxol, a polyoxygenated diterpene alkaloid approved by the FDA for use in the treatment of breast, ovarian and lung cancer, and of Kaposi's sarcoma—related to HIV. Due to the interest of this drug, several attempts have been made to find new sources to obtain Taxol, unsuccessfully. Only recently, the production of Taxol has been achieved by the submerged fermentation of one of the native strains of endophytic fungi of *Taxodium mucronatum* (Mendoza and Escamilla Silva 2018).

A useful approach to elucidate plant secondary metabolite pathways(s) is to clone and express the putative gene in a microbe (*Escherichia coli* or *Saccharomyces cerevisiae*) and determine its biochemical activity on pure substrates to assign its role in the plant pathway. Microorganisms have thus immensely contributed to the biotechnology of plant secondary metabolism by providing a model for its elucidation and as biosynthetic factories for the production of phytochemicals (Gandhi et al. 2015).

Biotechnology offers several choices through which secondary metabolism in medicinal plants can be altered in innovative ways to overproduce phytochemicals of interest, to reduce the content of toxic compounds or even to produce novel chemicals. Detailed investigation of chromatin organisation and microRNAs affecting biosynthesis of secondary metabolites as well as exploring cryptic biosynthetic clusters and synthetic biology options may provide additional ways to harness this resource (Gandhi et al. 2015).

3.9 Perspectives/Future Developments

As it can be seen from the various applications described in this chapter, the potential of plant biotechnology to revolutionise and transform the world is huge; it is one of our most important hopes to overcome some of the major challenges humanity is facing at present, such as food supply, clean-energy production or the creation of environmentally friendly products. Plant biotechnology tools can be applied for generating innovations in many areas and industries; this is a transversal technology that is becoming more important over the years, as it can impact many different fields, ranging from food production and healthcare to electricity production or even insurance.

One of the game changers and something that can truly lead to unleashing the full potential of plant biotechnology is legislation. European policies and regulations, specifically regarding genetic engineering technologies, have long been a source of polemics and social debate and are creating a lot of uncertainty (Ricroch et al. 2016; Davison and Ammann 2017). Even though the regulations for conventional GMOs are quite clear, new genome editing technologies are being developed rapidly, which brings regulatory uncertainty as legislators cannot catch up with the development pace, and there is a big political and social component on how these technologies should be regulated (Braun and Dabrock 2017; Ishii 2018). The recent decision of the European Court of Justice considering the products of genome edition (e.g. using the CRISPR/Cas9 system) as GMOs, against the general scientific opinion, will add to the confusion. How can a technology that leaves no trace be regulated? Indeed, it is impossible to demonstrate, or to disprove, that a plant variety with a specific change in its genome has been generated by genome editing and is not the result of a spontaneous or induced mutation, in which case it will not be legally considered as a GMO.

The European legislation is obviously limiting the development of commercial applications in the field of plant biotechnology, and it is undermining the European agriculture sector in favour of other regions where regulation is more permissive. Leaving aside this fact, there is a need of establishing a clear regulatory framework with a long-term vision, to give security to entrepreneurs, investors and stakeholders who are willing to create companies based on plant biotechnologies. Plant biotechnology-based start-ups usually require a large amount of money to bring a new product or service to the market, and investors, who bring in this capital, need stable and clear regulations since uncertainty means risk and risk is what all investors want to minimise (Blind 2012; McKinsey 2015). Therefore, entrepreneurs must be aware of regulatory frameworks and how they are being developed, as this is one of the main drivers that will shape the future of plant biotechnology.

Apart from legislation, there is a fact that entrepreneurs must take into consideration when starting their own plant biotechnology business: there are more opportunities of success when different people, with different backgrounds, knowledge and experience, work together; this is the way the most successful companies have been created and the most amazing innovations have been developed. Therefore, the rest of this section will be focused on how the combination of plant biotechnology tools with other technologies and areas of knowledge has the potential to disrupt many industries, with some examples of start-ups developing new products and services.

Energy production from plants has not been included explicitly in this chapter since this topic is dealt with in Chap. 4, although focusing on the commercial production of biofuels. Here, we will mention some recent developments in this area, using a completely different approach. Plants have been, and are still, our primary source of energy, as their decay has formed most of the fossil fuels we use today. Fossil fuels need to be replaced by sustainable sources of energy. Whereas many companies are trying to use wind, water or sun to generate energy, specifically electricity, there are a few initiatives trying to produce electricity directly from plants. For example, *Plant-e* (www.plant-e.com) and *Bioo* (www.biootech.com) are two companies using the decomposition of organic material released by plants to produce electricity. These two companies are using, indirectly, the result of photosynthesis to harvest energy that is not used by the plant. Plant photosynthesis is the most efficient way to produce energy from sunlight; as the result of millions of years of evolution, it has reached nearly 100% of quantum efficiency, meaning that through photosynthesis plants can convert almost every photon of sunlight that they capture into an electron. This efficiency is, by far, way higher than current solar panels, which generally operate at the 15%–17% range, reaching a maximum of 22% (Energysage 2018). The gap between the artificial solar panels and photosynthesis is enormous, which means that there is an excellent opportunity for entrepreneurs with knowledge in plant biotechnology, but also in nanotechnology, photonics and other fields, to develop innovations that could fill this gap (even only partly) and change the whole energy industry. *Plant-e* and *Bioo* are examples of what driven entrepreneurs in the field of plant biotechnology can achieve.

Even if this might not seem obvious, the clothing, or fashion, industry is one of the most promising areas for plant biotechnology. This industry is huge, valued at around 3×10^{12} US $ (https://fashionunited.com/global-fashion-industry-statistics) and with expectations to grow between 3.5 and 4.5% in 2019 (McKinsey 2018). One of the main challenges the industry is facing is the use of natural raw materials. Synthetic fibres have been largely used in the industry for many years, but now awareness is rising regarding the problems posed by these materials because of the release of microplastics, which end up in the oceans and enter the food chain. Microplastics have become a big environmental issue, and synthetic textiles are the main source of these particles, accounting for around 35% of the total (Boucher and Friot 2017).

On the other hand, the use of natural fibres also carries some problems as the farming of these fibres is chemical, water and land extensive; for example, 20,000 L of water are required to produce 1 kg of cotton, the quantity needed to produce one t-shirt and a pair of jeans (https://www.worldwildlife.org/industries/cotton). Moreover, processing the fibres to manufacture the desired textile is a process that involves the use of hazardous and contaminating chemicals such as different dyes. Plant biotechnology will play a central role in the race to find alternative fabrics and environmentally friendly processing methods, a race which has already started. This will include the development of less resource-demanding plants, the use of genetic engineering to obtain modified natural fibres with properties that nowadays can only be achieved by using synthetics or even to increase the yield of natural dyes production to meet the current demand of the industry. Biotechnology tools are already being used to deal with these problems, and there are companies such as *Pili Inc.* (www.pili.bio), which is using microorganisms as factories to produce dyes, or *Bolt Threads Inc.* (www.boltthreads.com), which is producing new fabrics in yeast. Similar approaches can be used to produce the same products in plants, and entrepreneurs who successfully enter the fashion industry will have access to capital and resources to drive the new wave of innovation in the clothing industry.

Table 3.5 Examples of companies using plant systems as biofactories

Industry	Company	What are they doing	Website
Pharma/ therapeutics	Leaf Systems	Production of recombinant antibodies, enzymes, virus-like proteins, etc., in plant-based expression systems	www.leafexpressionsystems.co.uk
	Icon Genetics (Denka Group)	Production of recombinant proteins for diagnostic and therapeutic use, in plant expression systems	www.icongenetics.com
	Protalix Biotherapeutics	Production of recombinant therapeutic proteins targeting Gaucher disease (commercial), Fabry disease, cystic fibrosis and immune and inflammatory diseases (on clinical trials)	www.protalix.com
Cosmetics	Vytrus Biotech	Development and production in plants of compounds for the cosmetic and dermatology industry	www.vytrus.com/
	Alternative Plants	Development of plant cell cultures to produce active ingredients for anti-ageing, skin protecting, anti-acne and skin whitening cosmetics	www.alternativeplants.eu
	Naolys	Production of active components for the cosmetic industry in plant cells, using a patented process	www.naolys.com
	LipoTrue	Use of a plant-based transient expression system for the production of recombinant proteins for cosmetic applications	www.lipotrue.com

As mentioned in the previous paragraph and described more extensively in Sect. 3.5, on transgenic plants, an area with tremendous potential is the use of plants as biofactories. In fact, there are already many entrepreneurs who have entered this field, to produce high added-value recombinant proteins in plant systems, with application, for example, in the pharmaceutical and cosmetic industries. A few examples are shown in Table 3.5.

Another field with great importance is, obviously, agriculture. The agriculture industry is under a revolution; the development of the 'Internet of things' (IoT), remote sensing, drones and the widespread use of Internet and communications is changing the process, methods and operations in agriculture. All these technologies are really beneficial as they provide more efficient ways of growing and harvesting plants while reducing the costs. However, plant biotechnology still plays a major role in agriculture, as shown in the different sections of this chapter. Application of molecular markers, mutagenesis, in vitro culture techniques, transgenesis and 'omics' technologies, to specific problems in plant breeding and agricultural production, will continue to provide business opportunities to entrepreneurs with background knowledge in plant biochemistry, molecular biology, genetics and genomics. The challenge will be to identify those 'specific problems' that can be addressed using the biotechnological tools available at present and to develop novel applications of plant biotechnology.

To finish this section, it is important to mention that entrepreneurs with a background in plant biotechnology must keep their eyes open on the developments in other fields of knowledge. There are two major technologies that will, most likely, change the world in the upcoming years: blockchain and artificial intelligence (AI). These two technologies are already being applied in many fields, and plant biotechnology will, for sure, be one of them. For example, AI systems, such as that developed by *Mötorleaf* (www.motorleaf.com), installed in commercial hydroponic greenhouses can predict events and establish the best growing conditions to achieve maximum yields. It is not far that we see commercial AI systems coupled with genomic or metabolomics tools, which will be able to make more precise forecasts to increase crop yields even more. This might sound weird, but these systems will also have an effect on the insurance industry, as more data will be available and more accurate predictions will be possible. Blockchain is still behind AI in terms of development, but there are already many commercial applications, and there are biotechnology companies starting to use this technology. *Macrogen* (www.macrogen.com), a South Korean DNA sequencing and genomics service provider, has announced that it will be using blockchain technology to share genetic information in a more secure way (Macrogen 2018). *Macrogen* is focused on biomedical biotechnology, but probably we will soon see applications of blockchain in the plant biotechnology industry.

The examples above provide a glimpse of what is coming and what can be achieved in the plant biotechnology field. Now it is the time for entrepreneurs to apply their knowledge and ideas to establish start-up companies to deliver novel products and services to customers, which can have a beneficial impact in society, apart from providing economic profits to the company.

References

Abraham J, Thomas TD (2017) Hairy root culture for the production of useful secondary metabolites. In: Malik S (ed) Biotechnology & production of anti-cancer compounds. Springer, New York, pp 201–230

Berlin J (1997) Secondary products from plant cell cultures. In: Rehm H-J, Reed G (eds) Biotechnology: products of secondary metabolism, vol 7. Wiley, New York, pp 593–640

Billeo JA (2005) The agony and ecstasy of "OMIC" technologies in drug development. Curr Mol Med 5 (1):39–52

Bino RJ, Hall RD, Fiehn O, Kopka J, Saito K, Draper J, Nikolau BJ, Mendes P, Roessner-Tunali U, Beale MH, Trethewey RN, Lange BM, Wurtele ES, Sumner LW (2004) Potential of metabolomics as a functional genomics tool. Trends Plant Sci 9:418–425

Blind K (2012) The impact of regulation on innovation. Nesta Working Paper 12/02

Boucher J, Friot D (2017) Primary microplastics in the oceans: a global evaluation of sources. IUCN, Gland

Braun M, Dabrock P (2017) Mind the gaps! Towards an ethical framework for genome editing. EMBO Rep 19 (2):197–200. https://doi.org/10.15252/embr.201745542

Brookes G, Barfoot P (2018) Farm income and production impacts of using GM crop technology 1996–2016. GM Crops Food 9(2):59–89. https://doi.org/10.1080/21645698.2018.1464866

Chemler JA, Koffas MAG (2008) Metabolic engineering for plant natural product biosynthesis in microbes. Curr Opin Biotechnol 19:597–605

Chiang L, Abdullah MA (2007) Enhanced anthraquinones production from adsorbent-treated *Morinda elliptica* cell suspension cultures in production medium strategy. Process Biochem 42:757–763

Davison J, Ammann K (2017) New GMO regulations for old: determining a new future for EU crop biotechnology. GM Crops Food 8:13–34. https://doi.org/10.1080/21645698.2017.1289305

Dunn WB, Ellis DI (2005) Metabolomics: current analytical platforms and methodologies. Trends Anal Chem 24:285–294

Endo T, Goodbody A, Misawa M (1987) Alkaloid production in root and shoot cultures of *Catharanthus roseus*. Planta Med 53:479–482

Energysage (2018) What are the most efficient solar panels on the market? Solar panel efficiency explained. https://news.energysage.com/what-are-the-most-efficient-solar-panels-on-the-market/

Financial Tribune (2015). Micropropagation: from laboratory to market. https://financialtribune.com/node/17287

Gandhi SG, Mahajan V, Bedi YS (2015) Changing trends in biotechnology of secondary metabolism in medicinal and aromatic plants. Planta 241:303–317

Gaosheng H, Jingming J (2012) Production of the useful secondary metabolites through the regulation of biosynthetic pathway in the cell and tissue suspension culture of medicinal plants. In: Gaosheng H, Jingming J (eds) Recent advances in plant *in vitro* culture. Intech, Rijeka

Giri A, Narasu M (2000) Transgenic hairy roots: recent trends and application. Biotechnol Adv 18:1–22

Herrera-Estrella L, Depicker A, Van Montagu M, Schell J (1983) Expression of chimaeric genes transferred into plant cells using a Ti-plasmid-derived vector. Nature 303:209–213

Horsch RB, Fry JE, Hoffmann NL, Eichholtz D, Rogers SG, Fraley RT (1985) A simple and general method for transferring genes into plants. Science 227:1229–1231

Hu ZB, Du M (2006) Hairy root and its application in plant genetic engineering. J Integr Plant Biol 48:121–127

Ibrahim R, Pasic M, Youse GM (2016) Omics for personalized medicine: defining the current we swim in. Exp Rev Mol Diagn 16:791–722

International Rice Genome Sequencing Project (2005) The map-based sequence of the rice genome. Nature 436:793–800

ISAAA (2017) Global status of commercialized biotech/GM crops in 2017: biotech crop adoption surges as economic benefits accumulate in 22 years. ISAAA brief no. 53. ISAAA, Ithaca, NY

Isah T, Umar S, Mujib A, Sharma MP, Rajasekharan PE, Zafar N, Frukh A (2018) Secondary metabolism of pharmaceuticals in the plant in vitro cultures: strategies, approaches, and limitations to achieving higher yield. Plant Cell Tissue Organ Cult 132:239–265

Ishii T (2018) Crop gene-editing: should we bypass or apply existing GMO policy? Trends Plant Sci 23:947–950
Ke M, Shen L, Luo S, Lin L, Yang J, Tian R (2016) Identification, quantification, and site localization of protein posttranslational modifications via mass spectrometry-based proteomics. In: Mirzaei M, Carrasco M (eds) Modern proteomics – sample preparation, analysis and practical applications, Advances in experimental medicine and biology, vol 919. Springer, Berlin, pp 345–381
Kennedy DO, Wightman EL (2011) Herbal extracts and phytochemicals: plant secondary metabolites and the enhancement of human brain function. Adv Nutr 2:32–50
Klein TM, Wolf ED, Wu R, Sanford JC (1987) High-velocity microprojectiles for delivering nucleic acids into living cells. Nature 327:70–73
Li Y-H, Zhou G, Ma J, Jiang W, Jin L-G, Zhang Z, Guo Y (2014) De novo assembly of soybean wild relatives for pan-genome analysis of diversity and agronomic traits. Nat Biotech 32:1045–1052
Liu C-Z, Guo C, Wang Y, Ouyang F (2003) Factors influencing artemisinin production from shoot cultures of *Artemisia annua* L. World J Microbiol Biotechnol 19:535–538
Lotina-Hensen B, King-Díaz B, Aguilar MI, Hernández Terrones MG (2006) Plant secondary metabolites. Targets and mechanisms of allelopathy. In: Reigosa MJ, Pedrol N, González L (eds) Allelopathy: a physiological process with ecological implications. Springer, Dordrecht, pp 229–265
Macrogen (2018) Press release. Macrogen-Bigstur builds big data distribution platform based on block chain. http://www.macrogen.com/ko/news/press_list.php?
Manzur JP, Oliva-Alarcón M, Rodríguez-Burruezo A (2014) *In vitro* germination of immature embryos for accelerating generation advancement in peppers (*Capsicum annuum* L.). Sci Hortic 170:203–210
McCallum CM, Comai L, Greene EA, Henikoff S (2000) Targeted screening for induced mutations. Nat Biotechnol 18(4):455–457
McKinsey (2015) The impact of regulation. https://www.mckinsey.com/business-functions/strategy-and-corporate-finance/our-insights/the-impact-of-regulation
McKinsey (2018) The state of fashion 2019: a year of awakening. https://www.mckinsey.com/industries/retail/our-insights/the-state-of-fashion-2019-a-year-of-awakening
Mendoza N, Escamilla Silva EM (2018) Introduction to phytochemicals: secondary metabolites from plants with active principles for pharmacological importance. In: Asao T, Asaduzzaman M (eds) Phytochemicals. Source of antioxidants and role in disease prevention. IntechOpen, London
Newman DJ, Cragg GM (2016) Natural products as sources of new drugs from 1981 to 2014. J Nat Prod 79:629–661
Nuffield Bioethics Council (2016). http://nuffieldbioethics.org/project/genome-editing/ethical-review-published-september-2016
Palazon J, Pinol MT, Cusido RM, Morales C, Bonfill M (1997) Application of transformed root technology to production of bioactive metabolites. Recent Research Developments in Plant Physiology 1:125–143
Pandit AA, Shah RA, Husaini AM (2018) Transcriptomics: a time-efficient tool with wide applications in crop and animal biotechnology. J Pharmac Phytochem 7:1701–1704
Potrykus I (1991) Gene transfer to plants: assessment of published approaches and results. Annu Rev Plant Physiol Plant Mol Biol 42:205–225
Prather KLJ, Martin CH (2008) De novo biosynthetic pathways: rational design of microbial chemical factories. Curr Opin Biotechnol 19:468–474
Rattan RS (2010) Mechanism of action of insecticidal secondary metabolites of plant origin. Crop Prot 29:913–920
Ricroch AE, Ammann K, Kuntz M (2016) Editing EU legislation to fit plant genome editing. The use of genome editing technologies in plant breeding requires a novel regulatory approach for new plant varieties that involves farmers. EMBO Rep 17:1365–1369. https://doi.org/10.15252/embr.201643099
Shahzad A et al (2017) Historical perspective and basic principles of plant tissue culture. In: Abdin M, Kiran U, Kamaluddin AA (eds) Plant biotechnology: principles and applications. Springer, Singapore
Sonnino A, Dhlamini Z, Santucci FM, Warren P (2009) Socio-economic impacts of non-transgenic biotechnologies in developing countries: the case of plant micropropagation in Africa. Food and Agriculture Organization of the United Nations (FAO), Rome
Srivastava S, Srivastava AK (2007) Hairy root culture for mass production of high-value secondary metabolites. Crit Rev Biotechnol 27:29–43
Suprasanna P, Mirajkar S, Bhagwat S (2015) Induced mutations and crop improvement. In: Bahadur B, Rajam MV, Sahijram L, Krishnamurthy KV (eds) Plant biology and biotechnology, vol 1. Springer, New Delhi, pp 593–617
The Arabidopsis Genome Initiative (2000) Analysis of the genome sequence of the flowering plant *Arabidopsis thaliana*. Nature 2000(408):796–815
Tisserat B, Berhow M (2009) Production of pharmaceuticals from papaver cultivars in vitro. Eng Life Sci 9:190–196
Van Emon JM (2015) The omics revolution in agricultural research. J Agric Food Chem 64(1):36–44
Wang W, Chao XC, Miclaus M, Xu J, Xiong W (2017) The promise of agriculture genomics. Int J Genomics 2017:9743749. https://doi.org/10.1155/2017/974374
Wise Guy Reports (2017) Agricultural biotechnology – global market outlook (2016–2022). Wise Guy Reports, Hadapsar Pune
Yue W, Ming Q-L, Lin B, Rahman K, Zheng C-J, Han T, Qin L-P (2016) Medicinal plant cell suspension cultures: pharmaceutical applications and high-yielding strategies for the desired secondary metabolites. Crit Rev Biotechnol 36:215–232

Creating Products and Services in Environmental Biotechnology

4

Calina Petruta Cornea, Catalina Voaides, Oana Alina Boiu Sicuia, Florentina Matei, and Narcisa Babeanu

Abstract

In the process of solving environmental protection problems biotechnology plays an essential role in providing alternative solutions to reducing pollution. The chapter approaches as a green alternative the phytoremediation of polluted environments, complete with microbial and vermiremediation as a clean-up alternative. Special attention is given to natural plant protection products, known as "biopesticides." Another aspect approached is the finding and development of new plants as a biomass source for energy production, which are objectives for start-ups, and have great business potential.

Keywords

Environmental biotechnology · Phytoremediation · Vermiremediation · Biopesticides · Energy crops · Start-up

4.1 Introduction

It is well known that the world is now experiencing the consequences of the overexploitation of natural resources by man and of technological development. The major concerns are related to the loss of biodiversity, to the extinction of many species with an impact on the good functionality of ecosystems, to the deterioration of the soil, water, and air quality, which have major economic implications and significant repercussions for the well-being of human populations (Leitão 2016). For example, contamination of ecosystems by xenobiotic compounds (organic petroleum hydrocarbons, agrochemicals such as pesticides, herbicides or other compounds, pharmaceutical products, heavy metals) causes serious environmental problems. Various measures have been proposed and, sometimes, adopted by governments, for preventing environmental degradation or for the reduction and cleanup of pollution produced by industrial, agricultural, and household waste and accidental spills, but the results are partially satisfactory. For example, in China, where the economy develops continuously and quickly, urbanization and industrialization are promoted, leading to serious environmental problems, a Bioindustry Development Plan was adopted that proposed that "priorities should be given to treating contamination of water, the atmosphere and organic substances, to treat and repair the impaired ecological system, vigorously develop biologically environmental protection materials and biological products with high performance, accelerate demonstration of the whole set of technological processes and equipment for efficient biological supervision, treatment, repair and waste

All authors have made equal contributions.

C. P. Cornea (✉) · C. Voaides · O. A. Boiu Sicuia · F. Matei · N. Babeanu
Faculty of Biotechnologies, University of Agronomic Sciences and Veterinary Medicine of Bucharest, Bucharest, Romania

F. Matei, D. Zirra (eds.), *Introduction to Biotech Entrepreneurship: From Idea to Business*,
https://doi.org/10.1007/978-3-030-22141-6_4

utilization, and expansion of industrial scale" (Wang et al. 2018). Attempts at remediating contaminated sites have usually used conventional but often costly approaches, such as "pump and treat," excavation and removal, soil vapor extraction, and other chemical treatments, but these methods are time-consuming, invasive, disruptive to natural ecosystems, and not always effective (Elekwachi et al. 2014).

Moreover, the most recent strategies in bioeconomy consider that, in addition to the terrestrial ecosystems, the marine environment is rapidly being polluted by human activities, and environmental biotechnology may provide important knowledge and tools that will help to protect the resource base upon which marine-related economic and social activities depend (Kalogerakis et al. 2015).

In a study performed by Gillespie in 2013, it was estimated by the European Environmental Agency (EEA) that in Europe there are over three million sites where potentially polluting activities have occurred, and more than 8% of them need to be remediated as they are highly contaminated with various pollutants. Moreover, it was estimated that the total number of contaminated sites could be increased by more than 50% by 2025 (Gillespie and Philp 2013).

For these reasons, in the process of solving these problems, environment protection biotechnology plays an essential role in providing alternative solutions to reducing pollution (Khan 2016). In this context, bioremediation has proven to be a safe, effective, low-cost, and environmentally friendly alternative for the sustainable remediation of environments contaminated by various pollutants. Bioremediation uses biological processes and naturally occurring catabolic activity realized by microorganisms (bacteria and fungi), green plants, or some animal organisms to eliminate, attenuate or transform contaminants into less hazardous products, such as carbon dioxide, water, inorganic salts, and biomass (Elekwachi et al. 2014).

Naturally occurring bioremediation and phytoremediation have been used empirically for many years. Processes such as desalinization of agricultural land by phytoextraction are applied in different world regions. More recently, the bioremediation technologies using microorganisms were used for treating the contaminated areas at the site (in situ) or after the removal of contaminated materials and their treatment elsewhere (ex situ) (Fig. 4.1).

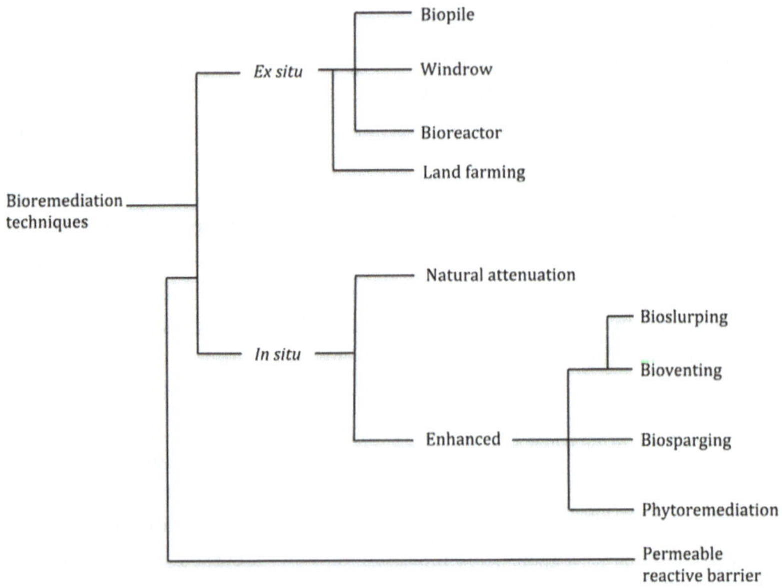

Fig. 4.1 Bioremediation techniques (after Azubuike et al. 2016)

Among the well-developed bioremediation technologies, the following could be mentioned:

- Bioventing is an in situ remediation technology that uses microorganisms to biodegrade organic constituents adsorbed on soils in the unsaturated zone. The technology enhances the activity of soil indigenous bacteria by introducing air/oxygen flow into the soil and, sometimes, by supplemental limited amounts of nutrients, resulting in the stimulation of the microbial biodegradation (http://www.cpeo.org/techtree/ttdescript/bioven.htm).
- Composting is a process that works to speed up the natural decay of organic material by providing the ideal conditions for detritus-eating microorganisms or other soil organisms to develop and act, the end-product of this concentrated decomposition process being nutrient-rich soil) (Ross 2018).
- Bioaugmentation is a process that involves the introduction into a polluted soil of microbial consortia selected from natural ecosystems or developed through successive adaptations under laboratory conditions, to enhance the degradation of toxic compounds (for example, oil spills) (Brown and Ulrich 2014).
- Biostimulation involves the addition in contaminated areas of limited amounts of nutrients to stimulate the growth of indigenous microorganisms and augment their catabolic activity for eliminating polluted compounds, mainly hydrocarbons (Sarkar et al. 2016).
- Phytoremediation is used to solve environmental problems caused by toxic elements by plant activities (Grison et al. 2015).
- Rhizofiltration is the adsorption onto or into plant roots (both terrestrial and aquatic) of various contaminants (heavy metals, radionuclides, etc.) from polluted aqueous sources (effluents discharged from industries and agricultural run-off, acid mine drainage, etc.) that surround the rhizosphere. Rhizofiltration decontaminates polluted water using plants grown in greenhouses in water from the sites instead of soil, acclimatizing the plants to the environment. The plants are then planted on the site of contaminated groundwater where the roots take up the water and contaminants; at the end of the process, when the roots are saturated with the contaminant, the entire plants are harvested (Abdullahi 2015).
- Landfarming is an ex situ waste treatment process that is performed in the upper soil zone or in biotreatment cells, using contaminated soils, sediments or sludges that are transported to the landfarming site, incorporated into the soil sur face, and, periodically, tilled to aerate the mixture. The aim of this procedure is the prevention of groundwater pollution by heavy metal, pesticides or other toxic compounds that could contaminate the upper soil layer (https://www.epa.nsw.gov.au/), etc.

In a global survey performed in 2014 by Elekwachi et al. the application of bioremediation technologies was examined in various regions and countries all over the world. It was shown that the use of low-cost in situ bioremediation technologies (such as monitored natural attenuation) (Table 4.1) are prominent in the developed economies (North America and Europe), whereas more expensive technologies, sometimes ex situ, are used in the developing regions.

The development of industries all over the world, the increased waste production, the increased use of pesticides (rodenticides, fungicides, algicides, acaricides or herbicides) for higher agricultural production, mining, petroleum extraction and its transport, are some of the "actors" involved in the increase in contamination by heavy metals, hydrocarbons, and other pollutants of agricultural land and freshwater sources. The consequences of such contamination lead to the erosion of soils, or even the phytotoxicity of soil systems, the migration of pollutants into soil–water systems, and the pollution of rivers, and could reduce the fertility of soils and contaminate agricultural and food products. The polluted resources are used by humans for food production and ultimately accumulate in the food chain (Yadav et al. 2018).

Table 4.1 Bioremediation technologies (adapted from Elekwachi et al. 2014)

Bioremediation technologies	
In situ	Ex situ
Monitored natural attenuation	Bioreactor technique
Bio-stimulation methods: Addition of fertilizers/nutrients; bioventing and air sparging; groundwater treatment and recirculation	Composting
Bio-augmentation methods: Enrichment cultures from the site; pure cultures specifically for the contaminant; commercial cultures/consortia; phytoremediation	Landfarming Biopile

4.2 Phytoremediation of Polluted Environments: A Green Alternative

Over time, phytoremediation process has had many definitions. In the first place, the term "phytoremediation" comes from associating two other terms—the Greek prefix *phyto* which means plant, and the Latin suffix *remedium*, meaning restoring balance or to correct or to remove something bad. Phytoremediation represents a group of technologies that use natural or genetically modified abilities of (superior) plants to clean up contaminated sites (Adriano et al. 2004; Pulford and Watson 2003; Robinson et al. 2009), cleaning-up being understood to mean the capacity to remove, degrade, detoxify or transform the contaminant from polluted environments—soil, sediments, groundwater, surface water, and/or atmosphere (Ying 2002). Likewise, phytoremediation has been defined as the employment of science and engineering to study problems and provide solutions involving plants and contaminated environments (Conesa et al. 2012).

This technology has been used to remove heavy metals, such as Hg, Cr, Cd, Cu, Ni, Zn, Pb, As, Mo, Se, Pd (Bolan et al. 2011; Shoji et al. 2008; Ayotamuno et al. 2009; Sampanpanish et al. 2006; Andreazza et al. 2013; January et al. 2008; Meeinkuirta et al. 2016), organic contaminants (alkylated polycyclic aromatic hydrocarbons, fungicides, pesticides, polychlorinated biphenyls) (White et al. 2005; Yavari et al. 2015), crude oil (Ayotamuno et al. 2009), some radioactive isotopes such as Cs, U (Schwitzguébel et al. 2002; Yavari et al. 2015).

Phytoremediation strategies utilize trees, shrubs, crop plants, aquatic macrophytes, and/or grasses from different species for treating contaminated air, soil or water. Some of these "green tools" are presented in Table 4.2.

The option to clean the contaminated environment with plants became more attractive to environmental scientists, as an alternative to the classic methods. These traditional technologies—excavation, chemical soil treatment, thermal treatment—proved to be expensive and destructive to the environment (Wenzel et al. 2004).

To use the most efficient plant for a given pollutant, sound studies are required, owing to the different potential of the plant species in different environments to remediate the problem (Andreazza et al. 2013). Consequently, because the implied factors differ from case to case, a unique phytoremediation scheme is difficult to apply. Each phytoremediation project has to be designed for a specific event, requiring certain approaches (Boroş et al. 2016).

Phytoremediation (Fig. 4.2) is based on different *phytotechnologies*, such as: phytovolatilization, phytoextraction/phytoaccumulation, phytodegradation/phytotransformation, phytofiltration, phyto-immobilization or phytostabilization.

Phytoextraction or *phytoaccumulation* represents the process in which the plant uptake translocates and accumulates the contaminants in harvesting plant parts, which can then be used or disposed of (Trapp and Karlson 2001; Rahman and Hasegawa 2011; Conesa et al. 2012). The aim of this process is to remove the polluted element from the site. The following techniques can be included in this category:

Table 4.2 Examples of plant species used in phytoremediation processes

Plant species	Process	References	Contaminant
Brassica juncea	Phytovolatilization and phytoextraction	Moreno et al. (2005), Ko et al. (2008), de Souza et al. (2000) cited by Bolan et al. (2011)	Heavy metals
	Reduction	Bolan et al. (2003) cited by Bolan et al. (2011)	
	Phytoimmobilization	Bolan et al. (2003) cited by Bolan et al. (2011)	
	Phosphate-induced desorption followed by plant uptake	Neunhauserer et al. (2001) cited by Bolan et al. (2011)	
	Chelation followed by uptake	Quartacci et al. (2006), Duqučne et al. (2009) cited by Bolan et al. (2011)	
	Accumulation	Salt et al. (1994), Kumar et al. (1995) cited by Sampanpanish et al. (2006)	
Brassica napus L.	Chelation followed by uptake	Zeremski-Škoric et al. (2010) cited by Bolan et al. (2011)	Heavy metals
	Phytoextraction	Marchiol et al. (2004) cited by Bolan et al. (2011)	
	Phytoremediation	Shams et al. (2009)	
	Accumulation	Kumar et al. (1995) cited by Sampanpanish et al. (2006)	
	Phytoremediation	Schwitzguébel et al. (2002)	^{137}Cs
Medicago sativa	Chelation followed by uptake	López et al. (2005) cited by Bolan et al. (2011)	Heavy metals
	Phytovolatilization	Duckart et al. (1992) cited by Bolan et al. (2011)	
	Phytoremediation	Xu et al. (2010) cited by Yavari et al. (2015)	Polychlorinated biphenyls
Pome fruit trees	Phosphate-induced desorption followed by plant uptake	Peryea (1991) cited by Bolan et al. (2011)	Heavy metals
Echinochloa crus-galli	Root exudates-enhanced phytoextraction	Kim et al. (2010) cited by Bolan et al. (2011)	Heavy metals
Raphanus sativus	Phytoextraction	Marchiol et al. (2004) cited by Bolan et al. (2011)	Heavy metals
Sedum alfredii	Chelation followed by uptake	Liu et al. (2008) cited by Bolan et al. (2011)	Heavy metals
Brassica rapa	Chelation followed by uptake	Meers et al. (2005) cited by Bolan et al. (2011)	Heavy metals
Cannabis sativa	Chelation followed by uptake	Meers et al. (2005) cited by Bolan et al. (2011)	Heavy metals

(continued)

Table 4.2 (continued)

Plant species	Process	References	Contaminant
Helianthus annuus	Chelation followed by uptake	Meers et al. (2005) cited by Bolan et al. (2011)	Heavy metals
	Accumulation	Zavoda et al. (2001) cited by Sampanpanish et al. (2006)	
	Hyperaccumulation	January et al. (2008)	Heavy metals
	Phytoremediation	Lotfy and Mostafa (2013) cited by Yavari et al. (2015)	Co
Zea mays L.	Improved metal uptake by plant growth regulators and EDTA	Hadi et al. (2010) cited by Bolan et al. (2011)	Heavy metals
	Phytoextraction	Murakami and Ae (2009) cited by Bolan et al. (2011)	Heavy metals
	Phytoremediation	Shams et al. (2009)	
	Phytoremediation	Ibrahim et al. (2013) cited by Yavari et al. (2015)	Atrazine (pesticide)
Oryza sativa L.	Phytoextraction	Murakami and Ae (2009) cited by Bolan et al. (2011)	Heavy metals
Glycine max [L.] Merr.	Phytoextraction	Murakami and Ae (2009) cited by Bolan et al. (2011)	Heavy metals
Solanum nigrum L.	Improved plant growth and Cd uptake by fungi and citric acid	Gao et al. (2010) cited by Bolan et al. (2011)	Heavy metals
Lolium perenne	Chelation followed by uptake	Duquc˘ne et al. (2009) cited by Bolan et al. (2011)	Heavy metals
Transgenic tobacco	Phytovolatilization	He et al. (2001) cited by Bolan et al. (2011)	Heavy metals
Lycopersicon esculentum	Phytovolatilization	Duckart et al. (1992) cited by Bolan et al. (2011)	Heavy metals
Festuca arundinacea	Phytovolatilization	Duckart et al. (1992) cited by Bolan et al. (2011)	Heavy metals
	Phytoremediation	Huang et al. (2005)	
Pteris vittata	Phytoremediation	Shoji et al. (2008)	Reduce As(V) to As(III)
Urtica dioica L.	Phytoremediation	Shams et al. (2009)	
Vetiveria zizanioides	Phytoremediation	Xia 2004 cited by Ayotamuno et al. (2009)	Heavy metals
Pennisetum purpureum	Phytoremediation		Crude oil
Amaranthus viridis	Phytoremediation	Sampanpanish et al. (2006)	Cr
Brachiaria decumbens	Phytoextraction and phytostabilization	Andreazza et al. (2013)	Cu
Kochia scoparia	Phytoremediation	Schwitzguébel et al. (2002)	^{137}Cs
Juniperus monosperma	Phytoremediation	Ramaswami (2001) cited by Schwitzguébel et al. (2002)	U
Eucalyptus camaldulensis	Phytostabilization	Meeinkuirta et al. (2016)	Cd
Lolium arundinaceum	Phytoremediation	White et al. (2005)	Alkylated polycyclic aromatic hydrocarbons
Cynodon dactylon	phytoremediation	White et al. (2005)	Alkylated polycyclic aromatic hydrocarbons

(continued)

Table 4.2 (continued)

Plant species	Process	References	Contaminant
Callitriche lusitanica	Phytoremediation	Favas et al. (2012) cited by Yavari et al. (2015)	As
Iris pseudacorus	Phytoremediation	Li et al. (2014) cited by Yavari et al. (2015)	Pesticide
Lemna minor and Spirodela polyrhiza	Phytoremediation	Dosnon-Olette et al. (2009) cited by Yavari et al. (2015)	Fungicides
Tagetes patula	Phytoremediation	Patil and Jadhav (2013) cited by Yavari et al. (2015)	Textile dye Reactive Blue 160

EDTA ethylenediaminetetraacetic acid

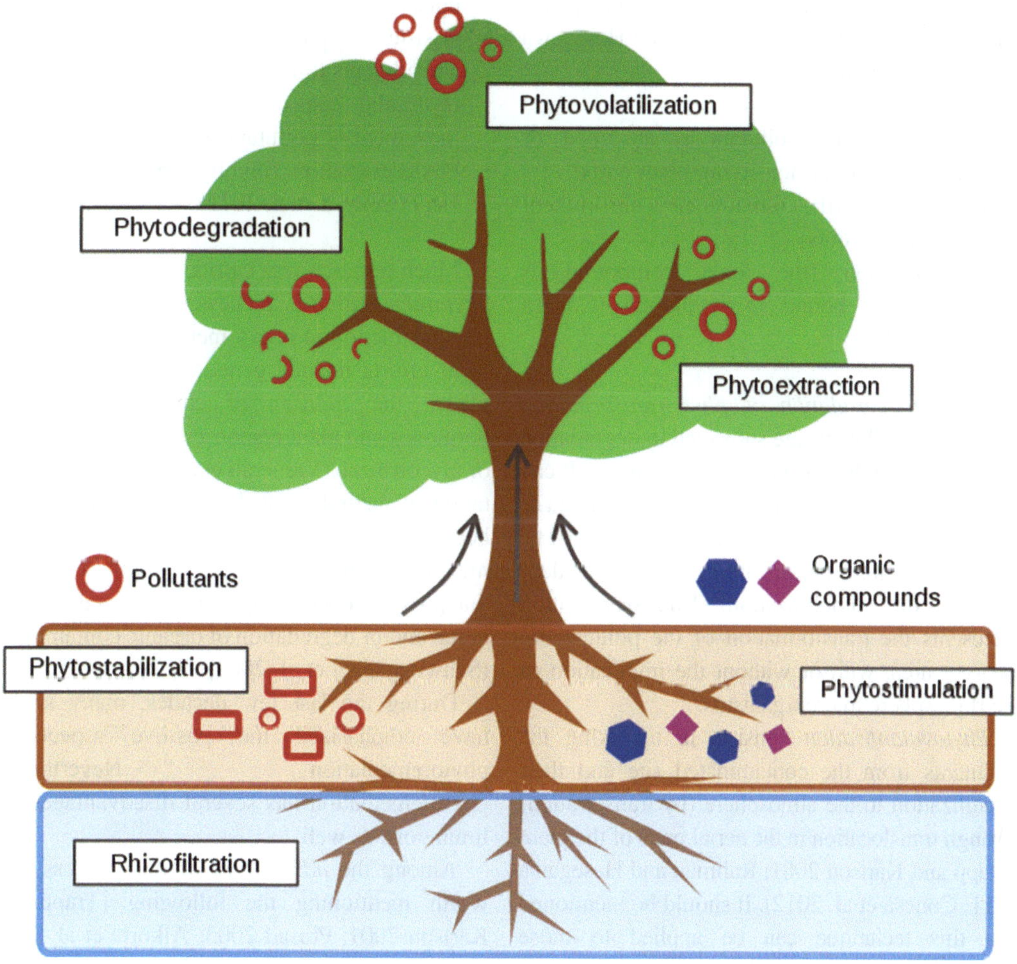

Fig. 4.2 Phytotechnologies (https://commons.wikimedia.org/w/index.php?curid=53861918)

- Bioaugmentation-assisted phytoextraction (including combination with mycorrhiza)
- Phytomining (obtaining economic profit from the metal accumulated by plants)
- Chelated-assisted phytoextraction (implies the adding of different chelants to the soil)

Phytofiltration is the capacity of plants to adsorb and absorb the pollutants from the contaminated environment, into roots or other plant parts. The precipitation of the pollutant in the root area is also a possibility. Usually, this technique is used to extract heavy metals or lipophilic compounds from water and is carried out by aquatic plants (Trapp and Karlson 2001; Rahman and Hasegawa 2011). Phytofiltration includes:

- Biosorption (the pollutant is absorbed or bound in living or non-living plant parts)
- Rhizofiltration (the contaminant is absorbed or bound in the roots)
- Blastofiltration (the toxic compound is absorbed or bound in seedlings) (Conesa et al. 2012)

In *phytodegradation* or *phytotransformation* the organic pollutants are subjected to degradation by plants through their metabolic processes. It can be considered a defense mechanism of the plant to the contaminant (Rahman and Hasegawa 2011), which results in its modification, inactivation, degradation or immobilization. Rhizodegradation represents the transformation of the pollutant in the root area, with or without the implication of the rhizosphere microorganisms.

Phytovolatilization consists in up-taking the pollutants from the contaminated site and their volatilization to the atmosphere (by transpiration) through translocation in the aerial parts of the plant (Trapp and Karlson 2001; Rahman and Hasegawa 2011; Conesa et al. 2012). It should be mentioned that this technique can be applied to those compounds that are volatile or those that can be transformed into volatile forms (chlorobenzene, trichloroethene, organically bound mercury, etc.) (Trapp and Karlson 2001). Transformation of the pollutant into volatile forms and releasing them into the atmosphere only displaces the pollution issue from one medium to another and is therefore seen as an improper process.

Phytostabilization is the process that can be applied to immobilize the pollutant (heavy metals or organic amendments) in soil through ab- or adsorption, accumulation in the roots of the plants or precipitation in the rhizosphere (transformation from a soluble form into a non-soluble one) (Andreazza et al. 2013). In fact, in this way, the mobility and the phytoavailability of the pollutant in the environment are reduced. This process includes:

- Phytoexclusion (use of plants with low metal uptake)
- Assisted phytostabilization (use of amendments to improve the process)
- Hydraulic control (to prevent leaching or movement of pollutants by water pumping)
- Phytorestoration (involves native plant species) (Conesa et al. 2012)

Much research was carried out to *improve the phytoremediation* process, especially the phytoextraction of heavy metals. These improvement efforts include genetic engineering of the plants, the addition of chelating agents or hormones and plant responses to them, formation of mycorrhizae (Vamerali et al. 2010), the exploitation of natural plant diversity, the interactions between plant roots and rhizosphere microorganisms, the use of endophytic bacteria that possess superior capacities for metal accumulation and/or degradation of organic contaminants (Schwitzguébel et al. 2002).

During the last few decades, many studies have emphasized the positive aspects of phytoremediation. Nevertheless, phytoremediation has several disadvantages and limitations as well.

Among the *advantages* of this process, it is worth mentioning the following (Trapp and Karlson 2001; Prasad 2003; Alkorta et al. 2004; Vasavi et al. 2010; Ekta and Modi 2018):

- It can be applied in situ (and ex situ as well)
- The plants can be easily monitored
- It reduces soil disturbance and the spread of pollutants

- The soil remains in place and is accessible for subsequent treatment
- It is solar driven
- It is considered inexpensive
- It costs less than 20% of conventional treatments
- There is no need for expensive equipment or highly specialized personnel
- It is a green tool, environmentally friendly
- It is aesthetically pleasing, socially accepted
- It is a low-tech alternative
- It maintains soil and stimulates soil life
- It can be combined with other methods of treatment
- The transfer of the contaminant is faster than natural remission
- It is considered to have fewer air and water emissions
- It is suitable for a wide variety of inorganic and organic pollutants
- It reduces the amount of waste
- It is possible to recover and re-use valuable metals
- It is easy to implement
- The plants represent a renewable resource, easily available
- It is capable of constantly treating a wide range of pollutants from different kinds of environments

Even if this green alternative has a number of notable advantages, its *limitations and disadvantages* must also be mentioned (Trapp and Karlson 2001; Prasad 2003; Alkorta et al. 2004; Ghosh and Singh 2005; Vasavi et al. 2010; Ali et al. 2013; Stephenson and Black 2014; Ekta and Modi 2018):

- Although faster than natural remission, it requires long time periods (several years)
- Even though it is a lengthy process, the contamination may still not be fully remediated
- It is slower than chemical and conventional treatments
- Only a few uses of the area are possible
- The phytotoxicity, ecotoxicity, and bioavailability of degradation products is unknown
- It is not applicable for all compounds
- It is limited in application to shallow soils, streams, and groundwater
- It is limited by the depth of the roots and both the solubility and the availability of the contaminant
- High concentrations of pollutant materials are toxic, even lethal to plants
- It is considered to be applicable to sites with low to moderate soil contamination over large areas, or to sites with large volumes of groundwater with low levels of contamination because plant growth is not sustained in heavily polluted environments
- Contaminants may accumulate in the groundwater because it is not possible to completely prevent leaching
- It is possible for pollutants to be transferred to another medium, the environment, and/or the food chain in the case of mismanagement and lack of proper care
- It is restricted to sites with low contaminant concentration
- Plant biomass from phytoextraction requires proper disposal as hazardous waste
- It is climate- and season-dependent, because unfavorable conditions can limit plant growth and biomass production, the result being decreased efficiency
- In the case of plant disease or attack by plant pests, effectiveness is lost as well
- The introduction of inappropriate, non-native or invasive plant species can affect biodiversity
- Some amendments and cultivation practices may have negative consequences for pollutant mobility
- Particularly in Europe, the limitation of phytoremediation is also associated with the potential use of genetically modified crops and the risk of their utilization to ecosystems. Consequently, its cost might be increased as sites require greater maintenance, monitoring, and disposal of genetically modified plant materials owing to the strict regulations

Regarding the phytoremediation process as a green alternative, it should be mentioned that results obtained in the field may be different from those obtained in the laboratory or at a greenhouse level. This is because the field is a real world, a real environment, where different factors act simultaneously. Factors that interfere with phytoremediation in the field also include variations in temperature, nutrients, precipitation and moisture, presence of plant pathogens, uneven distribution of pollutants, soil type, soil pH, and soil structure. Therefore, phytoremediation is an interdisciplinary domain and requires solid background knowledge in soil chemistry, plant biology, ecology, soil microbiology in addition to environmental engineering (Ali et al. 2013).

From an *economic* point of view, the dedicated literature maintains that phytoremediation will become feasible in the next few years. Approximately 20 years ago, in 2001, the US Environmental Protection Agency (EPA) published data regarding different completed soil remediation projects, a small number of which used phytoremediation (Vangronsveld et al. 2009). The costs varied as follows (Vangronsveld et al. 2009):

- Phytoremediation of a large site (USA) of contaminated soil with heavy metals—USD147–483/m^3
- Phytoremediation of a contaminated soil area with heavy metals (estimation)—USD13–131/m^3
- Phytostabilization (France) for soil contaminated with arsenic—minimum USD54/m^3
- Phytoremediation of a large groundwater site contaminated with heavy metals—USD4.8–6.9/m^3

Over 10 years ago, several commercial companies using phytoremediation technologies had been developed both in the USA and in Europe, for example: Phytotech (USA), Applied Natural Sciences (USA) (http://treemediation.com), Aquaphyte Remediation (Canada), BioPlanta (Germany) (http://www.bionity.com/en/companies/10451/bioplanta-gmbh.html), Consulagri (Italy), Earthcare (USA), Ecolotree (USA) (https://www.ecolotree.com/), Piccoplant (Germany) (https://www.piccoplant.de/en/), PhytoWorks (USA), Plantechno (Italy), Slater (UK) (http://www.slateruklimited.co.uk/), Thomas Consultants (New Zealand) (https://www.thomasconsultants.co.nz/), Verdant Technologies (USA), Viridian Resources (USA) (http://www.viridianresources.com/). As we noticed, some of them no longer exist.

Nowadays, according to Transparency Market Research (https://www.transparencymarketresearch.com/bioremediation-technology-services-market.html), the most important companies in the bioremediation technology and services market are: Altogen Labs (USA) (http://altogenlabs.com/), Aquatech International LLC (USA) (https://www.aquatech.com/about/), Drylet, LLC (USA) (https://www.drylet.com/), InSitu Remediation Services Limited (Canada) (http://irsl.ca/), Ivey International, Inc. (Canada) (https://www.iveyinternational.com/index.php), Probiosphere Inc. (Canada) (https://www.probiosphere.technology/about-us), Regenesis (USA) (https://regenesis.com/en/), Sarva Bio Remed, LLC (USA) (https://sarvabioremed.com/), Sevenson (USA) (https://sevenson.com/), Environmental Services, Inc. (USA) (www.esinc.cc), Soilutions Ltd, (UK) (https://www.soilutions.co.uk/), Sumas Remediation Services Inc. (Canada) (https://sumasrem.com/about-us/), Xylem Inc. (USA) (https://www.xylem.com/en-us/about-xylem/).

In their last report from 2017, the global market for bioremediation technology and services (including phytoremediation) was assessed to be worth USD32.2 billion (2016) and is estimated to reach USD65.7 billion by 2025.

4.3 Microbial and Vermiremediation: Clean-Up Alternatives

4.3.1 Microbial Bioremediation

Microorganisms have biosynthetic and biodegradative abilities, which proved very valuable in finding solutions for maintaining the quality of the environment or repairing the damaged ecosystems.

One of the main applications of microorganisms in environmental protection and bioremediation is the creating of cleaning technologies in oil-contaminated areas, but also in areas contaminated with polychlorinated biphenyl compounds (PCBs), hydrocarbons, dyes, pesticides, esters, heavy metals, or nitrogen-containing chemicals (Table 4.3) (Sharma et al. 2018). Compared with other methods, biological treatment using bacteria, fungi or microalgae is low in cost, highly efficient, and prevents secondary pollution.

4.3.2 Water Bioremediation

An important source of pollution is sewage water created by residences, hospitals, industrial establishments, farms, etc. Conventional sewage treatments are performed in treatment plants, in at least 12 phases, some of them involving aerobic biological processes realized by microorganisms, the technologies being well known and widely applied. For example, in the activated sludge many bacterial genera could be found such as *Pseudomonas*, *Flavobacterium*, *Alcaligenes*, *Bacillus*, *Acinetobacter*, and *Zooglea* spp., primarily involved in the biological treatment of municipal wastewater under aerobic conditions (the presence of organic matter supports the growth of heterotrophic bacteria able to degrade toxic compounds—nitrobenzene, tributyl phosphate, heavy metals, textile dyes, aliphatic and aromatic hydrocarbons, fatty acids, insecticides, etc.) (Shah 2017). In a synthesis performed in 2017, it was shown that the main microorganisms involved in efficient wastewater treatment include *Bacillus*, *Achromobacter*, *Pseudomonas stutzeri*, *P. putida*, *P. mendocina*, *Zooglea ramigera*, *Arthrobacter*, *Alcaligenes faecalis*, *Flavobacterium*, *Micrococcus*, *Rhodococcus species*, and lactic acid bacteria (*Lactobacillus casei*, *L. plantarum*, *Streptococcus* spp., *Rhodopseudomonas*) (Shah 2017).

In a case study carried out in China (Chengnan River) and published in 2018, a microbial product designated as HP-RPe-3 (national patent number: 2017114193785) composed of a large number (more than 100 types) of indigenous microorganisms (isolated from the Tibetan Plateau snow line—altitude 4650 m; species of *Bacillus*, *Micrococcus*, photosynthetic bacteria, nitrifying bacteria, denitrifying bacteria, lactic acid bacteria, yeasts, *Actinomycetes*, *Acetobacter*) and enzymes was used for the degradation of organic and inorganic matter, and toxic substances in water and sediments in the Chengnan River (Gao et al. 2018). The selected microorganisms have important properties (extreme cold resistance, high enzymatic activity, phage-resistant, and presented short cycles of development), but in the experiments microbial accelerating agents (enzymes, vitamins, amino acids, trace elements, and humic acid) were also used to stimulate the proliferation and activity of aerobiotic and facultative aerobic bacteria. The results obtained indicate that bioremediation technology, by adding microbial agents, improves water quality mainly by the degradation of NH3-N and elimination of the black-odor phenomenon of urban rivers (Gao et al. 2018).

4.3.3 Oil Spill Bioremediation

Regarding oil pollution, the accidental large-scale oil spills produced by Exxon Valdez in Alaska in 1989 and the BP Deepwater Horizon spill in the Gulf of Mexico in 2010 are well known: in these two environmental disasters 0.75 and 4.9 million barrels of crude oil were released respectively, which are still affecting some of the most productive and vulnerable marine ecosystems, and are having a high impact on terrestrial ecosystems too (Yavari et al. 2015). Another example is related to the Gulf War that occurred in 1991, when more than 700 oil wells were damaged, forming more than 300 oil lakes, and covering land areas in excess of 49 km^2 (Yateem 2014).

Microbial activity-based bioremediation processes used in situ in the field are classified as natural attenuation, bioaugmentation, and biostimulation (Table 4.4).

To establish a bioremediation technology based on the microbial degradative/biosynthesis activities it is necessary to isolate the

Table 4.3 Biological agents of bioremediation [adapted from Biswas et al. (2015)]

Microorganism		Toxic compounds used	
		Organic pollutants	Heavy metals
Bacteria	*Bacillus* spp.	Cresol, phenols, aromatics, long-chain alkanes, phenol, oil-based paints, textile dye (Remazol Black B), sulfonated di-azo dye Reactive Red HE8B, remazol navy blue dye	Cu, Zn, Cd, Mn
	Pseudomonas spp.	Benzene, anthracene, hydrocarbons, polychlorinated biphenyl compounds	U, Cu, Ni, Cr, Cd, Pb, Zn, As
	P. alcaligenes, *P. mendocina*, *P. putida*, *P. veronii*, *Acinetobacter*, *Achromobacter*, *Flavobacterium*	Petrol and diesel polycyclic aromatic hydrocarbons, toluene	
	Pseudomonas putida	Monocyclic aromatic hydrocarbons, e.g., benzene and xylene	
	Xanthomonas sp.	Hydrocarbons, polycyclic hydrocarbons	
	Nocardia sp.	Hydrocarbons	
	Streptomyces sp.	Phenoxyacetate, halogenated hydrocarbon, diazinon	
	Mycobacterium sp.	Aromatics, branched hydrocarbons benzene, cycloparaffins	
	Alcaligenes odorans, *B. subtilis*, *Corynebacterium propinquum*, *P. aeruginosa*	Phenol	
	Micrococcus luteus, *Listeria denitrificans*, *Nocardia atlantica*	Textile azo dyes	
	Acinetobacter sp., *Pseudomonas* sp., *Enterobacter* sp., *Photobacterium* sp., *Bacillus* spp., *Staphylococcus* spp.	Pesticides (chlorpyrifos, methyl parathion, malathion, endosulfan)	
	Rhodopseudomonas palustris, *Aerococcus* spp.		Pb, Cr, Cd
	Citrobacter sp.		Cd, U, Pb
	Lysinibacillus sphaericus		Co, Cu, Cr, Pb
Fungi	*Coprinellus radians*	Polyaromatic hydrocarbons, methylnaphthalenes, and dibenzofurans	
	Pycnoporus sanguineus, *Phanerochaete chrysosporium*, and *Trametes trogii*	Industrial dyes	
	A. niger, *A. fumigatus*, *F. solani*, and *P. funiculosum*	Hydrocarbons	
	Aspergillus versicolor, *A. fumigatus*, *Paecilomyces* sp., *Trichoderma* sp., *Microsporum* sp., *Cladosporium* sp.		Cd
	Saccharomyces cerevisiae		Pb, Hg, Ni
	Marasmiellus troyanus	Benzo[a]pyrene	
	Gloeophyllum trabeum	1, 1, 1-trichloro-2, 2-bis (4-chlorophenyl) ethane (DDT)	

(continued)

Table 4.3 (continued)

Microorganism		Toxic compounds used	
		Organic pollutants	Heavy metals
	Pleurotus ostreatus	Bisphenol A, hydrocarbons	
	Fomitopsis pinicola	1, 1, 1-trichloro-2, 2-bis (4-chlorophenyl) ethane (DDT)	
	Penicillium simplicissimum	Polyethylene	
	Rhizopus arrhizus		Ag, Hg
	Stereum hirsutum		Cd, Pb
Algae	*Chlamydomonas* sp.	Naphthalene	
	Dunaliella sp.	Naphthalene, DDT	
	Euglena gracilis	DDT, Phenol	
	Selenastrum capricornutum	Benzene, toluene, chlorobenzene, 1, 2-dichlorobenzene, nitrobenzene naphthalene, 2, 6-dinitrotoluene, phenanthrene, di-*n*-butylphthalate, pyrene	
	Chlorella sp.	Toxaphene	Au, Cu, Ni, U, Pb, Hg, Zn, As, Cd, Cr
	Cylindrotheca sp.	DDT	
	Zooglea sp.		Co, Ni, Cd
	Phormidium valderium		Cd, Co, Cu, Ni
	Volvariella volvacea		Cu, Hg, Pb
	Oscillatoria sp.		Ni, Cu, Co, Pb, Zn
	Tetraselmis chuii		Cu
	Spirogyra hyalina		Cd, Hg, Pb, As
	Lyngbya spiralis		Cd, Pb, Hg

DDT dichlorodiphenyltrichloroethane

Table 4.4 Microorganisms involved in oil bioremediation (adapted after Abatenh et al. 2017)

Microorganism	Oil type
Fusarium spp.	Oil
Alcaligenes odorans, Bacillus subtilis, Corynebacterium propinquum, Pseudomonas aeruginosa	Oil
Bacillus cereus	Diesel
Aspergillus niger, Candida glabrata, C. krusei, Saccharomyces cerevisiae	Crude oil
B. brevis, P. aeruginosa KH6, B. licheniformis, B. sphaericus	Crude oil
Pseudomonas aeruginosa, P. putida, Arthrobacter sp., *Bacillus* sp.	Diesel
Pseudomonas cepacia, B. cereus, B. coagulans, Citrobacter koseri, Serratia ficaria	Diesel

microorganisms able to perform, at the highest level, the decontamination of the environment.

The microbiological analysis of polluted sites (oil-contaminated soils or aquatic ecosystems) revealed the presence of a number of heterotrophic oil-utilizing bacteria or fungi (Table 4.4). Most studies revealed the presence in oil-contaminated sites of bacteria belonging to various genera such as: *Acinetobacter*, *Micrococcus*, *Rhodococcus*, *Pseudomonas*, *Bacillus*, *Staphylococcus*, *Kocuria*, etc. (Table 4.4). Moreover, the studies performed in oil-contaminated lakes from Kuwait allowed the identification of extreme halophilic archaea strains belonging to *Halobacterium*, *Haloferax* or *Halococcus* (Yateem 2014).

Recently, research into petroleum hydrocarbon microbial degraders in marine environments was conducted to identify the novel obligate hydrocarbon degraders typical of marine habitats such *Alcanivorax* and *Cycloclasticus* (Kalogerakis et al. 2015). These data suggest that the search for new efficient oil-degrading bacteria might still be the aim of many studies and could permit the selection of microbial consortia useful in specific technologies.

Microorganisms selected from specific polluted sites, such as consortia or as individual cultures, could be used in cleaning technologies of environments polluted with pesticides, heavy metals, different organic toxic compounds, etc. An example of bacteria useful in such approaches is *Pseudomonas putida*, a soil saprophytic Gram-negative bacteria able to produce a large diversity of enzymes for green chemistry applications and bioremediation. Until now, many strains of *P. putida* able to use aromatic hydrocarbons, trichloroethylene, indole, chlorophenols, nitrotoluenes, etc., as carbon sources were selected and studied from biochemical and genetic points of view. Many bacterial strains useful in bioremediation applications were subjected to legal protection by patent: the first patent for a biological remediation substance was recorded in 1974, and was a strain of *P. putida* capable of degrading petroleum (Biswas et al. 2015).

Moreover, many species of fungi and algae (eukaryotic microorganisms) are involved in biogeochemical transformations in both aquatic and terrestrial habitats. Fungi can mineralize xenobiotic compounds to CO_2 and H_2O generally through their non-specific ligninolytic and highly oxidative enzyme systems, which are also responsible for the degradation and decolorization of a wide range of dyes (Biswas et al. 2015). Various species of eukaryotic algae are able to produce organometallic complexes between algal peptides and heavy metals; the complexes are then included in vacuoles, thus neutralizing or preventing the toxic effects of metals. The transformations performed by them could influence plant productivity, the mobility of toxic elements, with important socio-economic relevance, especially in the mutualistic symbioses, lichens, and mycorrhizas. From a bioremediation point of view, fungal biotransformation has beneficial applications in environmental biotechnology, e.g., in metal leaching, recovery, and detoxification, and xenobiotic and organic pollutant degradation (Table 4.4) (Biswas et al. 2015; Majumder 2016).

The studies regarding the microbial mechanisms involved in the degradation of certain pollutants, such as aliphatic and aromatic hydrocarbons, revealed the presence of specific enzymes that contribute to the transformation of contaminants into less toxic final products, which are integrated into natural biogeochemical cycles (Peixoto et al. 2011). The efficiency of bioremediation depends on many factors such as: the chemical nature and concentration of pollutants, the physicochemical characteristics of the environment (type of soil, temperature, pH, the presence of oxygen or other electron acceptors, nutrients), the availability of xenobiotics to the microorganism, and the diversity of microbial consortia.

The biodegradation may occur under aerobic or anaerobic conditions, the processes being best studied in the case of hydrocarbons. Microbial strains involved in aerobic degradation are able to produce oxygenase enzymes that introduce oxygen atoms into hydrocarbons: for example, monooxygenases introduce one oxygen atom to a substrate whereas dioxygenases introduce two. Under anaerobic conditions, the hydrocarbon degradation is produced mainly by sulphate-reducing bacteria, by using different terminal electron acceptors (nitrate, sulphate, or Fe (III)). Similar studies were performed to establish the mechanisms involved in other microbial bioremediation processes, and more than 1000 different enzymes were described (Whiteley and Lee 2006).

Based on the results obtained in experiments performed with microorganisms and/or microbial enzymes, various products useful in practical applications, mainly in oil spill bioremediation, were developed. The US EPA has defined bioremediation agents (bioaugmentation agents or biostimulation agents) as "microbiological cultures, enzyme additives, or nutrient additives that significantly increase the rate of biodegradation to mitigate the effects of the discharge" (Zhu et al. 2004). Numerous bioremediation products (microbial cultures, enzyme additives, and nutrient additives) have been proposed and promoted by the manufacturers or vendors (Table 4.5).

In conclusion, the advantages of microbial bioremediation are related to the use of natural processes that cause less damage to ecosystems and take place underground, as additives and microbial cultures are introduced underground to clean up contaminants in ground water and soil. Bioremediation technologies are generally cheaper than most cleanup methods, and they does not require special equipment or labor. In a survey from 2012, it was encountered that bioremediation has been used to clean up more than 100 Superfund sites around the USA (https://www.investopedia.com/terms/b/bioremediation.asp).

The bioremediation could be considered a business domain, the large number of companies founded and the numerous products and technologies developed and commercialized being a proof of success. According to a study performed in 2018 by Transparency Market Research, the global market for bioremediation technology and services market was valued at USD32.2 billion in 2016 and is estimated to reach USD65.7 billion by 2025 at a compound annual growth rate of 8.3% from 2017 to 2025.

Based on technology developed, the bioremediation technology and services market is classified into phytoremediation, biostimulation, bioaugmentation, bioreactors, fungal remediation, and land-based treatments. Among these technologies, fungal remediation represents a major segment of the bioremediation technology and services market, as the use of mycelium to disintegrate contaminants from waterways, soil or even radioactive contaminated areas has increased. It was estimated that the use of fungi for the treatment of soils polluted by mercury and other heavy metals will increase by 2025.

Regarding the services, the market is divided into soil remediation, wastewater remediation, oilfield remediation, and others. Among the services offered by companies, wastewater remediation was accounted to hold the largest market share in 2016, whereas soil remediation services are likely to expand from 2017 to 2025. Owing to rapid industrialization, an increase in the disposal of pharmaceutical products, and a rise in the use of harmful insecticides, pesticides, petroleum hydrocarbons, chlorinated solvents, etc., were encountered, which is the reason why bioremediation technologies must be developed.

In the study it was shown that, at present, the major players in the bioremediation technology and services market are: Altogen Labs, Aquatech International LLC, Drylet LLC, InSitu Remediation Services Limited, Ivey International Inc, PROBIOSPHERE Inc, REGENESIS, Sarva Bio Remed LLC, Sevenson, Environmental Services Inc, Soilutions Ltd, Sumas Remediation Services Inc, and Xylem Inc.

Table 4.5 Bioremediation agents useful in oil spill bioremediation (adapted from Zhu et al. 2004)

Name of the product	Type of the product	Manufacturer
Bet Biopetro	Microbial culture	BioEnviro Tech, Tomball, TX
Inipol Eap 22	Nutrient additive	Societe, CECA SA, France
Land and Sea Restoration	Nutrient additive	Land and Sea Restoration LLC, San Antonio, TX
Oil Spill Eater II	Nutrient additive/enzyme additive	Oil Spill Eater International, Corporation, Dallas, TX
Oppenheimer Formula	Microbial culture	Oppenheimer Biotechnology, Inc, Austin, TX
Step one	Microbial culture	B & S Research, Inc, Embarrass, MN
Biosolve Pinkwater	Nutrient additive	The BioSolve Company, Lexington, MA, USA
Remediact™	Microbial culture	Chemtex, Inc., Cumberland, RI, USA
EcoPondSweep™	Microbial culture	Confluence Energy, Kremmling, CO, USA

4.3.4 Vermiremediation

Vermiremediation or vermicomposting is an effective, low-cost technology dedicated to recycling agricultural waste, city garbage, kitchen waste or even sewage sludge, by the activity of earthworms able to convert the organic waste materials into compost (Khan 2016). The beneficial role of earthworms (*Eisenia fetida*, *Lumbricus terrestris*, *Aporrectodea caliginosa*, *A. nocturna*, *Pheretima hawayana*, *Pontoscolex corethrurus*, *Dendrobaena veneta,* etc.) in the physical, chemical, and biological properties of soil (increasing soil fertility) is well known, but the use of these organisms in bioremediation technologies has been examined over the past decades. In this respect, the effect of earthworms on the removal of various contaminants, such as oil, PAHs, PCBs, pesticides, and heavy metals has been reported by many authors, both from scientific and practical points of view (in the laboratory or outdoors) (Rodriguez-Campos et al. 2014; Rorat et al. 2017). The results obtained demonstrated that more experiments are required so that the practical application of vermiremediation can be demonstrated on a large scale (soil remediation in extended areas). Additionally, the costs of the technologies may be too high to remediate large contaminated areas, because of the conditions needed for the survival and activity of the earthworms.

However, presently many products of the vermicomposting process are available on the market. Such products are obtained both in small-scale or home systems (from mixtures of fruit and vegetable waste, coffee grounds and filters, grains such as bread, crackers, and cereals, eggshells, leaves and grass, newspapers, paper) and in large-scale systems (generally using dairy cow or pig manure, brewery waste, other industrial and agricultural waste, grass clippings and wood, etc.) (Adhikary 2013).

In a report presented in 2019 it was shown that the vermicompost industry is very fragmented, manufacturers are mostly in the India and Southeast Asia, and products are manufactured and commercialized all over the world. The key manufacturers in the vermicompost market include: MyNOKE (the world leading manufacturer in global vermicompost market with the market share of 8.79% in 2015), NutriSoil, Davo's Worm Farms, Earthworm, Wormpower, Kahariam Farms, SAOSIS, Sri Gayathri Biotec, Jialiming, Dirt Dynasty, SLO County Worm Farm, Agrilife, and Suman Vermi Compost. Regarding the benefits, it was shown that compared with 2014, the vermicompost market managed to increase sales by 24.89% to USD38.09 million worldwide in 2015, which allows the conclusion that overall, the vermicompost performance is positive, despite the weak economic environment. For example, among the manufacturers, production in India accounted for less than 9.50% of the total value of global vermicompost in 2015 (http://www.qyresearchglobal.com/goods-1814370.html; http://www.marketsnresearch.com/global-vermicompost-market-report-2019-industry-analysis-size.html).

4.3.5 Microbial Biofertilizers for Bioremediation

Biofertilizers are microbially enriched products, containing latent or living cells of selected beneficial microorganisms that are able to improve soil quality and promote plant growth, mainly by increasing the uptake of nutrients. Biofertilizers accelerate certain bioconversion processes in the growing substrate and increase the bioavailability of nutrients for plants. They can be applied to the soil, seed or plant surface, enriching the microbial communities of the rhizosphere and colonizing the inner and external parts of the plants.

In sustainable agriculture, biofertilizers are cost-efficient supplements of plant nutrients that increase the efficacy of chemical fertilizers or reduce their application requirements.

Among the beneficial microorganisms used as biofertilizers are: mycorrhiza, several soil- and plant-inhabiting fungi, most of the plant growth-promoting bacteria, and some blue-green algae (Santra et al. 2015). Based on their function, they can be classified as nitrogen fixers, phosphorus solubilizers, phytohormone and enzyme producers, and others.

The nitrogen-fixing microorganisms can convert atmospheric nitrogen (unavailable for direct plant nutrition) into organic nitrogen compounds, which are available for plants. Such biofertilizers can substitute nitrogen fertilization in some cultivated plants. The microorganisms used as nitrogen-fixing biofertilizers include symbiotic bacteria and free-living or non-symbiotic microorganisms (bacteria, actinomycetes, and blue-green algae). Among the symbiotic bacteria, *Rhizobium* and related genera (*Azorhizobium, Bradyrhizobium, Sinorhizobium, Ensifer,* etc.) are able to fix nitrogen in leguminous plants, producing nodules on their roots. Worldwide, there are various commercialized biofertilizers based on such microorganisms, such as Rhizolife^{BJ}, Polarhizo, Effect Grow, Biobium, Rhizo-Enrich, Nitragin™ Gold, Biodoz™, Optimize™, Cell-Tech™, GlyciMax™, Nitrofix®, etc. Various other nitrogen-fixing microorganisms were found, such as both symbiotic and free-living bacteria and actinomycetes. In such cases *Acetobacter, Azotobacter, Azospirillum, Paenibacillus,* and *Frankia* were found. Commercially available biofertilizers based on such nitrogen-fixing bacteria are: Power Grain Booster, GreenAzoto, Azomax (containing *Azotobacter* strains), Sugar-Plus (containing endophytic *Acetobacter*), Abtech Azospirillum, and Azostim F9 PTS (containing *Azospirillum* strains).

Some biofertilizers contain mixed cultures of beneficial microorganisms such as Rhizodyne (containing *Azospirillum, Azotobacter, Rhizobium, Acetobacter* as NPK and Zn providers), Micosat (containing *Glomus mosseae* and *G. intraradices* mycorrhizal fungi and plant growth-promoting bacteria *Pseudomonas fluorescens* and *Bacillus subtilis*), BIO-NPK and Bharpur (containing a consortium of various bacterial strains providing NPK-balanced nutrition), and TagTeam™ LCO (which combines *Rhizobium* and phosphate-solubilizing inoculants with lipo-chitooligosaccharide molecules).

Among other beneficial microorganisms used as biofertilizers are phosphorus-solubilizing microorganisms, which increase phosphorus uptake from phytic acid and phytate organic phosphorus and improve the solubility of inorganic phosphates. Available commercial biofertilizers with phosphorus-solubilizing activity are VICI Routz GR soil probiotics, Rich Paddy biofertilizer, Rhizocell GC, and Rhizocell C. Some bioproducts are based on bacterial strains, such as BIOPHOS and GET-PHOS containing *Bacillus megaterium* var. *phosphaticum*, and others such as JumpStart® contain the soil fungus *Penicillium bilaii*.

Other biofertilizers found on the market are used to increase potassium accumulation (BioPotash and Potash-Cure based on *Frateuria aurantia*), sulfur solubilization (BIOSULF and SULF-CURE based on *Thiobacillus thiooxidans*), zinc solubilization (BIOZINC and ZINC-CURE), or silica (such as BioSilica and Silica-Cure containing strains of *Bacillus* spp.).

Mycorrhizal fungi are also very good soil fertilizers. Mostly, they are efficient in phosphorus uptake from insoluble sources, but because of their capacity for colonization, they improve plant nutrition with several other nutrients from sources generally unavailable to host plants. Moreover, they positively influence soil aggregation and water dynamics (Piotrowski et al. 2004). HPM Gold, Myco-Rise, Mycoxol, NutriVAM, BioVam, VAM Riches, Myconox, MycoStim, PlantSuccess, Myco-Win, Ecomax, Root care, and Glow Raja are some of the commercially available biofertilizers.

Regarding bioremediation with microbial biofertilizers, it has been noticed that bio-augmentation of hydrocarbon-polluted soils with nitrogen-fixing bacteria improves the soil decontamination process (Huesemann and Moore 1993). Several other authors maintain the fact that improved substrate fertilization stimulates contaminant degradation by microorganisms (Perez-Vargas et al. 2000; Santra et al. 2015).

4.4 Natural Plant Protection Products ("Biopesticides")

Plant protection products (PPPs) are mainly used to protect plants from harmful organisms such as pests, diseases or weeds. However, some plant protection products regulate plant growth by means other than nutrients or influence the shelf life of the harvest. PPPs contain one or more active substances responsible for the purposes mentioned. These active ingredients could be either chemical or natural. The latter category is low risk and includes microorganisms, insect pheromones, and plant extracts.

In the European Union (EU), the active ingredients of PPPs must be approved by the European Commission and the final product must be authorized before being marketed. Nowadays, around 25% of the active substances approved are natural products. This means that more than 70 natural active substances have been approved on the EU market (http://ec.europa.eu).

Among the microbial strains approved as plant protection products (Table 4.6), most have fungicidal and bactericidal activity (52.7%) and insecticidal effects (29.1%), the rest (18.2%) being elicitors, nematicides, and virus inoculants.

4.4.1 Microbial Biocontrol of Plant Pathogens

The spectrum of microorganisms used as biological control agents is relatively wide. In the EU, 22 species of fungi and bacteria have been approved for use as pesticide active substances, according to the data available in February 2019 (Table 4.7). However, worldwide, the spectrum of microbial biocontrol agents is much wider, especially in the USA, China, India, and the South American countries.

When searching for new microbial biological control agents (MBCAs), in addition to the efficacy and spectrum of activity, some other traits are also considered important for selection. As MBCAs are intended to be formulated, the preferred microorganisms are spore-forming fungi and bacteria, owing to their increased resistance and viability. Usually, in microbially based PPPs, in addition to the plant protective effects of the active ingredients, the self-replicating capacity of the formulated microorganisms is also exploited (Chattopadhyay et al. 2017). Other important issues are the adaptability and colonization capacity of the microorganisms. However, only strains that are neutral to non-target organisms, and safe for the environment, have been approved.

4.4.2 Entomopathogenic Fungi

The entomopathogenic fungi are highly efficient, as they produce insect-infecting spores that can induce pest death in 4–10 days, depending on the fungal strain. Moreover, the fungus continues its growth and sporulation on the body of the dead insect, continuing its biological control activity in the area of application. The main groups of entomopathogenic fungi occur in the phylum Zygomycota (mostly fungi of Entomophthorales order) and the phylum Ascomycotina, where the most efficient anamorphic genera known to have entomopathogenic activity are *Beauveria*, *Isaria*, *Metarhizium*, *Lecanicillium*, and *Purpureocillium.*

Most of these biological control agents are mass produced and formulated as PPPs. They are formulated as solid-state, emulsifiable suspension, oil dispersion, liquid suspension, dry flowable, wettable powder or water dispersible granules. Spore concentration depends on the formulation type. For example, powdery formulation has a lower concentration of colony-forming units (CFU)/g than liquid suspensions, in which the concentration is usually 10 times higher than in CFU/ml.

Although in the EU member states only 11 strains of entomopathogenic fungi have been approved, worldwide a wider spectrum of species and strains are used for insect, mite, and nematode control (Table 4.8).

4.4.3 Entomopathogenic Bacteria

Entomopathogenic bacteria are able to produce various toxins and virulence factors causing insect death after ingestion. Therefore, the insect must ingest the bio-pesticide, by feeding, which is less efficient than the entomopathogenic fungi, which act by contact. Moreover, entomopathogenic bacteria have a restrictive

Table 4.6 Selected microbial strains approved in the EU as active substances of plant protection products (http://ec.europa.eu)

No.	Active substance	Active substance ID	Category	Approval date	Expiration of approval	Rapporteur Member State (RMS)	Countries where authorized
1.	*Ampelomyces quisqualis* strain AQ10	959	FU	01/08/2018	31/07/2033	FR	BE, CY, DE, DK, EL, ES, FR, IE, IT, LU, NL, SI, SK, UK
2.	*Aureobasidium pullulans* (strains DSM 14940 and DSM 14941)	973	FU, BA	01/02/2014	31/01/2024	AT	AT, BE, DE, EL, ES, FR, HU, IT, NL, PL, PT, SI, SK
3.	*Bacillus amyloliquefaciens* (former *B. subtilis*) strain QST 713	986	BA, FU	01/02/2007	30/04/2020	DE	BE, CY, CZ, DE, DK, EE, EL, ES, FI, FR, IE, IT, LT, LU, LV, NL, PL, PT, SE, SI, SK, UK
4.	*Bacillus amyloliquefaciens* MBI 600	2325	FU	16/09/2016	16/09/2026	FR	
5.	*Bacillus amyloliquefaciens* strain FZB24	2324	FU	01/06/2017	01/06/2032	FR	
6.	*Bacillus amyloliquefaciens* subsp. *plantarum* D747	2252	FU	01/04/2015	31/03/2025	DE	BE, CY, EL, ES, FR, IT, SI, UK
7.	*Bacillus firmus* I-1582	2248	NE	01/10/2013	30/09/2023	FR	DK, EL, ES, FR, IT, NL, PT, SE, UK
8.	*Bacillus pumilus* QST 2808	2253	FU	01/09/2014	31/08/2024	IT	FR
9.	*Bacillus thuringiensis* subsp. *aizawai* strains ABTS-1857 and GC-91	988	IN	01/05/2009	30/04/2020	NL	AT, BE, CY, DE, DK, EL, ES, FI, FR, IT, LU, NL, PL, PT, SE, SI, UK
10.		989	IN	01/05/2009	30/04/2020	SE	

(continued)

Table 4.6 (continued)

No.	Active substance	Active substance ID	Category	Approval date	Expiration of approval	Rapporteur Member State (RMS)	Countries where authorized
	Bacillus thuringiensis subsp. *israeliensis* (serotype H-14) strain AM65-52						AT, DE, DK, ES, NL, UK
11.	*Bacillus thuringiensis* subsp. *kurstaki* strains ABTS 351, PB 54, SA 11, SA12 and EG 2348	990	IN	01/05/2009	30/04/2020	DK	AT, BE, BG, CY, CZ, DE, DK, EL, ES, FR, HR, HU, IE, IT, LT, LU, NL, PL, PT, RO, SE, SI, SK, UK
12.	*Bacillus thuringiensis* subsp. *tenebrionis* strain NB 176 (TM 14 1)	991	IN	01/05/2009	30/04/2019	IT	AT, DE, ES, FR, HR, HU, PL
13.	*Beauveria bassiana* IMI389521	2388	IN	19/02/2019	19/02/2029	NL	
14.	*Beauveria bassiana* PPRI 5339	2387	IN	20/02/2019	20/02/2029	NL	AT
15.	*Beauveria bassiana* strain 147	2311	IN	06/06/2017	06/06/2027	FR	FR
16.	*Beauveria bassiana* strain NPP111B005	2312	IN	07/06/2017	07/06/2027	FR	
17.	*Beauveria bassiana* strains ATCC 74040 and GHA	997	IN	01/05/2009	30/04/2020	DE	AT, BE, CY, DE, DK, EL, ES, FR, HU, IE, IT, NL, SI, UK
18.	*Candida oleophila* strain O	1074	FU	01/10/2013	30/09/2023	SI	AT, FR, NL, UK
19.	*Clonostachys rosea* strain J1446 (*Gliocladium catenulatum* strain J1446)	1435	FU	01/04/2005	31/07/2019	HU	AT, BE, CY, DE, EE, ES, FI, FR, IE, NL, PL, SE, SI, UK
20.		1156	FU	01/08/2017	31/07/2032	NL	

(continued)

Table 4.6 (continued)

No.	Active substance	Active substance ID	Category	Approval date	Expiration of approval	Rapporteur Member State (RMS)	Countries where authorized
	Coniothyrium minitans Strain CON/M/91-08 (DSM 9660)						AT, BE, CZ, DE, DK, EL, ES, FR, HU, IE, IT, LU, NL, PL, PT, SE, SK, UK
21.	*Cydia pomonella* granulovirus (CpGV)	1178	IN	01/05/2009	30/04/2020	DE	AT, BE, BG, CZ, DE, DK, EL, ES, FI, FR, HR, HU, IT, NL, PL, PT, RO, SE, SI, SK, UK
22.	*Fusarium* sp. L13	2389	FU	Pending	–	FR	
23.	*Isaria fumosorosea* Apopka strain 97 (formerly *Paecilomyces fumosoroseus*)	1653	IN	01/01/2016	31/12/2030	BE	BE, FI, FR, NL, SE
24.	*Lecanicillium muscarium* (formerly *Verticillium lecanii*) strain Ve6	1515	IN	01/05/2009	30/04/2020	NL	BE, DK, ES, FI, FR, NL, UK
25.	*Metarhizium anisopliae* var. *anisopliae* strain BIPESCO 5/F52	1559	IN	01/05/2009	30/04/2020	NL	AT, BE, DE, DK, EL, FR, IE, IT, LU, NL, PT, UK
26.	*Metschnikowia fructicola*	2457	FU	27/12/2018	27/12/2028	FR	
27.	*Paecilomyces fumosoroseus* strain Fe9901	1654	IN	01/10/2013	30/09/2023	PL	BE, ES, IT
28.	*Paecilomyces lilacinus* strain 251	1655	NE	01/08/2008	31/07/2019	HU	BG, CY, EL, ES, IT
29.	*Pasteuria nishizawae* Pn1	2460	NE	14/10/2018	14/10/2033	DK	
30.	*Phlebiopsis gigantea* (several strains)	1698	FU	01/05/2009	30/04/2020	EE	DK, EE, FI, FR, LT, LV,

(continued)

Table 4.6 (continued)

No.	Active substance	Active substance ID	Category	Approval date	Expiration of approval	Rapporteur Member State (RMS)	Countries where authorized
							PL, SE, UK
31.	*Pseudomonas chlororaphis* strain MA342	1786	FU	01/10/2004	30/04/2020	NL	AT, BE, DE, DK, ES, FI, FR, IT, LT, LU, NL, PT, SE, UK
32.	*Pseudomonas* sp. strain DSMZ 13134	1787	FU	01/02/2014	31/01/2024	NL	AT, BE, CY, CZ, DE, EL, FR, HR, IE, IT, NL, RO, SE, SI, SK
33.	*Purpureocillium lilacinum* PL 11	2391	NE	Pending	–	UK	
34.	*Pythium oligandrum* M1	1810	FU	01/05/2009	30/04/2020	SE	CZ, FR, HU, IT, PL, SK, UK
35.	*Saccharomyces cerevisiae* strain LAS02	2323	FU	06/07/2016	06/07/2031	FR	
36.	*Streptomyces* K61 (formerly *S. griseoviridis*)	1895	FU	01/05/2009	30/04/2020	EE	BE, CY, EE, FI, FR, HU, IT, LT, LV, NL, SE, UK
37.	*Streptomyces lydicus* WYEC 108	2256	FU, BA	01/01/2015	31/12/2024	NL	
38.	*Trichoderma asperellum* (formerly *T. harzianum*) strains ICC012, T25 and TV1	1979	FU	01/05/2009	30/04/2020	SE	DE, EL, ES, FR, IT, PL, PT
39.	*Trichoderma asperellum* (strain T34)	2066	FU	01/06/2013	31/05/2023	SE	BE, FR, IE, IT, NL, PT, RO, UK
40.	*Trichoderma atroviride* (formerly *T. harzianum*) strains IMI 206040 and T11	1980	FU	01/05/2009	30/04/2020	SE	EL, IT, SE
41.		2532	FU	Pending	–	FR	

(continued)

Table 4.6 (continued)

No.	Active substance	Active substance ID	Category	Approval date	Expiration of approval	Rapporteur Member State (RMS)	Countries where authorized
	Trichoderma atroviride AGR2						
42.	*Trichoderma atroviride* strain I-1237	1981	FU	01/06/2013	31/05/2023	IT	FR
43.	*Trichoderma atroviride* strain SC1	2329	FU	06/07/2016	06/07/2031	FR	AT, BE, FR, LU, PT
44.	*Trichoderma gamsii* (formerly *T. viride*) strain ICC080	1982	FU	01/05/2009	30/04/2020	SE	DE, EL, ES, FR, IT, NL, PL, PT
45.	*Trichoderma harzianum* strains T-22 and ITEM 908	1983	FU	01/05/2009	30/04/2020	SE	BE, DK, EL, ES, FR, HU, IE, IT, NL, PL, SE, UK
46.	*Trichoderma polysporum* strain IMI 206039	1984	FU	01/05/2009	30/04/2019	SE	SE
47.	*Verticillium albo-atrum* (formerly *Verticillium dahliae*) strain WCS850	2008	FU	01/05/2009	30/04/2020	SE	DE, DK, NL, SE, UK

spectrum of activity, usually being active against a reduced number of susceptible pest species.

Most common bacterial insecticides are based on *Bacillus thuringiensis (Bt)*. In the EU, several products are registered, all having as an active ingredient *Bt* ssp. *aizawai*, *israeliensis*, *kurstaki* or *tenebrionis*. *Bt* ssp. *aizawai* and *kurstaki* have restricted efficacy against lepidopteran larvae (caterpillars), *Bt* ssp. *israeliensis* is effective against some mosquito species, black flies, and a range of filter flies, and *Bt* ssp. *tenebrionis* is restricted to susceptible species of foliar feeding beetle larvae of Coleoptera.

Worldwide, other species have also been identified to act against insects, mites, and nematodes. Commercially available entomopathogenic bacteria include *Bacillus firmus* against cyst, lance, lesion, root-knot, sheath, spiral, sting, and stunt nematodes, *Lysinibacillus sphaericus* (sin. *Bacillus sphaericus*) for mosquito control (El-Bendary 2006; Glare et al. 2017), *Clostridium bifermentans* against mosquitos (Qureshi et al. 2014), *Paenibacillus popilliae* to control Japanese beetles (Kaya et al. 2008; Glare et al. 2017), *Pasteuria nishizawae* registered as a biological nematicide parasitic on cyst nematodes of genera *Heterodera* and *Globodera*, *Saccharopolyspora spinosa* against two-spotted spider mites (Sparks et al. 2012), *Streptomyces avermitilis* against Colorado beetles (Wang et al. 2011a, b), and Gram-negative bacteria such as *Pseudomonas alcaligenes* against locusts and grasshoppers (Ruffner et al. 2015), or bacteria in the genus *Serratia* to control beetle larvae.

The most successful bacterial pesticides to date are spore-forming *Bacillus* species, owing to their long-term stability and storage compared

Table 4.7 Commercial microbial plant protection products used as natural fungicides and bactericides (Woo Sheridan et al. 2014; Fravel et al. 1998; EFSA documents)

Active ingredients	Commercial names	Formulation type
Ampelomyces quisqualis	AQ10 Biofungicide	Water dispersible granules
Aureobasidium pullulans several strains	Botector, Blossom Protect	Wettable granules
Bacillus amyloliquefaciens several strains (some *formerly B. subtilis)*	CX 9030, TAE-022 WDG, TAEGRO, TAE-022 Technical, Subtilex®, BUEXP1780S, QST 713 Technical	Water dispersible granules, wettable powder
Bacillus pumilus	Ballad Plus, Sonata, Sonata® ASO, QST 2808 AS organic, QRD288 ASO, QST 2808 MUP, BAY2100	Suspension concentrate
Bacillus subtilis	Epic, Kodiak	Dry powder
Candida oleophila	Aspire	Wettable powder
Clonostachys rosea (formerly *Gliocladium catenulatum*)	Primastop biofungicide, Prestop Mix	Powder
Coniothyrium minitans	Contans	Aqueous biomass suspension
Fusarium oxysporum non-pathogenic	Biofox C, Fusaclean	Dust or alginate prill, microgranules
Gliocladium virens	GlioGard 12G	Granule
Metschnikowia fructicola	SHEMER	Water-dispersible granule
Phlebiopsis gigantea (several strains)	Rotstop	Spores in inert powder
Pseudomonas chlororaphis	Cerall	Flowable concentrate for seed treatment
Pseudomonas fluorescens	BlightBan A506, Conquer/Victus	Wettable powder, aqueous biomass suspension
Pseudomonas syringae	Bio-Safe 10, Bio-Safe 11	Wettable powder
Pythium oligandrum	Polygandron	Granule, powder
Saccharomyces cerevisiae	ALD1202	Water dispersible granule
Streptomyces griseoviridis	Mycostop	Powder
Streptomyces lydicus	Actinovate, Actinovate Soluble, Actinovate BioFungicid	Solid, water soluble powder
Trichoderma asperellum (several strains formulated as individual or mix cultures)	Ecohope, Ecohope-Dry, Quality WG, Trichodermax EC, Trichotech WP	Suspension, wettable powder, water-dispersible granules, emulsifiable concentrate
Trichoderma atroviride	Esquive WP, Trichopel, Trichodry, Trichospray, Vinevax Bio-dowel, Sentinel, Tenet	Wettable powder, pellet for soil incorporation
Trichoderma gamsii	Remedier WP	Wettable powder
Trichoderma harzianum (several strains)	TRIANUM-P, TRIANUM-G, Binab T wettable powder biorational fungicide, Rootshield WP biological fungicide, T-22 HC, T-22 WP, T-22 Granules, T-22 Planter Box, T-22 technical, T-Gro, Floragard, Trichodex, Supresivit	Wettable powder, granule, dust
Trichoderma polysporum	BINAB TF WP	Wettable powder
Trichoderma virens (several strains)	G-41 Technical, BW240 G, BW240 WPBiological Fungicide, Biocure F, Bio-Shield, Bioveer	Wettable powder, granules
Verticillium albo-atrum	Dutch Trig	Ultralow volume suspension

Table 4.8 Commercial biological acaricide, insecticides, and nematicides based on entomopathogenic fungi (Ruiu 2018; European Food Safety Authority, EFSA, documents)

Active ingredients	Commercial names	Formulation type
Beauveria bassiana	balEnce™ Fly Spray, BioCeres WP, Broadband, Bio-Power, BotaniGard® (22WP, ES, MAXX), Mycotrol® (ESO, WPO), Daman, Naturalis-L, Nagestra, Green Beauveria, Beauvitech-WP, Myco-B2, DuPont Benevia, Bb-Protec, Racer, Velifer	Liquid suspension, wettable powder, oil dispersion, emulsifiable suspension
Beauveria brongniartii	Bas-Eco	Liquid formulation, wettable powder
Hirsutella thompsonii	No-Mite, Hirsutellin	Liquid formulation, aqueous suspension, wettable powder
Isaria fumosorosea (formerly *Paecilomyces fumosoroseus*)	Preferal, Bioact WG, No-Fly-WP, Paecilomite	Water-dispersible granules, wettable powder
Lecanicillium muscarium (sin. *Lecanicillium lecanii* formerly *Verticillium lecanii*)	Mycotal, Bio Fire, VertiSoft, Vertiguard, Peak Victor, Verticon, Verti-Q, Vertici Power, Lecatech-WP, Varunastra, Bio-Catch, Mealikil, Bioline/Verti-Star	Wettable powder, liquid formulation
Mix of *Arthrobotrys oligospora*, *Hirsutella rhossiliensis*, and *Acremonium butyri*	Custon NC	Liquid formulation
Metarhizium anisopliae	Devastra, Kalichakra, Bio-Magic, Bio King, Metar-Q, Meta Power, Bio Storm, Emrald Dakshan, Biomet/Ankush, Novacrid, Met52/BIO1020 granular, Pacer	Liquid, granular, and powder formulations, emulsifiable concentrate
Metarhizium brunneum	Attracap	Granular formulation
Myrothecium verrucaria	DiTera	Dry flowable
Paecilomyces lilacinus (current name *Purpureocillium lilacinum*)	Bio-Nematon, Nematofree, MeloCon, BIONICONEMA, Mytech-WP, Paecilo, Wellpacilo, Agronema, Green Nemagon, Ecopal	Wettable powder, wettable granules, liquid formulation

with Gram-negative bacteria (Chattopadhyay et al. 2017).

Current goals include the identification and development of novel pathogens/strains and toxins that increase efficacy and extend the activity range.

Among other types of biological insecticides are entomopathogenic viruses such as *Spodoptera littoralis* nucleopolyhedrovirus, *Helicoverpa armigera* nucleopolyhedrovirus (HearNPV), and *Cydia pomonella* granulovirus (CpGV).

4.4.4 Plant Extracts Used in Plant Protection

Several plant extracts, decoctions, oils, and powders have been demonstrated to act as PPPs, being approved as fungicides, insecticides, acaricides, repellents or plant growth regulators.

When such natural products are approved as active substances in PPPs, the qualified presumption that they are safe is considered if these botanicals or botanical extracts are used as ingredients in food supplements or traditional herbal medicine.

Equisetum arvense L. is used as a natural PPP, prepared as a decoction in the water of dried edible sterile aerial stems. Its application as a fungicide is used to control foliar pathogens such as *Venturia inaequalis* and *Podosphaera leucotricha* in apple trees; *Taphrina deformans* in peach trees; downy and powdery mildew in grapevine; *Podosphaera xhanthii*, *Pythium* spp., *Alternaria solani*, and *Septoria lycopersici* in legumes (SANCO/12386/2013–rev. 5/2014).

Mustard seed powder of *Sinapis alba (Brassica alba)*, *Brassica juncea*, and *Brassica nigra* is approved in plant protection as a water-dispersible powder for slurry seed treatment. It is used against common bunt *Tilletia caries* and *Tilletia foetida* in different wheat species. It is recommended as a mix of 1.5 kg of mustard seed powder with 4.5 L water to create a slurry for treating 100 kg of seeds (SANTE/11309/2017–rev. 2/2017).

Salix spp. cortex is used as a natural PPP prepared as a water infusion. Its application as a fungicide is used to control peach leaf curl, foliar scab disease, and powdery mildews in apple trees, or powdery and downy mildews in grapevine (SANCO/12173/2014–rev. 4/2015).

Urtica dioica and *Urtica urens* extract is used as natural a PPP, prepared as a complex mixture for spray applications or soil-covering (mulch). Its function in plant protection is insecticidal, fungicidal, and acaricidal. As an insecticide, *Urtica* spp. Extract is used against a wide number of aphid species on fruit trees, bean, potato, leaf vegetables, and woody ornamental plants, against flea beetle and diamondback moths on cabbage, rapeseed, and radish, or against codling moth on apple and pear trees. Its acaricidal activity is against the two-spotted spider mite and red spider mite on beans and grapevine. Against fungi, *Urtica* spp. extract is used to prevent and control *Alternaria* sp. on Brassicaceae and Cucurbitaceae species and several fruit trees, powdery mildew in cucumber, brown rot blossom blight, gray mold and black bread mold of fruit trees, downy mildew of grapevine and potato blight (SANTE/11809/2016–rev.0.1/2017).

Several plant oils have also been approved as natural PPPs. Plant oils such as Citronella oil, clove oil, spear mint oil, and rapeseed oil have been approved for different purposes, such as herbicides, acaricides, insecticides, or repellants. Sunflower oil is used as a fungicide against tomato powdery mildew (SANTE/10875/2016), and onion oil is used as an insecticide against carrot root fly (SANTE/10615/2018–rev.1/2018).

4.5 Energy Crops in Europe: Feasible Resources for Biofuel Production

In the past decades, biofuels have attracted a lot of attention owing to the increasing demand on energy resources in addition to increasing concerns about greenhouse gas emissions due to the use of fossil fuels. Based on the type of the feedstock used, biofuels are classified into four generations. First-generation biofuels make use of edible biomass, which has caused controversy because it competes with global food needs. The second-generation biofuels are based on non-edible biomass, but there are some limitations related to the cost-effectiveness when scaling-up the production to a commercial level. The third-generation biofuels use the microorganisms as feedstock, while in the case of the fourth-generation biofuels the focus is on genetically modifying microorganisms able to achieve a preferable yield in the ratio hydrogen/carbon to eliminate or minimize carbon emissions (EC 2016).

Despite European efforts in the past decade, the USA is still the leader on the biofuel market with a target of substituting 20% of transportation fossil fuels with biofuel by 2022 (Saladini et al. 2016). The European Union 2020 Climate and Energy Package has committed to a 20% reduction of greenhouse gases, in addition to a target of a 20% renewable share in the energy market and a 20% increase in energy efficiency by 2020. Future ambitious targets are set for 2030 under the Climate and Energy Framework; these targets consist of a 40% reduction in emissions, with a 27% renewable share of the energy mix (Krol et al. 2019).

In the following paragraphs we are going to analyze the economic potential of the *second-generation biofuels*, which are based on renewable alternatives by utilizing inedible lignocellulosic biomass such as annual or perennial plants. We are not aiming to present here the drawbacks of the technical conversion into biofuel (thermal, biological, enzymatic or chemical processes), but we do focus on the non-edible feedstock production as a resource for the economic income of the farmers, because this biomass is considered to be an inexpensive, attractive biofuel resource (Westensee et al. 2018). Bioethanol can be produced from lignocellulosic biomass through hydrolysis and subsequent fermentation; this is why in *bioethanol* production the use of fermentative microorganisms is a must. Such examples of microorganisms can be yeast (*Saccharomyces*), bacteria (*Zymomonas*) or even molds.

Plants have been traditionally used for food, fiber, and feed applications. Their utilization for biofuels may require the breeding of novel phenotypes, or entirely new species. Scientists have provided different strategies for the genetic selection of plants as sources of biomass for biofuel production. Genetic modification of plants provides a wide range of options for improving the composition of biomass and for plant modifications to assist the fabrication of biofuels (Furtado et al. 2014). More references and solutions are provided in a special chapter included in this handbook, "Creating products and services in plant biotechnology".

In the past decades, most of the *biodiesel* was made from soybean, rapeseed, sunflower, and palm oils, whereas soybean oil was commonly used in the USA, about 80% of the EU's total biofuel production came from rapeseed and sunflower seeds (Demirbas et al. 2016). Because of socio-economic issues, nowadays biodiesel produced from edible vegetable oils is considered non-feasible and solutions have been proposed. Apart from different forms of agricultural waste, a wide variety of plants can be used as lignocellulosic biomass for biofuel production such as poplar trees, willow and eucalyptus, miscanthus, switchgrass, reed canary grass, camelina, Jatropha jojoba oil, etc.

The EU is one of the few global biofuel markets that explicitly addresses the sustainability impacts of biofuels; models of potential future biofuel systems have permanently focused on the resources available within certain EU regions and national areas (Tomei and Helliwell 2015).

When making the choice of which energy crop should be cultivated, apart from the plants' adaptability to different European climatic areas, it is important to have an already well-defined cultivation and harvesting technology. Generally, it is recognized that grass-plants (non-woody) are preferable in terms of cultivation technology because they can be employed in common agricultural techniques, which are not bringing complications to farmers. Still, some of the farmers consider that perennial crops are more simple to cultivate and harvest, being more profitable; in the latter case, high costs are involved only during the first year, when setting up the perennial plantation; costs are assumed to be 1.5 to 3 times higher than comparable costs of annual planting/seeding (OECD 2004), which is why incentives/subsidies from the governments are required.

In Europe, over the last few years, different trials have been conducted to establish feasible technologies for the cultivation and harvesting of suitable and *non-edible energy crops*, such as sorrel, red canary grass, camelina, miscanthus, ready to be implemented on a large scale by the farmers. The largest areas of energy crops reported in 2009 by Intelligent Energy Europe were found in Finland (reed canary grass), the UK (mainly willow and miscanthus), Sweden (reed canary grass, willow), Spain and Italy (miscanthus, poplar), and Germany (miscanthus, willow). According to the EU Energy Reference Scenario 2013, about 10.6% of land uses will change across the EU28 during the period 2010–2050. The largest countries producing energy crops are expected to be Poland, France, Germany, Spain, Romania and the UK, together accounting for about 83% of the total European acreage (Perpiña Castillo et al. 2015). In the following, some technological solutions reported in different EU countries are presented.

Among *annual herbaceous plants* cultivated in Europe for energy, the most frequently studied in the past few years, proving high potential, is camelina (*Camelina sativa* L. Crantz).

Camelina (*Camelina sativa*)

Camelina (*Camelina sativa* L. Crantz) is an annual plant member of the Brassicaceae family and returned to the attention of farmers relatively recently. The main advantage of the culture is that camelina is adaptable to many different environmental conditions and the only real limitations are heavy clay and organic soils. Camelina is a short-season crop (85–100 days) that is well adapted to production in the temperate climatic zone, germinating at a low temperature.

Camelina Company España (CCE) is the European reference company for the production of camelina. CCE develops camelina plantations for the production of camelina oil and meal. In cooperation with UASMV Bucharest, CCE has delivered a simple camelina cultivation technology for farmers with lands in a continental climate. It is recommended to plant camelina on plots that are sufficiently fertile and free of weeds, especially broad leaf weeds. Land with flooding or crusting problems must be avoided, in addition to shallow soils. The presence of residual herbicides affecting the Brassicaceae family must be avoided. The seeding rate is 6–8 kg/ha and the seeding depth is less than 1 cm. The seeding moment differs; it can be early autumn or spring. Regarding fertilization, the proposed scheme is (in fertilizing units/ha): N 50–60; P 30–40; the amount of the fertilizer is distributed between background and dressing fertilization. The optimal time for harvesting is when the crop becomes a yellow-cream color (Fig. 4.3). The yields vary according to environmental conditions and vary from 1200 to 2100 kg/ha in conditioned seeds.

It has been demonstrated that camelina can be cultivated as a double crop in Romania if the following requirements are met: sowing must be done at least 3–4 weeks prior to the sowing date of this trial; fertilization is a must, in addition to watering (Dobre et al. 2014).

A simple efficiency cost calculation has been proposed in Romania by CBM Biotehgen. To set up and harvest camelina seeds from a hectare, the estimated costs are 1800 Romanian Leu (RON)/ha (about 360 Euro/ha). If technology is respected, an average of 2 tonnes of dried seeds are obtained per hectare, the source of 600 l pressed oil. The costs of pressing 1 tonne of seeds is about 170 RON (34 Euro). The final cost for 1 l of pressed camelina oil is 3 RON (0.6 Euro/l).

A Case Study for Biojet Production: Camelina

An important example of scaling-up the biojet production on a European level was the FP7 funded project "Initiative towards sustainable kerosene for aviation" (ITAKA), coordinated by SENASA (Spain). The project has demonstrated the feasibility of using the biojet mixed into conventional airport fuel systems during conventional operation of the airport. Since the project at the end of 2015, all airplanes departing from

Fig. 4.3 Camelina culture during full-flowering (left) and at full maturity, ready for harvesting (right)

Gardermoen Oslo airport are partially using biojet fuel (below 3%), which would account for about 60,000 flights and about 6 million passengers, according to Avinor statistics from January to June 2016. Such results have been possible thanks to 100% EU-made biojet fuel, based on camelina oil produced in Spain and Romania and later refined in Finland. An important conclusion from ITAKA is related to the feedstock; camelina may be cultivated in a fallow rotation scheme (in dry land from Spain) or on polluted lands (demonstrated in Romania), bringing environmental and socio-economic benefits. It is estimated that in Eastern European countries there about 900,000 ha of polluted lands are available, ready to be used for the production of feedstock for biofuel (ITAKA sources). Meanwhile, in the ITAKA project, the camelina's cultivation protocol for cropping the plant under different European climate conditions has been developed and delivered to the farmers. During the project implementation, the consortium has learnt that even on this small scale, the availability of sustainable feedstock is a clear bottleneck: new crops, such as *Camelina sativa*, require a long time to expand and become significant.

As a consequence, UASMV Bucharest, in cooperation with CBM Biotehgen Romania, has been developing a new camelina variety adapted to heavy winters and resistant to strong winds (Matei et al. 2014); after 3 years of trials, the variety was patented in Romania under the name "Madalina" (Sauca et al. 2018).

In Europe, as energy crops are the preferred *perennial species*, woody or non-woody. The most frequently studied and successful species are sorrel, reed canary grass, willow, and miscanthus.

Sorrel (*Rumex* sp.)

Sorrel is a perennial herb with aerial parts of about half a meter and with roots that run deep into the ground. It can be propagated by seeds. As an energy crop a hybrid sorrel (*Rumex patientia* × *Rumex tianschanicus*) is actually used, known as sorrel of Uteush. Long-term trials have confirmed that the hybrid sorrel is one of the perennial energy crops with potential suitable for fuel biomass cultivation as a renewable source of energy in the European temperate climate (Ustak and Ustaková 2004). The studies have revealed that it is a plant with a high ecological plasticity, cold and winter resistance, and ia tolerant to salt stress and increased humidity (Kosakivska et al. 2008). According to Zhuang et al. (2005), hybrid sorrel has also been proven to be tolerant to heavy metals, and has potential in the phytoremediation of soils contaminated with heavy metals.

Nielsen (2008) has conducted trials in Norway on hybrid sorrel. The proposed technology for small parcels (for trials) was the following: sowing into rows30 cm apart, with a rate of 11 kg/ha; mineral fertilization with P, K, and 65 kg nitrogen/ha/year for the first 2 years, and then no fertilization in the 3rd year. On large-scale production, the seeding rate is recommended to be 5 kg/ha and the output is about 500 kg/ha of seeds.

Sorrel cultivation has been also tested in the Czech Republic as a pilot plot for commercial use; tests were done with the fodder sorrel of the variety Rumex OK 2, a hybrid of *Rumex patientia L.* and *Rumex tianschanicus A.Los*. REPROMO partnership (2003), coordinated by ENVIROS (Czech Republic) has drawn some conclusions. Fodder sorrel is productive for 10–12 years; the sowing rate is 5 kg/ha; it is highly tolerant to agrometeorological conditions and has low requirements for fertilization; it can be harvested in early July, the ripe biomass has low humidity (in dry weather 15–20%), so there is no need for additional drying; the dry biomass is 10–14 tonnes/ha since the 3rd year of cultivation.

Reed Canary Grass (*Phalaris arundinacea*)

Reed canary grass is a rhizomatous, perennial grass species that can be cultivated on the low-value areas and on marginal lands, which are in use for food production. It is mostly grown as a fodder crop in Europe, especially in the northern part, such as Norway, Denmark, or Sweden. The first crop can be harvested 2 years after sowing.

In a trial conducted by Nielsen (2008) in Norway, the culture was fertilized with 85 kg nitrogen per hectare per year in the form of mineral fertilizer also containing phosphorus and potassium. The culture was reported to have some problems because of the presence of weeds. The yield varied among the testing years from 6100 g/ha to 9000 kg/ha.

In Finland and Sweden reed canary grass is typically harvested in the spring, after the snow melts, when the water content of the biomass has decreased to the level enabling storage without additional drying. In the spring-harvested crop, the ash content is lower and the ash melting point higher than in an autumn-harvested crop (Intelligent Energy Europe 2009).

An example of good practice of the use of reed canary grass for biofuel production is the power plant of Kokkolan Voima Ltd. from Finland, which was commissioned in 2001.

Miscanthus (*Miscanthus* sp.)

Miscanthus is a perennial tropical plant, successfully adapted to temperate areas by European researchers. Its stems can be used as a heating biomass source or can be transformed into other useful products such as biogas, bioethanol or biodiesel. The sterile hybrid genotype *Miscanthus* × *giganteus*, obtained from *Miscanthus saccharifIorus* and *Miscanthus sinensis*, has attracted attention and is widely used in Europe. It has had perennial growth for 10–15 years, is considered a non-invasive plant, and the roots can reach up to 6 m deep to find water.

In Romania, a cultivation technology has been proposed by INMA Bucharest. The planting materials are rhizomes with a minimum of 3–4 viable buds; the optimal planting depth is 80–100 cm; and fertilization is compulsory only during the first year of cultivation. Harvesting starts in the 3rd year of cultivation by using harvesters or balers; every year before harvesting, the plants should be cut (in March–April) and the cuts should remain on the soil.

Harvestable *Miscanthus* yields (dry matter) have been reported to range from 15 to 44 tonnes/ha in Europe; higher values can be achieved in irrigated areas of southern Europe than in northern Europe owing to its higher average temperature and high accumulation of solar radiation (Brosse et al. 2012).

An example of cost calculation per hectare is provided by ARGE Mischantus Romania: the culture establishment cost is around 2400 Euros/ha of which about 90% is the cost of the rhizomes and the rest agricultural work); harvesting costs are about 50 Euros/ha; considering the lifetime of the culture is 25 years, of which 22 years will have full productivity, the total costs/year are 3600 Euros/ha; for an average yield of 20 tonnes of dry matter/ha/year after the 3rd year, the cost of a tonne of dry material is 7.95 Euros; if bracketing, additional costs of 20 Euros/tonne are considered.

Willow (*Salix* spp.)

Energetic willow (tree and shrub species) is characterized by a short rotation cycle and vegetative regeneration; it has rapid growth of up to 3–3.5 cm/day and a lifespan of 20–25 years. In 2–3 years, it can reach a shoots height of 6–7 m, with a shoots base diameter of 6–8 cm. Beginning with the 2nd/3rd year, a yield of at least 35 tonnes/ha/year (wet) biomass can be obtained in the form of raw biomass—bales, chips, or straight rods and pellets or briquettes. Clones of *Salix viminalis* are mainly used in energy forestry. Other species, such as *Salix dasyclados*, have also been cultivated, but to a much more limited extent.

Willow plantations are established from cuttings during spring. Planting material should be bought from a well-known source to ensure a good starting quality. Willow is usually supplied as 2- to 3-m long branches that are cut between December and March when the buds are fully dormant. The branches can be planted immediately or stored in cool conditions (−2 to −4 °C) until they are used. It is compulsory to protect the planting material from moisture loss prior to planting. Weed control during the first year is very important. The plantations are very demanding of water and nutrients, generally requiring 3–5 mm of water per day during the growing season. The demand for nutrients varies according to age of the plantation and the stage of crop development. For example, no nitrogen fertilization is recommended in Sweden during the year of establishment, but 45 kg nitrogen per hectare should be applied during the second year (i.e., the first harvest) and 100–150 kg nitrogen during the 3rd and 4th years (Intelligent Energy Europe 2009). Willow is better harvested in the winter when the ground is frozen and the moisture content in the biomass is lowest. A constant biomass quantity can be harvested for 20–25 years. By applying adequate technology, yields of over 60 tonnes/ha can be obtained. Without irrigation, over last the 5–6 years, average yields reached at the most 30–35 tonnes/ha/year on well-administered plantations (Jovicic 2016).

An example of cost breakdown for willow cultivation (Table 4.9) is proposed by Jovicic (2016) after information provided by REBINA Group Romania (group of German–Austrian–Romanian companies promoting non-conventional and renewable energy sources).

According to an Intelligent Energy Europe report (2009), a *good practice case* is the multi-functional willow plantation in Enköping (Sweden), which is an example of large-scale energy farming. The biomass of willow is supplied in chips to the ENA Energy's CHP plant in Enköping. The original concept comprises about 80 ha of willow plantation, an irrigation system, and three ponds connected to the municipal wastewater treatment plant.

Another example of good practice in the willow sector is SalixEnergi Europa, which is a privately owned incorporated company with its head office located in southern Sweden. They have conducted plant breeding for over 25 years and delivered new willow varieties that provide superior yields and characteristics. The company is offering planting project management, from concept development, planning and preparation to site management. Planting projects of 2–2000 ha have been implemented across Europe, for private and public investors and end users.

Table 4.9 Cost breakdown (Euros/ha) for willow cultivation (after Jovicic 2016)

Action	Estimated costs
Planting material (cuttings)	1400
Soil preparation	1000
Irrigation system	1000
Harvest, maintenance, land lease/ annually	300

References

Abatenh E, Gizaw B, Tsegaye Z, Wassie M (2017) The role of microorganisms in bioremediation – a review. Open J Environ Biol 2(1):038–046

Abdullahi MS (2015) Soil contamination, remediation and plants, Chapter 18. In: Hakeem KR, Sabir M, Öztürk M, Mermut AR (eds) Soil remediation and plants. Academic, Cambridge, MA

Adhikary SD (2013) Vermicompost, the story of organic gold: a review. Agric Sci 3(7):905–917

Adriano DC, Wenzel WW, Vangronsveld J, Bolan NS (2004) Role of assisted natural remediation in environmental cleanup. Geoderma 122:121–142

Ali H, Khan E, Sajad MA (2013) Phytoremediation of heavy metals–concepts and applications. Chemosphere 91:869–881

Alkorta I, Hernandez-Allica J, Becerril JM, Amezaga I, Albizu I, Garbisu C (2004) Recent findings on the phytoremediation of soils contaminated with environmentally toxic heavy metals and metalloids such as zinc, cadmium, lead, and arsenic. Rev Environ Sci Bio-Technol 3:71–90

Andreazza R, Bortolon L, Pieniz S, Camargo FAO, Bortolon ESO (2013) Copper phytoextraction and phytostabilization by *Brachiaria decumbens* Stapf. in vineyard soils and a copper mining waste. Open J Soil Sci 3(6):273–282. https://doi.org/10.4236/ojss.2013.36032

AVINOR statistics, January–June 2016

Ayotamuno JM, Kogbara RB, Agoro OS (2009) Biostimulation supplemented with phytoremediation in the reclamation of a petroleum contaminated soil. World J Microbiol Biotechnol 25:1567–1572. https://doi.org/10.1007/s11274-009-0045-z

Azubuike CC, Chikere CB, Okpokwasili GC (2016) Bioremediation techniques–classification based on site of application: principles, advantages, limitations and prospects. World J Microbiol Biotechnol 32(11):180

Biswas B, Sarkar B, Rusmin R, Naidu R (2015) Bioremediation of PAHs and VOCs: Advances in clay mineral microbial interaction. Environ Int 85:168–181. https://doi.org/10.1016/j.envint.2015.09.017

Bolan NS, Park JH, Robinson B, Naidu R, Huh KY (2011) Phytostabilization: a green approach to contaminant containment. Adv Agron 112:145–204. https://doi.org/10.1016/B978-0-12-385538-1.00004-4

Boroş MN, Micle V, Flămînd L, Sur IM (2016) Phytoremediation planning in the case of former industrial sites. Studia Ubb Ambientum LXI(1–2):5–14

Brosse N, Dufour A, Meng X, Sun Q, Ragauskas A (2012) Miscanthus: a fast-growing crop for biofuels and chemicals production. Biofuels Bioprod Biorefin. https://doi.org/10.1002/bbb

Brown LD, Ulrich AC (2014) Bioremediation of oil spills on land. In: Handbook of oil spill science and technology. Gulf, Houston, TX, pp 395–406

Chattopadhyay P, Banerjee G, Mukherjee S (2017) Recent trends of modern bacterial insecticides for pest control practice in integrated crop management system. 3 Biotech 7:60. https://doi.org/10.1007/s13205-017-0717-6

Conesa HM, Evangelou MWH, Robinson BH, Schulin R (2012) A critical view of current state of phytotechnologies to remediate soils: still a promising tool? Sci World J. https://doi.org/10.1100/2012/173829

Demirbas A, Bafail A, Ahmad W, Sheikh M (2016) Biodiesel production from non-edible plant oils. Energy Explor Exploit 34(2):290–318

Dobre P, Jurcoane S, Matei F, Cristea S, Farcas N, Moraru AC (2014) *Camelina sativa* as a double crop using the minimal tillage system. Roman Biotechnol Lett 19 (2):9190–9195

EC (2016). https://ec.europa.eu/clima/news/articles/news_2016100401_en

EFSA (European Food Safety Authority) (2017) Technical report on the outcome of the consultation with Member States and EFSA on the basic substance application for mustard seeds powder from *Sinapis alba* (*Brassica alba*), *Brassica juncea* and *Brassica nigra* for use in plant protection as fungicide. EFSA supporting publication 2017:EN-1169. 35 p. doi:https://doi.org/10.2903/sp.efsa.2017.EN-1169

Ekta P, Modi NR (2018) A review of phytoremediation. J Pharmacogn Phytochem 7(4):1485–1489

El-Bendary MA (2006) *Bacillus thuringiensis* and *Bacillus sphaericus* biopesticides production. J Basic Microbiol 46:158–170

Elekwachi CO, Andresen J, Hodgman TC (2014) Global use of bioremediation technologies for decontamination of ecosystems. J Bioremed Biodeg 5:225. https://doi.org/10.4172/2155-6199.1000225

EU Energy, Transport and GHG Emissions – Trends to 2050. http://ec.europa.eu/energy/observatory/trends_2030/index_en

Fravel DR, Connick WJ, Lewis JA (1998) Formulation of microorganisms to control plant diseases, Chapter 5. In: Burges HD (ed) Formulation of microbial biopesticides, 1st edn. Springer Science+Business Media, Dordrecht, pp 187–202. https://doi.org/10.1007/978-94-011-4926-6

Furtado A, Lupoi JS, Hoang NV, Healey A, Singh S, Simmons BA, Henry RJ (2014) Modifying plants for biofuel and biomaterial production. Plant Biotechnol J 12:1246–1258

Gao H, Xie Y, Hashim S, Akhtar Khan A, Wang X, Xu H (2018) Application of microbial technology used in bioremediation of urban polluted river: a case study of Chengnan River, China. Water 10(5):643

Ghosh M, Singh SP (2005) A review on phytoremediation of heavy metals and utilization of its byproducts. Appl Ecol Environ Res 3(1):1–18

Gillespie IM, Philp JC (2013) Bioremediation, an environmental remediation technology for the bioeconomy. Trends Biotechnol 31(6):329–332

Glare TR, Jurat-Fuentes J-L, O'Callaghan M (2017) Basic and applied research: entomopathogenic bacteria, Chapter 4. In: Microbial control of insect and mite pests from theory to practice. Academic, Cambridge, MA, pp 47–67. https://doi.org/10.1016/B978-0-12-803527-6.00004-4

Grison C, Escande V, Biton J (2015) Ecocatalysis – new integrated approach to scientific ecology. ISTE Press Elsevier, London

http://ec.europa.eu/assets/sante/food/plants/pesticides/lop/index.html#inline_content

http://ec.europa.eu/food/plant/pesticides/eu-pesticides-database/public/?event=activesubstance.selection&language=EN

http://www.cpeo.org/techtree/ttdescript/bioven.htm

https://www.certiseurope.co.uk/products/insecticides/

https://www.epa.nsw.gov.au/~/media/EPA/Corporate%20Site/resources/clm/140323landfarmbpn.ashx

Huang XD, El-Alawi Y, Gurska J, Glick BR, Greenberg BM (2005) A multi-process phytoremediation system for decontamination of persistent total petroleum hydrocarbons (TPHs) from soils. Microchem J 81:139–147

Huesemann MH, Moore KO (1993) Compositional changes during landfarming of weathered Michigan crude oil-contaminated soil. J Soil Contam 2:245–246

Intelligent Energy Europe (2009) Energy from field energy crops – a handbook for energy producers. Jyväskylä Innovation Oy & MTT Agrifood Research Finland

January MC, Cutright TJ, Van Keulen H, Wei R (2008) Hydroponic phytoremediation of Cd, Cr, Ni, As, and Fe: Can *Helianthus annuus* hyperaccumulate multiple heavy metals? Chemosphere 70:531–537

Jovicic B (2016) Energy willow *Salix viminalis* – biomass where you want it. Blakan Green Energy News. 26 Apr 2016

Kalogerakis N, Arff J, Banat IM, Broch OJ, Daffonchio D, Edvardsen T et al (2015) The role of environmental biotechnology in exploring, exploiting, monitoring, preserving, protecting and decontaminating the marine environment. New Biotechnol 32(1): 157–167

Kaya HK, Klein MG, Burlando TM (2008) Impact of *Bacillus popilliae*, *Rickettsiella popilliae* and entomopathogenic nematodes on a population of the scarabaeid, *Cyclocephala hirta*. Biocontrol Sci Tech 3:443–453

Khan FA (2016) Biotechnology fundamentals, 2nd edn. CRC, Boca Raton, FL

Kosakivska D, Klymchuk V, Negretzky D, Bluma D, Ustinova A (2008) Stress proteins and ultrastructural characteristics of leaf cells of plants with different types of ecological strategies. Gen Appl Plant Physiol 34(3–4):405–418

Krol DJ, Jones MB, Williams M, Ni Choncubhair O, Lanigan GJ (2019) The effect of land use change from grassland to bioenergy crops Miscanthus and reed canary grass on nitrous oxide emissions. Biomass Bioenergy 120:396–403

Leitão A (2016) Bioeconomy: the challenge in the management of natural resources in the 21st century. Open J Soc Sci 4:26–42. https://doi.org/10.4236/jss.2016.411002

Majumder DR (2016) A multifaceted bioremediation of xenobiotics using fungi. In: Sangeetha J, Thangadurai D, David M, Abdullah MA (eds) Environmental biotechnology: biodegradation, bioremediation, and bioconversion of xenobiotics for sustainable development. CRC, Boca Raton, FL

Matei F, Sauca F, Dobre P, Jurcoane S (2014) Breeding low temperature resistant *Camelina sativa* for biofuel production. New Biotechnol 31:S93

Meeinkuirta W, Kruatrachueb M, Pichtelc J, Phusantisampand T, Saengwilaib P (2016) Influence of organic amendments on phytostabilization of Cd-contaminated soil by *Eucalyptus camaldulensis*. Sci Asia 42:83–91

Nielsen HK (2008) Sorrel and reed canary grass in southern Norway. Asp Appl Biol Biomass Energy Crops III 90:75–79

OECD (2004) Biomass and agriculture sustainability, markets and policies. doi:https://doi.org/10.1787/9789264105546-en

Peixoto RS, Vermelho AB, Rosado AS (2011) Petroleum-degrading enzymes: bioremediation and new prospects. Enzyme Res. doi:https://doi.org/10.4061/2011/475193

Perez-Vargas I, Poggi-Varaldo HM, Calva-Calva G, Rios-leal E, Rodriguez-Varquez R, Ferrera-Cerralto R, Esparza-Garcia F (2000) Nitrogen-fixing bacteria capable of utilizing kerosene hydrocarbons as a sole carbon source. Water Sci Technol 42:407–410

Perpiña Castillo C, Baranzelli C, Maes J, Zulian G, Lopes Barbosa A, Vandecasteele I, Mari Rivero I, Vallecillo Rodriguez S, Batista E Silva F, Jacobs C, Lavalle C (2015) An assessment of dedicated energy crops in Europe under the EU Energy Reference Scenario 2013 Application of the LUISA modelling platform – Updated Configuration 2014. EUR 27644. doi:https://doi.org/10.2788/64726

Piotrowski J, Denich T, Klironomos JM, Graham J, Rillig M (2004) The effects of arbuscular mycorrhizas on soil aggregation depend on the interaction between plant and fungal species. New Phytol 164:365–373. doi: https://doi.org/10.1111/j.1469-8137.2004.01181.x

Prasad MNV (2003) Phytoremediation of metal-polluted ecosystems: hype for commercialization. Russ J Plant Physiol 50(5):686–700

Pulford ID, Watson C (2003) Phytoremediation of heavy metal-contaminated land by trees – a review. Environ Int 29(4):529–540

Qureshi N, Chawla S, Likitvivatanavong S, Lee HL, Gill SS (2014) The Cry toxin operon of *Clostridium bifermentans* subsp. malaysia is highly toxic to *Aedes* larval mosquitoes. Appl Environ Microbiol 80:5689–5697

Rahman MA, Hasegawa H (2011) Aquatic arsenic: phytoremediation using floating macrophytes. Chemosphere 83:633–646

Robinson BH, Banuelos G, Conesa HM, Evangelou MWH, Schulin R (2009) The phytomanagement of trace elements in soil. Crit Rev Plant Sci 28:240–266

Rodriguez-Campos J, Dendooven L, Alvarez-Bernal D, Contreras-Ramos SM (2014) Potential of earthworms to accelerate removal of organic contaminants from soil: a review. Appl Soil Ecol 79:10–25

Rorat A, Wloka D, Grobelak A, Grosser A, Sosnecka A, Milczarek M et al (2017) Vermiremediation of polycyclic aromatic hydrocarbons and heavy metals in sewage sludge composting process. J Environ Manag 187:347–353

Ross R (2018) The science behind composting. https://www.livescience.com/63559-composting.html

Ruffner B, Péchy-Tarr M, Höfte M, Bloemberg G, Grunder J, Keel C, Maurhofer M (2015) Evolutionary patchwork of an insecticidal toxin shared between plant-associated pseudomonads and the insect pathogens Photorhabdus and Xenorhabdus. BMC Genomics 16:609

Ruiu L (2018) Microbial biopesticides in agroecosystems. Agronomy 8:235. https://doi.org/10.3390/agronomy8110235

Saladini F, Patrizi N, Pulselli FM, Marchettini N, Bastianoni S (2016) Guidelines for energy evaluation of first, second and third generation biofuels. Renew Sustain Energy Rev 66:221–227

Sampanpanish P, Pongsapich W, Khaodhiar S, Khan E (2006) Chromium removal from soil by phytoremediation with weed plant species in Thailand. Water Air Soil Pollut Focus 6:191–206. https://doi.org/10.1007/S11267-005-9006-1

SANCO/12173/2014–rev. 4/2015. Final review report of the Standing Committee on Plants, Animals Food and Feed at its meeting on 29 May 2015 in view of the approval of *Salix* spp. cortex as basic substance in accordance with Regulation (EC) No 1107/2009

SANCO/12386/2013–rev. 5/2014. Final review report of the Standing Committee on the Food Chain and Animal Health at its meeting on 20 March 2014 in view of the approval of *Equisetum arvense* L. as basic substance in accordance with Regulation (EC) No 1107/2009

SANTE/10615/2018–rev.1/2018. Final review report of the Standing Committee on Plants, Animals, Food and Feed at its meeting on 20 July 2018 in view of the approval of onion oil as basic substance in accordance with Regulation (EC) No 1107/2009

SANTE/10875/2016. Final review report of the Standing Committee on Plants, Animals, Food and Feed at its meeting on 7 October 2016 in view of the approval of sunflower oil as basic substance in accordance with Regulation (EC) No 1107/2009

SANTE/11309/2017–rev. 2/2017. Final review report of the Standing Committee on Plants, Animals, Food and Feed at its meeting on 6 October 2017 in view of the approval of mustard seeds powder as basic substance in accordance with Regulation (EC) No 1107/20091

SANTE/11809/2016–rev. 0.1/2017. Final Review report of the Standing Committee on Plants, Animals, Food and Feed at its meeting on 24 January 2017 in view of the approval of *Urtica* spp. as basic substance in accordance with Regulation (EC) No 1107/2009

Santra SC, Mallick A, Samal AC (2015) Biofertilizer for bioremediation. In: Pati BR, Mandal SM (eds) Recent trends in biofertilizers. I K International Publishing House Pvt. Ltd., pp 205–234

Sarkar J, Kazy SK, Gupta A, Dutta A, Mohapatra B, Roy A, Sar P (2016) Biostimulation of indigenous microbial community for bioremediation of petroleum refinery sludge. Front Microbiol 7:1407

Sauca F, Jurcoane S, Dobre P, Matei F, Podgoreanu E, Moraru A, Cristea S (2018) Patent ISTIS 504/2018. New variety of *Camelina sativa* – Mădălina

Schwitzguébel JP, van der Lelie D, Baker A, Glass DJ, Vangronsveld J (2002) Phytoremediation: European and American trends successes, obstacles and needs. J Soils Sediments. https://doi.org/10.1065/jss2002.03.37

Shah MP (2017) Waste water pollution. J Appl Biomater Biomech 3(1):262–263

Shams KM, Tichy G, Fischer A, Sager M, Peer T, Bashar A, Filip K (2009) Aspects of phytoremediation for chromium contaminated sites using common plants *Urtica dioica*, *Brassica napus* and *Zea mays*. Plant Soil. https://doi.org/10.1007/s11104-009-0095-x

Sharma B, Dangi AK, Shukla P (2018) Contemporary enzyme based technologies for bioremediation: a review. J Environ Manag 210:10–22

Shoji R, Yajima R, Yano Y (2008) Arsenic speciation for the phytoremediation by the Chinese brake fern, *Pteris vittata*. J Environ Sci 20:1463–1468

Sparks T, Dripps J, Watson G, Paroonagian D (2012) Resistance and cross-resistance to the spinosyns – a review and analysis. Pestic Biochem Physiol 102:1–10

Stephenson C, Black CR (2014) One step forward, two steps back: the evolution of phytoremediation into commercial technologies. Biosci Horiz. https://doi.org/10.1093/biohorizons/hzu009

Tomei J, Helliwell R (2015) Food versus fuel? Going beyond biofuels. Land Use Policy 56:320–326

Trapp S, Karlson U (2001) Aspects of phytoremediation of organic pollutants. J Soil Sediment I:37–43

Ustak S, Ustaková M (2004) Potential for agricultural biomass to produce bioenergy in the Czech Republic. In: Parris K (ed) Biomass and agriculture: sustainability, markets and policies. OECD, Paris, pp 229–239

Vamerali T, Bandiera M, Mosca G (2010) Field crops for phytoremediation of metal-contaminated land. A review. Environ Chem Lett 8:1–17. https://doi.org/10.1007/s10311-009-0268-0

Vangronsveld J, Herzig R, Weyens N, Boulet J, Adriaensen K, Ruttens A, Thewys T, Vassilev A, Meers E, Nehnevajova E, van der Lelie D, Mench M (2009) Phytoremediation of contaminated soils and groundwater: lessons from the field. Environ Sci Pollut Res 16:765–794. https://doi.org/10.1007/s11356-009-0213-6

Vasavi M, Usha R, Swamy PM (2010) Phytoremediation – an overview review. J Ind Pollut Control 26(1):83–88

Wang XJ, Wang JJ, Wang JD, Zhang J, Xu MD, Xiang WS (2011a) Two new doramectin analogs from *Streptomyces avermitilis* NEAU1069: fermentation, isolation and structure elucidation. J Antibiot 64: 591–594

Wang XJ, Zhang J, Wang JD, Huang SX, Chen YH, Liu CX, Xiang WS (2011b) Four new doramectin congeners with acaricidal and insecticidal activity from *Streptomyces avermitilis* NEAU1069. Chem Biodivers 8:2117–2125

Wang R, Cao Q, Zhao Q, Li Y (2018) Bioindustry in China: an overview and perspective. New Biotechnol 40:46–51

Wenzel WW, Lombi E, Adriano DC (2004) Biogeochemical processes in the rhizosphere: role in phytoremediation of metal-polluted soils. In: Prasad M, Hagemeyer J (eds) Heavy metal stress in plants: from biomolecules to ecosystems. Springer, Berlin, pp 273–303

Westensee DK, Rumbold K, Harding KG, Sheridan CM, Van Dyk LD, Simate GS, Postma F (2018) The availability of second generation feedstocks for the treatment of acid mine drainage and to improve South Africa's bio-based economy. Sci Total Environ 637:132–136

White Jr PM, Wolf DC, Thoma GJ, Reynolds CM (2005) Phytoremediation of alkylated polycyclic aromatic hydrocarbons in a crude oil-contaminated soil. Water Air Soil Pollut 169: 207–220

Whiteley CG, Lee DJ (2006) Enzyme technology and biological remediation. Enzym Microb Technol 38 (3–4):291–316

Woo SL, Ruocco M, Vinale F, Nigro M, Marra R, Lombardi N, Pascale A, Lanzuise S, Manganiello G, Lorito M (2014) *Trichoderma*-based products and their widespread use in agriculture. Open Mycol J 8(Suppl-1, M4):71–126

Yadav KK, Gupta N, Kumar A, Reece LM, Singh N, Rezania S, Khan SA (2018) Mechanistic understanding and holistic approach of phytoremediation: a review on application and future prospects. Ecol Eng 120:274–298

Yateem A (2014) Bioremediation strategies for the clean-up of oil contamination in an arid environment: an insight on the Kuwaiti experience. In: Velazquez-Fernandez JB, Muniz-Hernandez S (eds) Bioremediation: processes, challenges and future prospects. Nova, Hauppauge, NY, pp 269–284

Yavari S, Malakahmad A, Sapari NB (2015) A review on phytoremediation of crude oil spills. Water Air Soil Pollut 226(8):226–279

Ying OY (2002) Phytoremediation: modeling plant uptake and contaminant transport in the soil-plant-atmosphere continuum. J Hydrol 266(1–2):66–82

Zhu X, Venosa AD, Suidan MT (2004) Literature review on the use of commercial bioremediation agents for cleanup of oil-contaminated estuarine environments EPA/600/R-04/075

Zhuang P, Ye ZH, Lan CY, Xie ZW, Shu WS (2005) Chemically assisted phytoextraction of heavy metal contaminated soils using three plant species. Plant Soil 276:153–162

Medicinal Biotechnology for Disease Modeling, Clinical Therapy, and Drug Discovery and Development

5

Robin Duelen, Marlies Corvelyn, Ilaria Tortorella, Leonardo Leonardi, Yoke Chin Chai, and Maurilio Sampaolesi

Abstract

Over the past decades, stem cell technology has revolutionalized medical biotechnology due to the unlimited self-renewal ability and differentiation capacity of stem cells to generate cells and tissues of the entire human body. Many efforts have focused on providing cutting-edge stem cell therapies in order to repair or replace damaged cells or tissues, hoping to ultimately cure devastating diseases. Undoubtedly, this novel technology guarantees a serial entrepreneur's confidence in the future prospects of stem cell-based products and services. Here, we describe the state of the art of several applications of adult stem cells, as well as of embryonic and induced pluripotent stem cells in biotechnology that represent entrepreneurial opportunities. Although the contribution of stem cells to medical research is enormous, several hurdles still have to be overcome, including ethical and regulatory issues, functional maturation of stem cell progenitors, stringent manufacturing guidelines, immune rejection, and tumorigenicity.

Nevertheless, key studies applying microfluidic technology, "organ-in-a-dish" and 3D bioprinting have been published, reporting the successful development of human pluripotent stem cell-based healthy and disease models for deciphering pathological mechanisms, drug discovery and toxicity screening, and regenerative medicine. Interestingly, because of the increasing amount of newly identified targets, assistance from computational chemistry and bioinformatics became indispensable to reduce the quantity of molecules that need to be tested in vitro or in vivo. In the past years, a boom in companies and start-ups all over the world occurred, focusing on bioinformatics and machine learning. Furthermore, biotechnological applications are highly applied in the veterinary medicine nowadays, and stem cell-based

R. Duelen · M. Corvelyn · I. Tortorella
Translational Cardiomyology Laboratory, Stem Cell Biology and Embryology Unit, Department of Development and Regeneration, KU Leuven, Leuven, Belgium

L. Leonardi
Department of Veterinary Medicine, University of Perugia, Perugia, Italy

Y. C. Chai
Translational Cardiomyology Laboratory, Stem Cell Biology and Embryology Unit, Department of Development and Regeneration, KU Leuven, Leuven, Belgium

Department of Development and Regeneration, Stem Cell Institute Leuven, KU Leuven, Leuven, Belgium

M. Sampaolesi (✉)
Translational Cardiomyology Laboratory, Stem Cell Biology and Embryology Unit, Department of Development and Regeneration, KU Leuven, Leuven, Belgium

Human Anatomy Unit, Department of Public Health, Experimental and Forensic Medicine, University of Pavia, Pavia, Italy
e-mail: maurilio.sampaolesi@kuleuven.be

F. Matei, D. Zirra (eds.), *Introduction to Biotech Entrepreneurship: From Idea to Business*,
https://doi.org/10.1007/978-3-030-22141-6_5

biotechnology is opening an exciting era in human therapeutics. In conclusion, scientists with strong entrepreneur mind-set are crucial to generate economic value in medicinal biotechnology. Thus, we need to educate next generation of scientists in entrepreneurship and work directly with institutions and funding agencies to guarantee a successful translational process in hiring and training our next generation of students.

Keywords
Stem cell biotechnology · Entrepreneurship · Regenerative medicine · Disease modeling · Computer-aided drug discovery and toxicity · Organ-in-a-dish · Veterinary medicine

5.1 State of the Art in Adult Stem Cell Biotechnology

Every tissue and organ in the human body is made up of different types of cells. Cells that give rise to skin, for example, are different from those that generate skeletal muscles. This makes it impossible for cells that generate one tissue or organ to be used for repairing damage to another tissue or organ. However, in the very beginning of the development, all cells derived from the same source, the zygote, which is a fertilized oocyte. This cell will divide and assemble all tissues in the development of every human or animal being. Thus, during the embryonic development, the cells adopt different capacity to become certain cell types.

Embryonic cells within the first couple of cell divisions after fertilization generate a globular solid mass named morula and until probably the 4–16-cell stage are totipotent (Fig. 5.1). This means that they can still differentiate into all kinds of tissues, both embryonal and extra-embryonal. After 5 to 6 days after fertilization, a blastocyst is formed from the morula, dividing the cells in the inner cell mass (ICM) and a single layer surrounding the blastocyst, the trophectoderm. The trophectoderm cells will give rise to tissues in the placenta, while the ICM will lead to development of the embryo itself. From this point on, ICM cells are not totipotent anymore, since they lost the capacity to give rise to the extra-embryonal tissues, and are now pluripotent (Rossant 2007). The embryonic cells are scientifically very useful since they could give rise to any kind of mature cells; however, they are also very controversial for further applications. The major ethical and policy issue concerning stem cells is the objection to the derivation and use of embryonic stem cells for research. A vocal minority of people objects to the destruction of embryos that occurs when stem cells are derived. During further development, the capacity of these cells keeps on narrowing down. After birth, still adult stem cells can be found in many human tissues, such as blood, brain, intestine, skin, and muscle. These cells are multi- or even unipotent, meaning that they are restricted to only one **germ layer (ecto-, endo-, or mesoderm)** or one specific cell type respectively. They are responsible for the repair and regeneration of tissues in the body during the entire life (Maitre 2017).

> **Germ Layer (Ecto-, Endo-, or Mesoderm)**
> During embryogenesis, the cells become predefined to a certain germ layer. These germ layers eventually will give rise to all the tissues and organs of the organism, through a process called organogenesis.
>
> The **ectoderm** will primarily form the nervous system and epidermis, the **endoderm** the gastrointestinal and respiratory tracts, and the **mesoderm** blood cells and muscle tissue.

Stem cells have two important features that make them different from other types of cells. On the one hand, they are unspecified, or undifferentiated, and can renew themselves through **symmetrical cell divisions** to maintain their cell pool. On the other hand, under certain biochemical cues they can be made to differentiate into distinct cell types (Khanlarkhani et al. 2016).

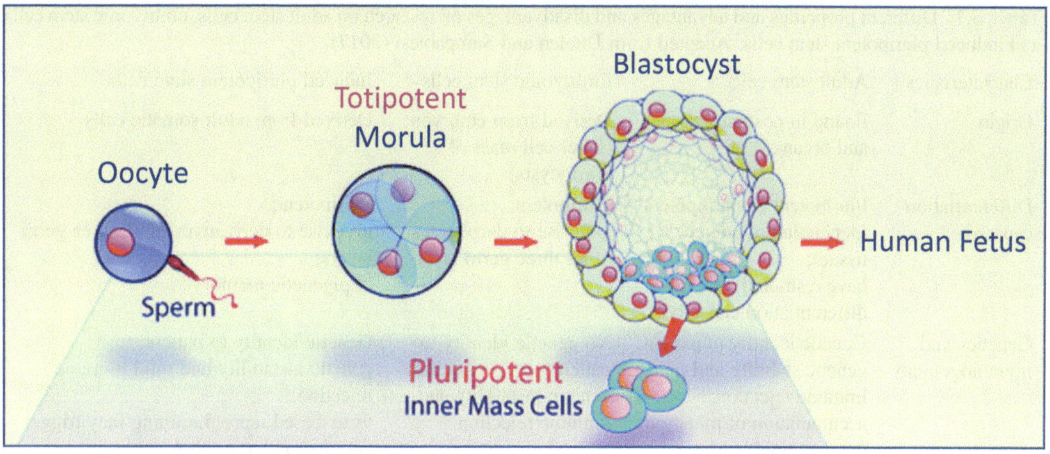

Fig. 5.1 Potency phases of (stem) cells during embryonic development. Adapted from: Mike Jones

Symmetrical Cell Division
These divisions will give rise to daughter cells of equivalent fates, keeping their original differentiation potential. This is in contrast to asymmetrical cell division, which leads to two distinct daughter cells: one copy of the original stem cell as well as a second daughter cell programmed to differentiate into a non-stem cell fate.

These different stem cell types withhold a lot of therapeutic potential, as well as they can be used as valuable tool for research and drug screening. However, it needs to be kept in mind that stem cell research also can lead to controversy. All advantages and disadvantages of adult stem cells, embryonic stem cells, and induced pluripotent stem cells (will be explained in more detail in the next subchapter) are listed in Table 5.1 (Duelen and Sampaolesi 2017).

Adult stem cells have the feature to mimic the situation in vivo, in a patient itself. Therefore, these cells can be very valuable in gaining insights into the physiology of the tissue, toxicity screens, drug testing, etc. Based on the capacity of these cells to repair damage to the organ where they are situated, it was not surprising that the industry showed a lot of interest into this field. In the last decade, cell-based therapy approaches are moving from bench to clinical trials. Moreover, companies (also such as the previously mentioned Moraga Biotechnology Corporation) are aiming to provide a stem cell bank for **HLA-haplotype-matched** engraftment to treat as many patients as possible.

HLA-Haplotype-Matched
Human leukocyte antigen (HLA) molecules are expressed on almost all nucleated cells and they are the major molecules that initiate graft rejection. The immune system uses these markers to recognize which cells belong in your body and which do not.

The most well-known application is the use of blood stem cells in bone marrow transplants, which are already performed since the late 1960s (Ferrebee and Thomas 1960; Barnard et al. 2006). Bone marrow contains immature blood-forming stem cells known as hematopoietic stem cells (HSCs). HSCs have the potential to multiply through cell division and either remain stem cells or differentiate and mature into many different kinds of blood cells such as white blood cells, platelets, or red blood cells. However, in bone marrow cancer (leukemia) or blood disorders (such as thalassemia) HSCs are impaired and can be replaced by HSCs from a healthy donor (allogeneic transplantation). In case of planning for chemotherapy or radiation, patient's own stem cells can be harvested before

Table 5.1 Different properties and advantages and disadvantages on research on adult stem cells, embryonic stem cells, and induced pluripotent stem cells. Adapted from Duelen and Sampaolesi (2017)

Characteristics	Adult stem cells	Embryonic stem cells	Induced pluripotent stem cells
Origin	Found in postnatal tissues and organs	Derived from embryos (inner cell mass of blastocysts)	Derived from adult somatic cells
Differentiation capacity	Pluripotent or multipotent (depending of source tissue); have restricted differentiation capacity	Pluripotent; give rise to derivatives of all three germ layers	Pluripotent; give rise to derivatives of all three germ layers; "Epigenetic memory"
Genetics and immunogenicity	Genetic identity to patient; genetic stability and mild immune rejection; accumulation of mutations due to aging	No genetic identity to patient; genetic instability and immune rejection	Genetic identity to patient; genetic instability and mild immune rejection; virus-based reprogramming may trigger antiviral and anti-DNA antibody-mediated immunity
Cell accessibility	Hard to access and advanced purification strategies needed	Depends on ethical issues	Depends on reprogramming efficiency
Ethics	No ethical issues	Ethical and legislative issues	No ethical issues

starting invasive treatments and transplanted back afterward. When patient's own cells are returned to the body, this is called autologous transplantation (Hardy and Ikpeazu 1989). More recently, a certain type of skeletal muscle stem cell, mesoangioblasts, was discovered which was able to migrate through the blood stream and ameliorate the phenotype of muscular dystrophy (Sampaolesi et al. 2006).

Additionally, the previously mentioned databases of adult stem cells would be of great value for drug developers who wish to pursue the **pharmacogenomics** and **toxicogenomics** of a particular disease category. Using the cell-based assay system, the drug developer can further optimize their lead compounds prior to entering costly clinical trials with a prospective drug.

Pharmacogenomics and Toxicogenomics
Determination of the potential response of an individual by studying gene activity within a particular cell or tissue in response to exposure to therapeutic or even toxic drugs.

Moreover, the industry focused on the discovery of different kinds of adult stem cells. Quite recently, the scientists from Moraga Biotechnology Corporation discovered and isolated a very primitive population of stem cells from adult tissues. These so-called blastomere-like stem cells (BLSCs) resemble totipotent cells, which are normally found in the developing embryo before the blastocyst stadium. They have the ability to differentiate into most, if not all, tissues and organs of the body, which was surprising for cells found postnatally. Furthermore, they also developed a specific serum-free culture medium that allows only symmetrical divisions, ensuring a homologous population of BLSCs that does not lose its broad differential capacity. They patented both the isolation process of these BLSCs and the serum-free culture medium.

A very recent application of adult stem cells is for the production of exosomes. Exosomes are extracellular vesicles, originated from the inward budding of the cell membranes followed by formation of multivesicular bodies (Fig. 5.2). When these fuse with the plasma membranes, exosomes are released. Currently, exosomes have been verified to be secreted from various cells including platelets, tumor cells, and stem cells, among others. Exosomes are characterized by their diameters ranging from 30 to 150 nm. Moreover, it was discovered that exosomes are enriched with various nucleic acids including mRNAs,

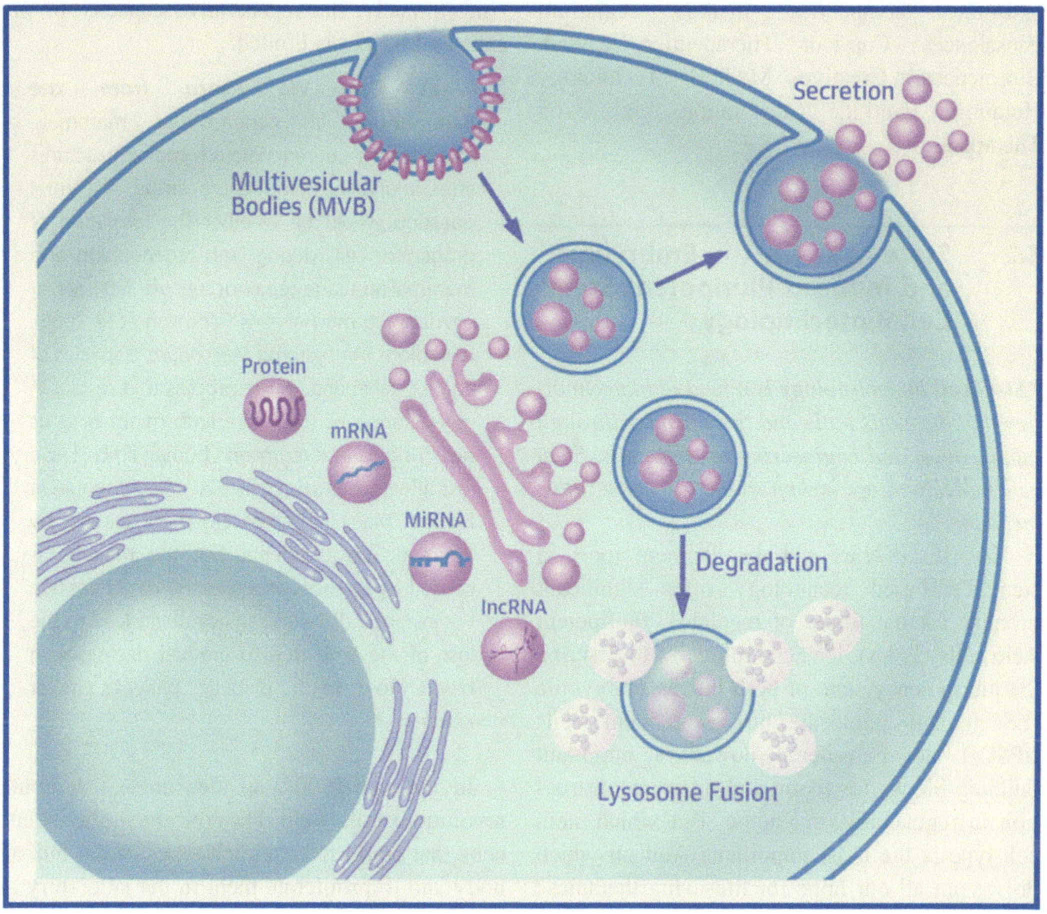

Fig. 5.2 Exosome production derived from multivesicular bodies. Adapted from Marban (2018)

microRNAs (miRNAs), and other noncoding RNAs. These RNAs can be taken up by neighboring cells or remote cells, subsequently modulating recipient cells; on the other hand, RNAs are protected from degradation after being packed into the exosomes or microvesicles, which altogether results in increased attention to exosomes and the carried RNAs. These exosomes have their function in intercellular communication for modulating cellular processes (Han et al. 2016). The adult mesenchymal stem cells (MSCs) are the most prolific producers of exosomes compared with other cell types. Moreover, extensive studies demonstrated that exosomes derived from MSCs could repair injured tissues. For example, in acute myocardiac injury animal models, the secreted exosomes have shown to be cardioprotective and preserved cardiac performance, avoiding the possible immune response associated with traditional cell therapy approaches (Bian et al. 2014). Not only do exosomes have enormous potential as vehicles for drug delivery, they can also be used as biomarkers in noninvasive tests and active agents in cosmeceutical products. Therefore, the industry showed a lot of interest in them and more broadly in manufacturing technologies and tools to support exosome applications and research. Over the last 5 years, more than 200 million dollars were used for funding in this field. Among others, big life science companies like ThermoFisher and Qiagen jumped in, launching tools and systems to support exosome research. Companies leading the charge in the area of exosome therapeutics include Anjarium Biosciences, Capricor Therapeutics, Codiak Biosciences, Creative Medical Technology Holdings, Everkine Corporation, and Evox Therapeutics.

5.2 State of the Art in Embryonic and Induced Pluripotent Stem Cell Biotechnology

*"**Stem cell biotechnology** is a field of biotechnology that develops tools and therapeutics through modification and engineering of stem cells. Stem cell biotechnology is important in regenerative medicine."*

New discoveries and development tools in stem cell-based technology offer significant promise for the future of medicine. Pluripotent stem cells (PSCs), the scientific term to describe the pluripotency state of both human embryonic (ESCs) and induced pluripotent stem cells (iPSCs), are considered nowadays important building blocks for tissue and organ reconstruction in regenerative medicine. But which stem cell type is the most important? And on which do we put all our hope for lifesaving therapies? Both will be crucial in the upcoming years since regenerative medicine is aiming at repairing the entire scale of diverse tissues and organs in the human body. Organ transplantation is one way to do so, although its applicability is severely limited. The majority of the people who need a donor organ never find a compatible donor, closely matching to the given patient. If cells and tissues from the adult organ could regenerate, and had the ability to reconstitute entire organs, we would simply regrow amputated arms and legs, as well as heart tissue would be able to regenerate by proliferating cardiomyocytes after a heart attack or the damage in our brain would be repaired by remaining brain cells due to a stroke. Unfortunately, the regenerative capacity of our cells and tissues is limited.

▶ **What Can We Learn from the Salamander?** In contrast to mammals, salamanders can regenerate complex structures after injury, including entire limbs. A central question would be whether the formation of progenitor cells during limb regeneration and mammalian tissue repair occurs via separate or overlapping mechanisms (Morrison et al. 2006). In Axolotls (*Ambystoma Mexicanum*, a species of aquatic salamander), the process that results in regeneration of the entire limb might give us new insights into regrowing human limbs. Limb loss affects approximately two million people in Europe. Many limb loss cases are caused by traumatic car accidents, but the majority is caused by diseases affecting the blood vessels. For instance, diabetes can eventually lead to the loss of the limb due to gradual decreases in blood flow to a diabetic patient's lower extremities.

In the mid-1990s, an idea arose that could revolutionize the field. The concept was to isolate cells that could form all cell types of the human body and to proliferate them in the laboratory in culture dishes as cell lines, in order to expand them into millions and millions of cells. Afterward, these cell lines could be used as powerful tools for differentiation to any kind of cell of interest of the human body, by turning on the molecular switches that would cause them to become anything we needed to make, like pancreatic cells to secrete insulin for diabetic patients, blood cells for leukemia patients, or cartilage for the treatment of arthritis. In 1998, the first human ESCs were derived by Thomson and colleagues. They successfully isolated human ESCs from the inner cell mass of blastocysts (Thomson et al. 1998). Because ESCs exhibit unlimited self-renewal and can differentiate into any cell type present in the adult organism, they are considered as a promising cell source to achieve regeneration.

▶ **Case Study** ESCs are interesting candidates as a renewable cell source to achieve heart regeneration, for example after myocardial infarction. The first transplantation reports in pigs and guinea pigs of human ESC-derived cardiomyocytes (ESC-CMs) have demonstrated their potential to function as biological pacemakers in electrophysiologically silenced or atrioventricular blocked hearts (Kehat et al. 2004). One of the initial technical challenges in the differentiation of ESCs toward the cardiovascular lineages was to obtain a high purity and large yield of differentiated cells. However, as we gradually increased our knowledge of the mouse embryonic heart development, mouse, and human ESC-CM differentiation became more efficient by manipulating the cardiac-specific signaling pathways (Sumi et al. 2008). Several strategies, including specialized cell culturing methods, treatments with biological and chemical factors, or genetic modifications, have been conducted to purify and scale up homogeneous and functional ESC-CMs (Schwach and Passier 2016). Recently, human ESC-CMs, generated on a large scale, were able to engraft and repair damaged areas of the heart tissue in a primate model with myocardial infarction (Chong et al. 2014). These results seemed promising since only in a few cases different cell types showed efficacy in large animal models. However, the clinical use of human ESC-CMs has been hampered by crucial limitations, such as potential immunogenic and tumorigenic properties, genetic instability, as well as ethical issues (Robertson 2001).

Human ESCs are pluripotent stem cells derived from the inner cell mass of pre-implantation stage blastocysts, thereby sacrificing human embryos. As a consequence, the ethical and legislative debate arose surrounding the involvement of ESCs for research purposes (Thomson et al. 1998). In 2006, Yamanaka and Takahashi developed a new method to induce a pluripotent state in mouse somatic cells, referred to as iPSCs, by introducing a defined combination of transcription factors through retroviral transduction. The cocktail of transcription factors used for cellular reprogramming was composed of Krüppel-like factor 4 (Klf4), sex determining region Y-Box 2 (Sox2), V-Myc avian myelocytomatosis viral oncogene homolog (cMyc), and POU class 5 homeobox 1 (Oct3/4) (Takahashi and Yamanaka 2006). One year later, in 2007, the first iPSCs were successfully generated from human origin, using the same set of transcription factors (KLF4, SOX2, cMYC, and OCT3/4) (Takahashi et al. 2007) or mediated via SOX2, OCT3/4, Lin-28 homolog A (LIN28), and Nanog homeobox (NANOG) (Yu et al. 2007). The introduction of iPSCs has lit a fire under the scientific community. Like ESCs, iPSCs have wide differentiation ability, giving rise to all the cell derivatives of the three germ layers (Lee et al. 2014). Although iPSCs have several characteristics in common with ESCs (Gherghiceanu et al. 2011), genome-wide analyses revealed significant differences between both cell types regarding methylation signatures, gene expression profiles, and microRNA patterns (Chin et al. 2009). However, outweighing these differential landscape marks, iPSCs circumvent the ethical problems and can be generated identical to the patient to prevent transplant rejection in regenerative therapy (Takahashi et al. 2007; Yu et al. 2007).

5.3 Current Obstacles for Use of Pluripotent Cells in Stem Cell Biotechnology

A major reason for the fairly slow progression in the field of PSCs for biotechnology-oriented applications is due to various difficulties, associated with intrinsic properties of PSCs. *"We can readily drive the cells from the undifferentiated state to the differentiated state. However, getting those cells to pause anywhere in the middle of this continuum to yield progenitor cells is incredibly challenging"* (citation from

Dr. James Peyer, Managing Partner at Apollo Ventures).

- *Teratomas and Karyotypic Abnormalities:*
  A major concern that has to be considered in every clinical therapy based on PSCs is their tendency to form tumors or teratomas. Inefficient differentiation or purification techniques of undifferentiated PSCs can cause undesirable tumors or teratomas after transplantation. PSCs can show karyotypic instability with long-term culturing in vitro, raising concerns about potential neoplastic transformation and dysregulation of gene expression. Chromosomal abnormalities are highly variable and depending on the cell line and culturing conditions (Taapken et al. 2011).
- *Genetics and Immune Rejection:*
  Transplantation of (allogeneic) human ESC-CMs can be identified as foreign, triggering an immune response. Therefore, in preclinical models, immunodeficient or immunosuppressed animals are used. Human iPSCs could have an important clinical advantage over human ESCs because they can be created from the patient, leading to genome matching iPSC-derived tissues.
- *Immature Phenotype:*
  The maturation phenotype of human PSC-derived tissues must be addressed, before their clinical application in cell-based therapies since these de novo generated CMs should replace the damaged heart tissue. For example, PSC-CMs exhibit features of an early and immature phenotype, rather resembling fetal than adult CMs (Karakikes et al. 2015). In case the CM maturation issue cannot be solved using the current available cardiac differentiation protocols, alternative methods should be developed to enhance maturation. Research groups are exploring the beneficial effect on the maturation of CMs mediated by electrical stimulation, cyclic stretch-induced mechanisms (Tulloch et al. 2011), chemical manipulation (Pillekamp et al. 2012), and 3D tissue engineering techniques (Yang et al. 2014). Nowadays, robust protocols to increase the maturation of human PSC-CMs are still lacking, and therefore, functional maturation is one of the critical hurdles to overcome before PSC-CMs can be used for clinical applications.

5.4 Moving Toward Biotech Applications of Induced Pluripotent Stem Cells in Regeneration

The discovery of iPSCs sparked widespread enthusiasm for the development of new stem cell-based models of human disorders, extraordinary platforms for drug discovery and screening, as well as more widespread use of autologous cell-based therapies. Axiogenesis (Cologne, Germany) is a biotech company that develops cell lines derived from iPSCs for preclinical approaches. Studies using directed differentiation of iPSCs frequently unravel disease-specific phenotypes at cellular levels, caused by (monogenic) mutations. Efforts have been made to study diseases at the level of tissues and organs by the development of more complex 3D and multicellular systems. Moreover, organoids and human-rodent chimeras, which represent more accurately the diverse cellular ecosystems of complex tissues, are being applied to stem cell-based disease models to recapitulate the pathobiology of a broad spectrum of human disorders, like genetic and infectious diseases, as well as cancer (Rowe and Daley 2019). BioTalentum Ltd. (Gödöllö, Hungary) is an RTD research-intensive SME, a leading technology provider in Central and Eastern Europe, which specializes in stem cell research and is working on stem cell research and services, focusing on animal and medical biotechnologies. BioTalentum's mission is to develop innovative human cellular systems and animal models for biomedical research and drug testing.

- *Organoids derived from iPSCs:*
  A spectrum of studies, starting from 2008, reported disease-related phenotypes in models derived from patient-specific iPSCs (Robinton and Daley 2012). Although drug screening arrays on these disease-specific iPSC-based

Table 5.2 Important reports of disease modeling using organoids from human iPSCs

Disease model	Organ (location)	Cell type(s) analyzed	Disease genotype	Disease phenotype	Refs.
Microcephaly	Neural (cerebrum)	Neurons, NPCs, RG, retina, choroid plexus, meninges	*CDK5RAP2*	Premature neuronal differentiation	Lancaster et al. (2013)
Autism	Neural (cerebrum)	Neurons, NPCs, RG	NA	Increased GABAergic neuron fate	Mariani et al. (2015)
Glioblastoma	Neural (cerebral)	NPCs	*MYC* and others	Organoid overgrowth	Bian et al. (2018)
Cystic fibrosis	Gastrointestinal (liver)	Cholangiocytes	*CFTR*	Dysfunctional epithelial transport	Ogawa et al. (2015)
Colon cancer	Gastrointestinal (colon)	Colonocytes	*APC*	Epithelial proliferation	Crespo et al. (2017)
Polycystic kidney disease	Kidney (glomeruli and proximal tubules)	Podocytes, epithelial and endothelial cells	*PKD1* and *PKD2*	Cyst formation	Freedman et al. (2015)
Cystic fibrosis	Lung (airway)	Epithelial cells	*CFTR*	Impaired forskolin-induced swelling	Mccauley et al. (2017)
Surfactant deficiency	Lung (alveolus)	Type 2 epithelial cells	*SFTPB*	Decreased surfactant production	Jacob et al. (2017)

NPCs neural progenitor cells, *RG* radial glia, *NA* not applicable

models identified potential candidate small molecule therapies, to date unfortunately, very few human clinical trials have satisfied early hopes and expectations, after extrapolating these in vitro cell-based screening results. An important advance in disease modeling arose with the generation of iPSC-derived organoids, which are 3D multicellular aggregates that differentiate and self-organize to recapitulate structural properties and inter-cellular interactions to mimic those in adult mature tissues (Dutta et al. 2017). Refined guidance of organoid differentiation from iPSCs led to the generation of increasingly complex "tissues in a dish." Table 5.2 summarizes some key studies describing advanced iPSC-derived organoid systems for disease modeling, providing invaluable insights into certain disease pathogenesis.

In the last few decades, innovative advances have been made in the development of human iPSC-derived organoids, ultimately aiming for autologous iPSC-based therapies. In many of these studies, disease-causing mutations in patient-specific iPSCs have been corrected to reverse disease phenotypes at the level of the organoids, providing proof of principle that this approach could be applied in autologous therapies in the future. If progressive advances will be obtained in the assembly of organoids, then even more complex and functional engineered tissues and organoids will be exploited to increase further our knowledge in disease pathogenesis at the tissue and organ levels (Liu et al. 2018; Workman et al. 2017; Taguchi and Nishinakamura 2017; Turco et al. 2018).

iPSC-based models for host–pathogen interactions:
The earliest iPSC-based studies modeled genetic disorders (Ebert et al. 2009; Moretti et al. 2010) However, it has been documented that cells terminally differentiated from human iPSCs are susceptible to infection with human pathogens, which opens a new frontier for

Table 5.3 Key studies modeling infectious disorders using human iPSCs

Pathogen	Target	Phenotype	Refs.
HIV	T cells	iPSC-derived NK cells killing of target cells	Ni et al. (2011, 2014)
HIV	Monocytes	HIV resistance by introducing CCR5del32 mutation	Ye et al. (2014)
Zika	Neural organoids	Impaired organoid growth, decreased ventricular zone thickness, increased ventricle size	Garcez et al. (2016), Cugola et al. (2016), Gabriel et al. (2017), Nowakowski et al. (2016), Wells et al. (2016)
Helicobacter Pylori	Gastric organoids	Increased proliferation of epithelial cells	Mccracken et al. (2014)
Salmonella	Intestinal organoids	Epithelial cell invasion	Forbester et al. (2015)

iPSC induced pluripotent stem cell, *NK cells* natural killer cells

studying interactions between host and pathogens (Table 5.3). The advantage of using human iPSC systems is reduced species specificity of infectious pathogenicity and inflammatory responses that limited the extrapolation of results from model organisms to a human setting.

Modeling host–pathogen interactions with human iPSC-based systems is amenable, although further sophistication of organoid setups, by including immune and inflammatory cells, could result in significant broadening of the repertoire of human iPSC infectious disease models.

Human–animal chimeras:

Organoid systems limit proper analysis of interactions with the nervous and immune system, the blood circulation, as well as hormones and other mediators that are present in vivo (Liu et al. 2018). Moreover, organoids rely on self-organization growth in in vitro cultures, lacking many morphogenetic cues present in vivo (Dutta et al. 2017). Recently, these restrictions have been tackled by innovations in the xenotransplantation of cells and organoids differentiated from human iPSCs. Given the high amount of ongoing and starting cell-based therapies under investigation in preclinical or clinical settings for many diseases nowadays (Bollard and Heslop 2016; Maude et al. 2018; Mandai et al. 2017), iPSC-based chimeric models to study disorders by xenotransplantation have emerged as a trustworthy tool to fuel these progressions by modeling human diseases. For example, in order to model adoptive T cell therapies in infectious disorders and cancer, human iPSCs have been used, providing an unlimited source of customized T lymphocytes for immunotherapy. The use of these chimeric antigen receptor T cells (CAR T cells) has been approved for treating B-ALL (Maude et al. 2018). CD19-targeted CAR T cells derived from human iPSCs have been demonstrated to be effective against $CD19^{+}$ B-ALL cells in a chimeric model of leukemia (Themeli et al. 2013). Moreover, natural killer (NK) cells differentiated from human iPSCs can inhibit HIV replication in vivo (Li et al. 2018). Together, these studies have shown the strength of preclinical models of adoptive immunity to identify hopeful immune cell-based therapies.

5.5 Companies in Stem Cell-Based Biotechnology

In the last few years, healthcare group Bayer (Leverkusen, Germany) and Versant Ventures (USA Healthcare investor) joined forces, spending approximately 200 million euros, to launch BlueRock Therapeutics in Toronto (Canada), a stem cell-based regenerative medicine company, and Fujifilm (Madison, WI, USA) invested more than 270 million euros to acquire the iPSC company Cellular Dynamics. The ultimate goal for stem cell therapies is to regenerate injured or diseased tissues, as for instance described in heart failure, diabetes, or blindness. So far, many of the first human clinical trials with PSCs are focusing on the eyes to restore the loss of the retinal epithelium, like age-related macular degeneration. Holostem Terapie Avanzate, a spin-off company from the University of Modena and Reggio Emilia (Italy), was the first company that put a step toward this goal. Nowadays, after 20 years of intense research, the biotech company has developed Holoclar (London, United Kingdom), one of the rare companies that succeeded in entering the European market. Holoclar is specialized in treating patients who lost eyesight due to burn injuries by using stem cells.

Over the past decades, much progression has been made in the development of disease models from human iPSC lines, including sophisticated 3D organoid and chimeric systems, which are being exploited for the discovery of novel therapeutic applications. Although successes have been achieved in differentiating patient-derived iPSCs toward target cells of interest with measurable phenotypic disease characteristics, the question remains whether the biotech field advanced to a point in which these stem cell-based models are of sufficient accuracy to support the identification of new therapies to counteract disease onset and/or progression of a wide range of disorders.

5.6 Toxicity Screening and Drug Discovery

As a result of the many advances that have been made in the field of biotechnology, pharmacology underwent significant remodeling. A shift occurred from a predominant trial-and-error approach (known as forward pharmacology) to more accurate methods, using the latest discoveries of molecular biology in order to discover new pharmaceutical entities (known as reverse pharmacology).

In forward pharmacology, also called phenotypic drug discovery (PDD), compounds are screened in cellular or animal disease models to identify molecules that have beneficial effect: only after an active drug has been identified, an attempt is made to identify the biological target of the drug (Moffat et al. 2017).

In reverse pharmacology, also known as target-based drug discovery (TDD), a biological target is hypothesized to be disease modifying. A high-throughput screening (HTS) of compound libraries against the purified protein target is completed, identifying hit compounds that are then optimized and, differently than the forward pharmacology, tested for in vivo efficacy in the final drug discovery stages (Saeidnia et al. 2016).

This new approach satisfied the demands related to the increased complex knowledge of biological systems enlightened by the new "omics" sciences, which enhanced our capability to link diseases to their causes and therefore led to an exponential rise in drug targets. In light of the new discoveries and approaches in pharmacology and molecular biology, biotechnologies became the leading force of drug discovery and one of the main research fields of new start-ups and companies all over the world (Morrison 2018).

5.6.1 Drug Discovery

Drug discovery is the process by which new pharmaceutical compounds are discovered and brought to the market. It is a process made up of several stages that takes an average of 15 years to be completed (see Fig. 5.3). The first step is to decide which pathology to study and identifying and validating the target that might be disease modifying. Afterward, the exploratory research starts: with the first large screening tests it is possible to identify HIT molecules (chemical entities that have a promising affinity with the target). After more accurate investigations, there is the selection of a molecule that binds specifically and selectively to the target and is able to modify its normal mechanism of action, the LEAD compound. The latter is rationally modified in order to improve biological activity and ADME (absorption, distribution, metabolism, and excretion): if a promising compound is

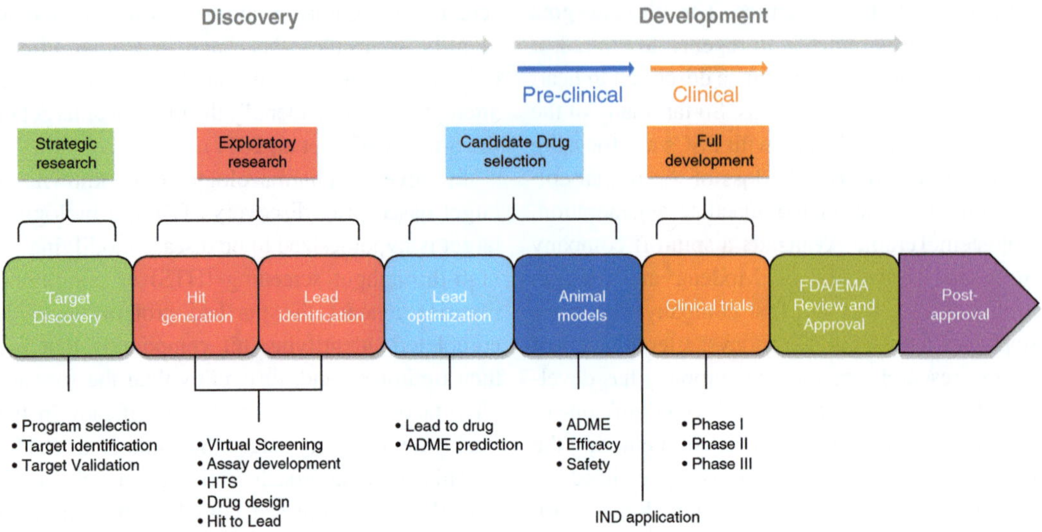

Fig. 5.3 Schematic representation of the drug discovery process. The two main phases, discovery and development, are articulated in sub-phases: for each sub-phase the major strategies and aims are listed (*IND* investigational new drugs)

found during the screening, the preclinical and clinical phase will be started. Upon finalization of the clinical trials, an approval of the drug by the Food and Drug Administration (FDA) or the European Medicines Agency (EMA) has to be given before the drug can be brought to the market: during the years of its distribution, the safety of the drug will be continuously monitored thanks to pharmacovigilance.

5.6.2 Computer-Aided Drug Discovery (CADD)

Because of the high amount of new targets identified, and consequently the increasing data to be analyzed, it is necessary to develop new approaches that could handle the increasing costs of R&D (research and development), especially in pharmacology (Paul et al. 2010). For that purpose, assistance from computational chemistry and bioinformatics became indispensable in order to reduce the quantity of molecules that need to be tested in vitro or in vivo, making possible to minimize the number of trial experiments by the order of tens of thousands of possible combinations. As a proof of this need, in the past years there has been a boom in companies and start-ups all over the world that focus on bioinformatics and machine learning (Table 5.4) (source: https://www.biopharmatrend.com/post/72-2018-ai-is-surging-in-drug-discovery-market/).

With the advent of more sophisticated computational machine the in silico approach became the leading force of all the discovery and preclinical phases of drug discovery, allowing an improved rational design of small molecule with pharmaceutical activity in combination with a significant gain in time and cost during several stages of the drug discovery process: target identification and validation, lead optimization, and preclinical tests (Buchan et al. 2011).

The term computer-aided drug design (CADD) refers to all the processes through which it is possible to identify a molecule with a selective pharmaceutical activity against the target of interest. Using CADD it is possible to predict biological activity, ADME, and toxicity and to do drug repurposing before performing any in vitro assay: the major role is to screen out big compound libraries, cluster them into small groups of active compounds (HIT molecules), and allow the optimization of LEAD candidates with improved biological properties (Yu and Mackerell 2017).

Table 5.4 List of companies focused on machine learning and bioinformatics

Company	Country	Research field
BenevolentAI	London, UK Cambridge, UK New York, USA Antwerp, Belgium	Machine learning for drug discovery and development process
Atomwise	San Francisco, USA	Neural network for bioactivity prediction in structure-based drug discovery
Cyclica	Tortonto, Canada	Cloud-based proteome screening platform
NMD Pharma	Aarhus, Denmark	Small molecule drug discovery for treatments of neuromuscular disorders
OWKIN	New York, USA Paris, France	Building mathematical models and algorithms that can interpret biomedical images, genomics, and clinical data to discover new biomarker
Verge Genomics	San Francisco, USA	Use of machine learning and AI to develop therapeutics against Alzheimer's and Parkinson's disease

AI artificial intelligence

CADD is a comprehensive term that includes different strategies that can be divided in two main branches:

1. *STRUCTURE-BASED drug design:* Relies on knowledge of the 3D structure of the target or of the target–ligand complex, obtained by X-ray crystallography, nuclear magnetic resonance (NMR) spectroscopy, or homology modeling.

   Therefore, the design of molecules is based on the study of the interaction that occurs between the binding site and the ligand.
2. *LIGAND-BASED drug design:* In the absence of a known target structure, known ligands are studied in order to correlate their physiochemical properties to a biological activity. This information is then used for guiding the design of new drugs with improved activity or the optimization of existing drugs.

Software use force fields to estimate energies and forces associated with a drug–target complex; then virtual screening is performed in order to identify molecules (from large libraries) that are likely to bind to the target. This identification is carried out by simulating ligand–target interactions (docking) of every compound in the database with every target conformation. A score is given to every simulation and compounds are ranked from the ones with more affinity (HIT compounds) to the ones with less/none affinity. Once the size of the library is minimized, few compounds (LEAD compounds) will be analyzed better in order to optimize the structure of molecules and to have a compound with promising biological effect and ADME. One of these will likely become a drug candidate ready to be tested in vitro and in vivo (Yu and Mackerell 2017).

5.6.3 Toxicity

Toxicology is a very important step of drug design. Determining toxicity is essential to identify adverse effects of new potential compounds both during molecules optimization and after the approval and commercialization with pharmacovigilance. Discovering inappropriate drug candidates early in the drug development process allows to save money and design risk factor during the LEAD optimization stage. Many factors

Table 5.5 List of methods for testing the most common types of toxicity observed in drug development

Toxicity	Methods
Hepatotoxicity	• Cell death: membrane integrity assays (LDH release), dead cell-specific protease activity, DNA-binding dye • Cell viability: metabolic activity (MTT assay), ATP production quantification, oxidative stress (ROS production)
Cardiotoxicity	• Patch-clamp assay • hERG Predictor assay
Genotoxicity	• AMES assay • Micronucleus test (MNT)
Immunotoxicity	• ELISA test • T cell proliferation and activation after treatment
Phospholipidosis	• Fluorescent probes (fluorescent phospolipid analogues), microscopy techniques

LDH lactate dehydrogenase, *MTT* 3-(4,5-dimethylthiazol-2-yl)-2,5-diphenyltetrazolium bromide, *ROS* reactive oxygen species, *hERG* human Ether-à-go-go-Related Gene, *ELISA* enzyme-linked immunosorbent assay

contribute and determine toxicity of chemicals, such as physical-chemical properties, ADME properties, route of administration, dose, and exposure time, so it is clear that evaluating toxicological profile of molecules is not an easy task (Raies and Bajic 2016).

One of the major challenges has always been to find a correct model that could mimic the human response to drugs, and this has been done via three different approaches: in vivo, in vitro, and in silico.

In Vivo Usually toxicity is tested in vivo with animal models. However, ethical concerns arose to prevail and the scientific community began to question whether animals were really the best option for testing organ and dose-specific effects for investigational compounds.

In addition to being expensive and time-consuming, performing experiments with animals does not allow to predict human response to drugs due to the many species-specific differences in terms of physiological responses and living conditions.

Animal testing is still needed for the complete assessment of safety and activity of new drugs, but a lot of efforts are still ongoing in order to reduce as much as possible the use of animal models.

In Vitro Progress made in HTS and cell culturing allowed using cellular models to predict toxicity (Table 5.5). Since 2D cellular assays have quite obvious limitations, in terms of representing the human system, currently 3D systems are preferentially used.

3D cultures (with or without scaffold materials) can better mimic the interactions that happen within tissues and therefore have a metabolic competence that resemble more the in vivo situation when compared to the classical 2D cultures.

The newest cell culturing techniques, such as organ-on-a-chip models, have revolutionized the way in which in vitro toxicology testing is conducted. These models, thanks to the use of microfluidic systems and tissue engineering, are able to mimic physiological conditions, providing an accurate representation of the toxicological effects of compounds. Even if the organ-on-a-chip technology can be used to improve preclinical safety and efficacy testing, progress still has to be made to improve the scalability of these systems in order to use them in pharmaceutical industries.

Instead of using human primary cells from donors, which are difficult to obtain and have variable background, to date stem cells are considered as a more useful tool for doing bioassays. Using them allows to develop basically every cell line, without the need of many human donors; using iPSCs permits the creation of cell lines that are difficult to obtain and maintain, such as neurons (important for evaluating neurotoxicity) (source: E-book in vitro toxicology https://www.

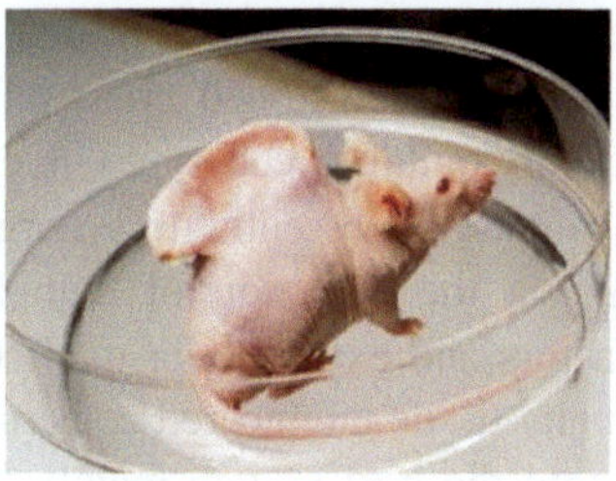

Fig. 5.4 The Vacanti mouse – a human ear grown on mouse? In 1997, Vacanti C. et al. reported on the successful growth of human ear-like structure on the back of a mouse by implanting synthetic biodegradable polymer embedded with bovine-derived cartilage cells (Cao et al. 1997). This study was originally dedicated to plastic and reconstructive surgery, but since then has caused an international surgc of tissue engineering research and applications (picture source: https://en.wikipedia.org/wiki/Vacanti_mouse

admescope.com/material-portal/e-books/in-vitro-toxicology.html).

In Silico Most of the available computer software packages (as the ones developed by the leading company Schrodinger) for molecular modeling and drug design are able to predict toxicological effect. Through different strategies, it is possible to evaluate ADME profile as much as off-targets and adverse effects.

In addition to prediction tools, one of the most recent approaches is to correlate metabolic transformation predictions to liquid chromatography-mass spectrometry data (LC-MS) in order to rapidly identify the formation of possible toxic metabolites of drugs (as the software Mass-MetaSite by Molecular Discovery).

Computational methods are, to date, the most promising tools for understanding biological activities and adverse effects of drugs. Using in silico analysis methods allows minimizing the need for animal testing and reducing costs and time during the entire drug discovery pipeline. They have the advantage of being able to evaluate the activity and toxicity of small molecules even before they are synthesized and make possible to analyze very large quantity of data in a short time (Raies and Bajic 2016).

In conclusion, toxicity screening and drug discovery will rely more and more on innovative technologies to enable rapid synthesis and high-throughput screening of large libraries of compounds. Many ongoing data-generation and data-sharing programs are gaining momentum thanks to the input from the European government, and together automated data management approaches will simplify and accelerate chemical toxicity studies. Then, it is possible to predict that modern toxicology research will be an attractive field for young investigators to better predict the systemic effects of compounds on animals and humans.

5.7 "Organ-in-a-Dish" for Disease Modeling, Drug Development, and Regenerative Medicine

5.7.1 From 2D to 3D Cell Culture Models

Transferring preclinical data obtained from animal models to clinical setting represents a major challenge in bench-to-bedside translation (Fig. 5.4). This is mainly due to the fact that the use of animal testing suffers from several limitations, including (1) lengthy experimental time and high costs associated with purchasing, housing, and maintaining of the animals; (2) the ethical issues related to the number and procedures on animal trials; and (3) the unpredictability of animal results for human diseases. For instance, animal models have been essential in cancer research due to practical and ethical concerns associated with human experimentation. However, the average rate of successful translation from animal models to clinical cancer trials is less than 8%. This is mainly attributed to the limited ability of animal models in mimicking extremely complex process of human carcinogenesis, physiology, and progression (Mak et al. 2014). In fact, animal models often do not possess the same mechanisms as in human diseases and have been increasingly criticized for their limited ability to predict new chemical entities' (NCEs) efficacy, safety, and toxicity in humans (Mcgonigle and Ruggeri 2014).

On the other hand, despite two-dimensional (2D) cell culture models have demonstrated additional values in biomedical research, they are lacking three-dimensionality and insufficiently reconstitute the multicellular functionality and organ niche as compared to in vivo tissues. As a result, 2D cell culture tests sometimes provide misleading and unpredictable results for in vivo responses. This is reflected by merely 10% of the drug candidates progressing successfully to clinical development, largely due to the lack of clinical efficacy and/or unacceptable toxicity where a portion of these failures is attributed to data collected from the 2D culture studies (Edmondson et al. 2014). To overcome these limitations with the conventional 2D cell cultures, a more complex system such as 3D cell cultures has been developed in recent years and often combining the advances in tissue engineering, 3D biomaterials, stem cell biology, and 3D printing or additive manufacturing technologies. The ultimate goal is to produce tissue constructs that behave similar to the living organs, in which the 3D structures, mechanical properties, and biochemical microenvironment cues are tightly and dynamically regulated. This approach has been shown to more accurately present the actual microenvironment where cells reside in tissues, resulting in enhanced expression of differentiated cell functions and tissue organization (Pampaloni et al. 2007). In the past decade, tremendous scientific efforts have been put into the development of a variety of 3D culture systems for drug discovery, cancer and stem cell biology, cell-based assays, tissue engineering, and regenerative medicine (Birgersdotter et al. 2005; Justice et al. 2009; Reininger-Mack et al. 2002). However, designing of such 3D construct requires full understanding on how tissues form and function, as well as their pathophysiology.

In fact, 3D culture construct still lacks most of the vital biological cues that reconstitute the features of living organs, such as spatiotemporal gradients of growth factors and oxygen, tissue-to-tissue interfaces to resume organ functionality, and biomechanical active milieu (Huh et al. 2011). Therefore, the complexity of 3D cultures needs to be enhanced and most often requires co-culturing multiple cell types in one system in order to promote cell-to-cell cross talks. Cross talks between multiple cell types relevant to a tissue of interest are essential to synergize tissue formation via the secretion of complementing growth factors and extracellular matrices. Nonetheless, many tissue-engineered constructs fail to integrate into the host system mainly due to the lack of vascularity. A step forward has been made by the scientists to facilitate implant integration, mainly by incorporating blood vessel networks into the construct design via micropatterning and bioprinting of endothelial cells in architecture resembles human vasculature (Zhu et al. 2017). Implanting pre-vascularized tissue-engineered constructs promotes vasculature connection to the host circulation and thus enhances tissue growth and functional restoration instead of cell death due to hypoxia condition at implantation site (Levenberg et al. 2005; Muscari et al. 2014; Ben-Shaul et al. 2019).

Decellularization followed by recellularization of harvested tissues or whole organs [e.g., liver (Uygun et al. 2010), heart (Ott et al. 2008), lungs (Petersen et al. 2010), and kidney (Nakayama et al. 2010)] represents another promising alternative source of animal testing for preclinical R&D development and ultimately functional organ transplantation in clinics. Decellularization is performed by exposing the organ to selected non-physiological chemical or biological agents (e.g., detergents, enzymes), physical forces, as well as perfusion techniques to lyze and remove cellular components without disrupting the extracellular matrix compartment (Crapo et al. 2011). This procedure will render the decellularized allogeneic or xenogeneic organ non-immunogenic upon transplantation, which in turn serves as a highly biological-relevant template for organ regeneration by recellularization with stem cells. Therefore, the host tissue response following implantation of these scaffold materials is highly dependent on the efficacy of the selected decellularization method. Unfortunately, the shortcomings of this technique are that it is highly dependent on the organ availability and requires an optimized decellularization method before its potential beneficial effects in the field of tissue engineering and regenerative medicine can be realized. Nonetheless, various decellularized tissues derived from human/animal dermis and pericardium and from porcine small intestine, urinary bladder, and heart valve have been commercialized for clinical applications (Crapo et al. 2011; Gilbert et al. 2006).

5.8 Organ-in-a-Dish: The Future of Medicine?

In spite of the increasing organ transplantation used in medical treatment, which has saved thousands of lives and greatly improved the patient's quality of life, the severe shortage of donors for organs remains a major constraint. It has resulted in a significant number of patient deaths. In the European Union (EU), the organ donor rate is recorded at only 21% between 2008 and 2015, which is widely different across the EU and cannot be explained by the public attitude or mortality rates. In 2017, a total of 34,000 organ transplants have been carried out in EU, and 60,000 patients remain on waiting lists for organ transplants (Fig. 5.5). Kidney transplant recorded the highest demand (62.1%), followed by liver (23.5%), heart (6.4%), and lung (5.9%). Other organ transplants include pancreas, small bowel, hand, and face. These data not only suggested a huge demand in organ donors, but also highlighted the need to improve the health of the EU population, potentially through the development of innovative approaches that enable the effective diagnosis and treatment of diseases related to organ dysfunctions and failure.

Recent emergence of organoid technology has revolutionized the field of 3D tissue cultures and provided sophisticated alternative organ-in-a-dish models to animal testing (Table 5.6). With this technology, the scientists could study human development and disease as well as validate their results in a representative "organ niche" microenvironment (Eisenstein 2018). Organoids are in vitro generated 3D cellular structures, which acquired correct identity, functions, and spatial organization that resembles an organ of interest in a relatively simplified form. Under tightly regulated culture regimes with growth factors, chemokines and/or small molecules that

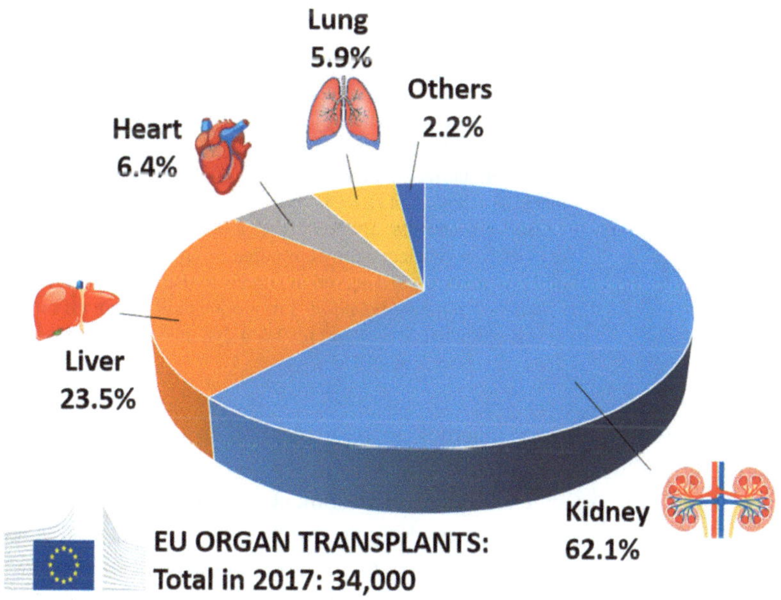

Fig. 5.5 Organ transplants trend in EU. In 2017, only 34,000 organ transplants have been performed across all member states of the EU, which is corresponding to 36% of patients who required organ transplantation. A total of 60,000 patients are still on waiting list by the end of 2017. Between 2008 and 2015, organ donation was recorded at 21%, in which 35% of organs received from living donation and 16% from donation after death. Kidney transplantation is the highest demand, followed by liver, heart, and lung (Source: https://ec.europa.eu/health/blood_tissues_organs/organs_en)

Table 5.6 Types of organoids and examples of their potential field of applications

Type of organoid	Example of potential field of application
1. Brain organoids 	• *Neurological disorders:* e.g., schizophrenia (Moslem et al. 2018), Alzheimer's (Raja et al. 2016), and Parkinson's (Schwamborn 2018) • *Disease modeling:* Modeling neurodegenerative microenvironment using cortical organoids (Yan et al. 2018) • *Zika virus infection:* Brain organoids were used as a model to study how Zika virus infects brain and halts neurogenesis, which leads to microcephaly in newborns (Garcez et al. 2016) • *Human brain evolution:* Cerebral organoid was used as novel platform to reveal molecular pathways underlying remarkable specializations of the human brain as compared to chimpanzee, giving new insights into human brain development and evolution (Pollen et al. 2019)
2. Liver organoids 	• *Acute liver failure:* Transplantation of liver organoids into acute liver failure (ALF) mouse model improved the survival rate (Nie et al. 2018b) • *Assessment of drug clearance:* Liver organoids demonstrated relevant biotransformation of pharmaceutical in fish (Baron et al. 2017) • *Hepatitis infection:* Infection of liver organoids with hepatitis viruses robustly triggers an antiviral response, which provided basis for understanding the pathogenesis (Wang et al. 2018a; Nie et al. 2018a) • *Liver inflammation:* Liver organoids were used to study how inhibition of miRNA leads to liver inflammation, necrosis, steatofibrosis, and insulin dysregulation (Sendi et al. 2018) • *Liver regeneration:* Liver organoids were embedded in liver-derived extracellular matrix hydrogel and developed into liver-like functional tissue structure with potential used for liver tissue regeneration (Saheli et al. 2018); self-assembled liver organoids into hepatobiliary formation (Vyas et al. 2017)
3. Heart organoids 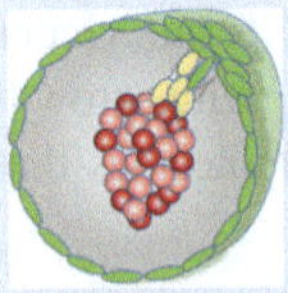	• *Heart regeneration:* Using specific growth factor cocktails (via BMP and Wnt signaling), cardiac organoids could develop two heart fields with cardiac chambers (Andersen et al. 2018) • *Environmental toxin screening:* 3D heart organoids were exposed to environmental toxins (leads, mercury, thallium, and glyphosate) and the toxicity concentrations could be identified (Forsythe et al. 2018)
4. Blood vessel organoids 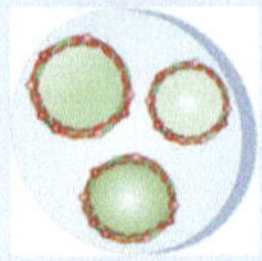	• *Disease modeling:* Human blood vessel organoids were generated as model to study vasculopathy related to diabetes (Wimmer et al. 2019)
5. Lung organoids 	• *Disease modeling:* Branching lung airway structures were created in lung organoids and used as 3D cellular model to study lung bronchiolitis and fibrosis due to respiratory syncytial virus infection (Chen et al. 2017) • *Lung regeneration:* By combining a bioartificial microporous polymer scaffold, lung organoids were successfully engrafted in vivo and exhibited improved cell differentiation, airway structures, and secretion similar to the native adult human lung (Dye et al. 2016) • *Modeling parasite infection:* Lung organoids were used to study how a protozoan parasite infects epithelial tissues and revealed dynamic regulation of the parasite transcriptomes to its life cycle (Heo et al. 2018).

specifically activating or inactivating particular signaling pathways, these "mini-organs" grown in a dish within a specific 3D culture environment using one or a few cell types deriving from tissues, embryonic stem cells, or induced pluripotent stem cells (Fig. 5.6). Numerous organoid systems have been developed for the small intestine (Sato and Clevers 2013; Sato et al. 2009), liver (Hu et al. 2018; Fiorotto et al. 2019), heart (Andersen et al. 2018; Nugraha et al. 2019), lung (Miller et al. 2019; Barkauskas et al. 2017), and brain (Lancaster et al. 2013). They offer great promise as alternative human cognate models to animal models for assessing infection, toxicity, and efficacy screening of pharmaceutical molecules and for personalized medicine,

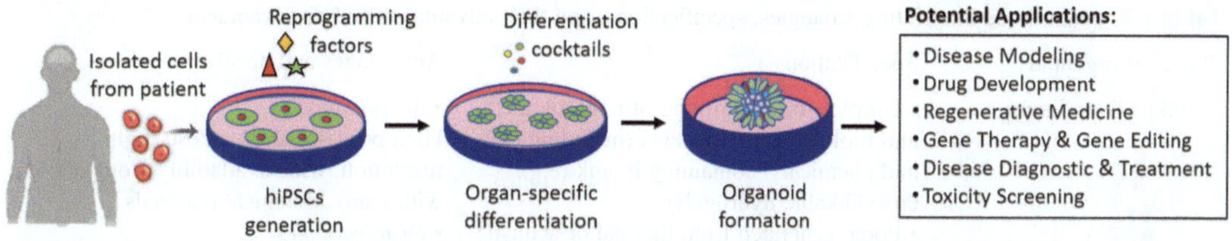

Fig. 5.6 Schematic showing the process from human-induced pluripotent stem cell (hiPSC) generation and organ-specific differentiation to organoid formation and their potential biomedical applications

regenerative medicine, as well as organ transplantation. In fact, organoids are highly expandable, thus representing an unlimited cell source. They can be made to reflect the healthy cells/tissues metabolisms, thus allowing human-relevance evaluation of drugs pharmacokinetics and pharmacodynamics. Additionally, organoids are human origin models and therefore overcome ethical issues with animal testing. They also allow standardization of assays for pharmaceutical compounds testing, thus ensuring reliable safety and efficacy testing. It is conceivable that the field of organoid research may potentially pave the way toward generating functional organ parts (and eventually the whole organ) in vitro and thus fulfilling the acute demand of organ transplantation.

Despite organoids representing powerful emerging tools for precision medicine, these exciting tools are still works in progress (Joachim and Tooze 2018). The main caveat is that these organoids lack vasculature that allows perfusion of blood through the structure, and the absent of immune system as well as elimination system for the removal of metabolic waste products. The organoids could suffer from inefficient nutrient, oxygen, and waste exchange in particular at the center of the organoid when they reach critical size—a scientific challenge that must be overcome. Certainly, this technology needs to be continually evaluated for its technical and biological variations, and the findings need to be interpreted with careful assessment of its limitations before reaching full utility. Additionally, as organoid technology matured, ethical questions started to arise (Yeager 2018): (1) is it ethical or acceptable by the patients/donors to implant human organoids into animals. (2) At what point would we be concerned about organoids developing perception or sensation? (3) Could organoids ever develop something like human consciousness or intelligence? In any case, experiments in animal models should not be abandoned, and the use of human embryonic tissue or cells should go in parallel as benchmarking.

5.9 3D Bioprinting of Human Organs: Promises Hold by Current Organoids Technologies

Since the 1980s, industrial 3D printing has been around and was actively used for rapid prototyping and manufacturing. The emergence of stereolithography, fused deposition modeling, laser sintering or melting, and electron beam sintering has resulted in the coinage of the term additive manufacturing. Recent advance in additive manufacturing toward bioprinting (Table 5.7) has then opened up a whole new realm of possibilities to 3D print human tissues in combination with biomaterials, cells, as well as biomolecules like growth factors or drugs. Undoubtedly, it has been a dream for scientists working in the field of tissue engineering and regenerative medicine to 3D printing replacements of human organs. With the accumulating promises seen nowadays in various organoid technologies that highly resemble the in vivo human organs, the ability to print whole organs may one day become a reality. In this context, the organoids can be used as biological

Table 5.7 Main tissue bioprinting strategies, specifications, and their advantages and disadvantages

Types of organoid	Specifications	Advantages and disadvantages
1. Inkjet Bioprinting 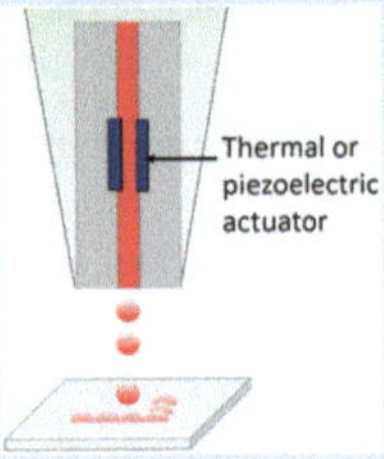	• Droplets-based printing of nonbiological and biological components (including cells and chemicals) containing bioink (e.g., crosslinkable hydrogels) • Force generated from thermal or acoustic is used to eject drops of bioink onto a substrate on which desired 3D anatomically-relevant structure is built gradually	• *Advantages:* High print speed; low cost; high resolution; wide availability; compatible with many biological materials • *Disadvantages:* Risk of exposing cells and materials to thermal and mechanical stress; frequent clogging of the nozzle; requires solidification of liquid form in time to enable 3D droplet formation and 3D tissue build-up; risk of cell toxicity by cross-linkers; low cell density is required to avoid nozzle clogging
2. Microextrusion 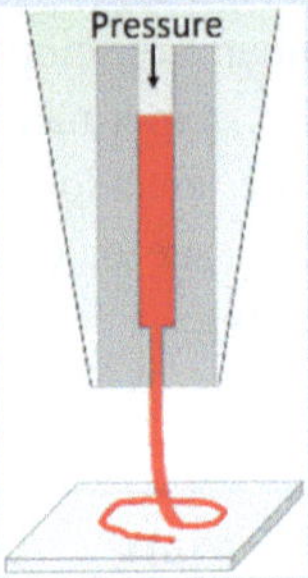	• Microextrusion of temperature-controlled bioink slurry onto a substrate using pneumatic or mechanical dispensing system • It can yield continuous beads or treads in *x*, *y*, and *z* axis and thus via layer-by-layer technique forms 3D structure	• *Advantages:* Compatible to a wide range of bioink properties including high viscosity; capable of depositing high cell densities (e.g., using microspheroids or organoids) and rapid 3D structure formation. • *Disadvantages:* Low cell viability due to extrusion pressure or non-physiological temperature; use of large nozzle size (to increase cell viability) results in loss of printing resolution and speed
3. Laser-assisted 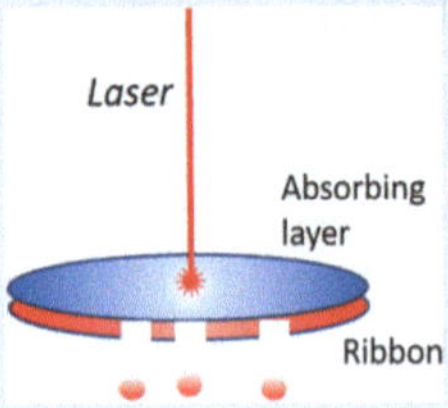	• Based on the principles of laser-induced forward transfer technology • Use focused laser pulses to generate a high-pressure bubble that propels cell-containing bioink onto the collector substrate, thus offering a high-throughput and rapid printing process	• *Advantages:* No clogging of nozzle; compatible with a range of viscosities; high cell viability; capable of depositing high cell density and at single cell resolution. • *Disadvantages:* Requires rapid gelation kinetics to achieve high shape fidelity; preparation is time-consuming when multiple cell types and/or materials are used; high cost

building blocks for printing 3D organ, thus resulting in 3D printed structures with high cell mass in addition to their structural and functional properties.

In general, there are two central approaches for 3D bioprinting (Murphy and Atala 2014):

1. *Bioinspired 3D Bioprinting:* Specific cell types and extracellular components (e.g., bioinks, growth factors, molecular sensors, extracellular matrices, and signaling biomolecules) are printed sequentially or in mixtures to reproduce identical structures and organizations similar to a tissue or organ of interest. For example, printing vascular networks that are similar to in vivo blood vessel architecture, deposition of soluble or insoluble factors in biomimicry gradients to induce embryogenesis and organogenesis, or layering of hydrogels with different mechanical properties that resembles tissue interface (e.g., joint-to-bone interface) as optimized biomechanical microenvironment for cell differentiation and tissue maturation.
2. *3D Cells Self-Assembly, Mini-Tissues, or Organoids Bioprinting:* An approach that initially relies on the intrinsic cell signaling and autonomous tissue organization and patterning behavior to establish the desired organ micro-architecture and function via microspheroid or organoid formation (Fig. 5.7). Through high-throughput generation techniques, high number of these microspheroids or organoids are then used as biological and functional building

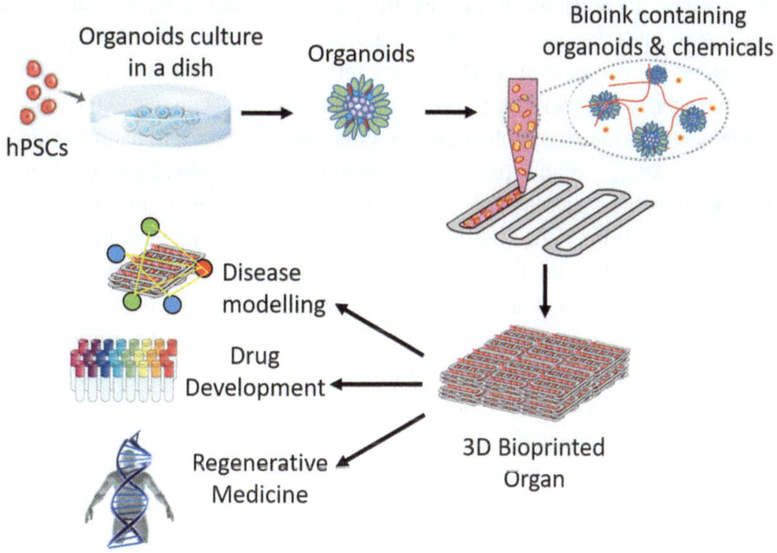

Fig. 5.7 Bioprinting of 3D organ for disease modeling, drug development, and regenerative medicine applications. Human pluripotent stem cells (hPSCs) differentiated in vitro into specific tissue lineage to form organoids. The organoids are then encapsulated into bioink (e.g., hydrogel) containing chemicals that stimulate further maturation of the organoids and printed into 3D organ structures via layer-by-layer method

blocks to fabricate 3D organ structure with the aid of high-resolution bioink printing and assembly. The combinations of the above two strategies will facilitate bioprinting of complex 3D biological structure with multiple functions and physicochemical properties.

Nevertheless, 3D bioprinting still faces a number of challenges before its potential could be fully realized in regenerative medicine: (1) the bioprinter technology needs to be scaled up and be compatible with more biomaterials at physiological-relevant conditions; (2) more biomaterials need to be explored as printable bioink to suit different cell types at the right biological and biomechanical properties; (3) enables printing of vasculature networks as well as innervation within the tissue constructs; (4) controls fabrication timing and facilitates nutrient-waste transport within the construct during the course of printing. Encouragingly, despite these challenges the companies remain determined to build a 3D bioprinting industry. However, the business models and product targets have shifted from emphasizing on printing functional organ to delivering different printing technology and bioinks and providing small-scale engineered tissues (e.g., skin wound healing products, scaffold implants, or blood vessel grafts) for research and development as well as drug testing. The goal is to achieve tangible results in a time frame that fits to the investment and profit gained. Stringent manufacturing guidelines and regulatory requirements are the major hurdles faced by these products before they can be commercialized.

5.10 Microfluidic Enabling Biotechnologies for Biomedical Exploration

5.10.1 When Microengineering Meets Developmental and Stem Cell Biology

The intrinsic variation in biological responses represents a major challenge in obtaining controllable, predictable, and reproducible biological results. At the cellular level, individual cell reacts relatively different to a stimulus within an "on/off" threshold margin depending on its genetic and cellular states.

At the tissue and organ levels, the physiological and functional states (largely depending on genetic and epigenetic background) of each subject influence the consistency of experimental data. Nevertheless, implementing a more controlled experimental environment using high precision quantitative microengineering system could minimize the biological variation, hence producing downstream results with higher predictability and reliability (Peela et al. 2017; Folch and Toner 2000; Keatch et al. 2002). Most remarkably, integration of cellularized constructs within microengineered platform has enabled the recapitulation of the physiological and pathological conditions of complex tissues and organs at micro-scale precision that greatly influences cellular signaling events (Perestrelo et al. 2015). To achieve this, microfabrication techniques, such as photolithography, replica molding, micromilling, and microcontact printing, are essential tools to create structures with defined dimensions and structures at micrometer scale, including microfluidic devices.

Microfluidics is a multidisciplinary field combining biotechnology, microtechnology, physics, chemistry, analytical physicochemistry, and fluid dynamics to generate precision devices with controlled parameters and process environment at micro- to milli-scale for accurate qualitative and quantitative assays (Sonnen and Merten 2019). Conventional as well as novel biological analysis are replicated in the microfluidic devices whereby samples are subjected to microchannel flow system (in the range of tens to hundreds of micrometers) to manipulate small fluid volumes ($10^{-9} - 10^{-18}$ L) (Whitesides 2006). Such systems are capable of processing low volumes of samples (in particular when sample volume is limited) using lower amounts of expensive reagents in order to detect the molecule of interest (higher detection sensitivity) at faster rate, for example, via ligand-antibody binding or fluorescent-based measurement via a high-throughput, high-resolution, and sensitive manner. In fact, miniaturization of an assay device facilitates the reduction in the scale of assay (including sample volume, size of the detection system, faster analysis, and reduced response time), thus imposing lowered production and operation costs in particular when combining semiconductor technology to streamline manufacturing process. It is also unique over conventional system, due to its compact size, disposable nature, and decreased procedural times and reduces hands-on manipulation of samples within a controlled field of operation at micro- to nano-precision.

These properties are essential to reduce sample reagents and to allow multiplexing screening or detection in biomedical diagnostics, drug discovery, and regenerative medicine. Additionally, the introduction of polydimethylsiloxane (PDMS) (and other improved polymeric substrates) and soft lithography techniques boosted the growth of this field enormously due to their cost-effectiveness and rapid microfabrication technology (Duffy et al. 1998; Unger et al. 2000). Recent advances in manufacturing methods such as lithography (Dendukuri et al. 2007), xurography (Neuville et al. 2017), laser machining (An et al. 2008), and 3D printing (Waheed et al. 2016) have enabled the production of compact microfluidic system at lower cost. Moreover, the emergence of droplet microfluidics allows compartmentalizing immiscible fluids within pico- or nano-liter (Teh et al. 2008), whereas the invention of paper microfluidics replaces hollow microchannels with woven microfibers of papers that move fluids via capillary action, thus circumventing the need of using additional pumps (Li et al. 2012). These peripheral improvements render microfluidic technology highly versatile and open new avenues for unprecedented exploitation in development of various biotechnological assays in broad application sectors: biotechnology, consumer health, cosmetics, chemical, food, agrochemicals, government agencies, academic researchers, personalized health, and pharmaceutical industries. In fact, it has paved the way toward the emergence of an innovative and contemporary industrialization of "lab-on-chip" biotechnologies. A database containing emerging, innovative companies in microfluidics-related technologies has been created to promote the microfluidic circle (http://circle.ufluidix.com/microfluidic-companies/).

Today, many lab-on-a-chip technologies are being developed rapidly and commercialized for targeted biotechnological applications, including genetic mutation/cancer screening via blood or urine samples, glucose monitoring, and diagnosis

Table 5.8 Main advantages and disadvantages of microfluidic-based biotechnologies compared to conventional technologies

Advantages	Disadvantages
1. *Low assay volume:* Only small amount of sample and less reagents needed for analysis, thus reducing cost of analysis or increasing cost-effectiveness for more analysis at the same cost 2. *User-friendly, compact design, and high parallelization:* Easy handling of complex operations integrated on a chip design for simultaneous analysis in highly parallelized manner. On-site analysis is possible 3. *Prompt analysis and real-time process control and monitoring:* Fast response-time for swift changes of analysis parameters at micro-scales (i.e., temperature, molecular diffusion rate, in situ mixing of analytes, and high biochemical reactions) 4. *Automation with minimal human errors:* Automated diagnosis and samples handling reduces human errors compared to classical analytical procedures in the laboratory	1. *Signal/noise ratio:* Increase of signal-to-noise ratio intrinsic to miniaturization of the assay system results in poorer results than classical assays 2. *Dependency of specific equipment:* Requires specific equipment, e.g., low-pulsatile pump or high precision pressure/vacuum generator, to work properly. This increases system cost and size 3. *Human misconducts and ethics:* Unsupervised usage of microfluidic devices may lead to untrained public diagnosis, privacy violation, and criminal activities, etc.

of bacterial or viral infection. In future, this technology is predicted to be expanded to governmental system as well as domestic monitoring of health, in combination with the advances of telecommunication technology due to popularity of World Wide Web networking and popularity of personal smart phone usage. It is conceivable that in the near future, lab-on-a-chip devices may revolutionize the conventional practices in health and medical sectors, for instance, immediate diagnosis during clinical consultation or emergency cases, thus increasing survival chance of patients, and in reducing social-economical burden of patients due to lower cost. Table 5.8 summarized the main advantages and disadvantages of microfluidic-based biotechnologies as compared to conventional fluidic handling technologies.

5.11 Microfluidic-Based Biotechnology Applications

5.11.1 Microfluidics for DNA and Protein Detection and Analysis

Microfluidics has been integrated into a range of DNA-based analysis such as capillary flow direct polymerase chain reaction (PCR) assay (Zhang et al. 1999) and shown to accelerate hybridization time in few minutes (Chang et al. 2013; Pjescic et al. 2011) and recently in less than 1 min (Samuel 2016) using extreme PCR. This is possible due to the minimization of sample processing and handling up to nanoliter (He et al. 2001) and potentially at individual molecules (Foquet et al. 2002) and at single cell level (Zhu et al. 2012). For this, construction of fluidic channels of <1 μm and the control of flow rate by electrical field are essential to allow the detection of each DNA molecule in mixture per several milliseconds. The hybridization times for targeted DNA binding to the probe elements (from hours to <1 min) have been improved significantly by using microfluidic devices with, for instance, a microarray design (Wang et al. 2003), integrated pump system (Lenigk et al. 2002), or based on cavitation microstreaming technology by inducing vibration of air bubbles using a sound field to enhance circulatory flow and mixing mechanism (Liu et al. 2003). Detection of gene fragments carrying low-abundant point mutation was also reported to be highly feasible when a low-density array was implemented in microfluidic channels (Wang et al. 2003). Combining biobarcode (fluorescent-labeled single-stranded barcode DNA with on-chip microcapillary electrophoresis provides rapid, sensitive, multiplex, and accurate biological agent identification for bioterrorism or biological warfare (Cho et al. 2014). As the technology advances, more sophisticated system such as micro-qPCR and digital microfluidic are developed to perform PCR with lower detection limit and higher efficiency and specificity.

Integrating microfluidics with mass spectrometry enables automated protein or peptide analysis (Figeys et al. 1998; Figeys and Aebersold 1998) up to pico-molar level (Lion et al. 2003) within a controlled microenvironment. Generic profiling of peptides from protein samples using microfluidic combined with combinatorial peptidomics approach enables the identification of peptides from proteolytically digested samples through chemical cross-linking via amino acid side chains (Soloviev et al. 2003). Purification of proteins from cell lysates is feasible by subjecting the samples to microfluidic channels linked to membranes imprinted with corresponding enzymes (e.g., trypsin); thus, protein digestion, separation, and identification are achieved at high throughput (in minutes) and low sample amount (in nanograms) manners (Gao et al. 2001). To meet the demand of discovering proteins or small molecules of pharmacological value at low sample volume but high assay sensitivity and short hybridization time, various high-throughput microarray methods based on fluorescent detection principle or waveguide technology in combination with microfluidic system are developed to quantitate protein profiling efficiently (Scrivener et al. 2003; Barry et al. 2003; Pawlak et al. 2002). Other applications of microfluidics include rapid forensic DNA analysis (Hopwood et al. 2010), handling of mammalian embryos and reproduction (Glasgow et al. 2001; Beebe et al. 2002), and isolation of motile spermatozoa (Schuster et al. 2003).

5.11.2 Microfluidics for Organs-on-Chip Biotechnologies

Despite animal models being indispensable for preclinical drug discovery and development, various issues such as ethical considerations and species differences have impeded animal experiments as useful models to study human diseases. Mismatching in the pharmacokinetic predictions caused by species differences between humans and experimental animals has halted many candidate compounds toward clinical trials (Lin 1995), which hugely jeopardized the cost-effectiveness in new drug development. This is compounded by the recently implemented European Union (EU) legislation on Animals Used for Scientific Purposes (Directive 2010/63/EU; http://ec.europa.eu/environment/chemicals/lab_animals/index_en.htm), whereby animal testing for cosmetic development has been completely prohibited since January 1, 2013. Among others, firmly anchoring to the three Rs principle (replace, reduce, and refine) in the use of animals improves animal welfare. In fact, in collaboration between the European Commission, European Trade Associations, and companies from seven industry sectors, the European Partnership for Alternative Approaches to Animal Testing (EPAA) has been initiated since 2006. The association aims to accelerate the development, validation, and acceptance of alternative approaches to animal use in regulatory testing (https://ec.europa.eu/growth/sectors/chemicals/epaa_en).

Over the past decades, human stem cell-based therapeutic approaches are being pursued actively to overcome the unavoidable limitations of animal studies. This includes pluripotent stem cells (PSCs) derived from human embryos or induced via genetic reprogramming (namely, induced PSCs) or multipotent stem cell (MSC)-derived from human adult tissues (e.g., bone marrows, fat tissue, umbilical cord, urine, and skins). Nonetheless, cultivated cells in two-dimensional (2D) or even in 3D often fail to retain their original organ functions, morphologies, and blood circulation, rendering the cell culture models inaccurate in predicting drug efficacy, toxicity, and organ interactions. Incorporating organ cultures in microfluidics provides unprecedented potential in biotech and medicine, which more precisely imitates in vivo conditions at miniature fashion in vitro. Through biochemical engineering and microfabrication technologies, it is possible to mimic the in vivo microenvironment in vitro, thus replicating the cellular function, morphology, and organ interactions. It will help in elucidating how tissue responds to new drugs, as future alternative to traditional animal testing which is essential for ADMET (adsorption, distribution, metabolism, elimination, and toxicity) characteristics of a drug. Animal testing is expensive and cumbersome, and bearing an intrinsic unavoidable huge difference in the drug uptake between animal models and humans. Through the use of human cells, these interspecies differences can be eliminated. So far, functions of various vital organs such as the liver, kidney, lung, and gut have been successfully replicated as clinically relevant in vitro models. However, it is essential to produce safe novel therapeutics for human.

Recent innovations in microfluidic and "organ-in-a-dish" technologies have enabled the creation of "organ-on-a-chip" platforms (Table 5.9). These systems integrate the advances of 3D tissue engineering, stem cell biology, and biomaterials with microfluidic networks, with an aim to overcome the indispensable shortcomings of in vitro 2D cell culture models as well as preclinical animal models. Indeed, the organ-on-chips mimic basic organ-to-organ coupling, enabling ADMET testing and drug screening all done in vitro instead of through the use of animal models. Recently, there is an increase in demand for organ-on-a-chip in drug screening. Organ-on-a-chip simulates perfused chambers of living cells in vitro with function at the organ level in their physiologically relevant "organ" niche. For instance, vasculature and interstitial fluid flow are simulated on chip in order to emulate the in vivo physiological conditions for studying bacteria/viral infections, stem cell differentiation, cancer metastasis, environmental toxicants, etc. Use organ-on-a-chip culture devices to monitor and assess drug toxicity and effectiveness in various organs of the body (i.e., pharmacodynamics (PD) and pharmacokinetics (PK) of drugs). In 2018, the European Organ-on-Chip Society (EUROoCS) was officially launched (https://etp-nanomedicine.eu/european-organ-on-chip-society-launched/), with the purpose of encouraging and developing organ-on-a-chip research as well as providing opportunities to share and advance knowledge and expertise in the field toward better health for all. Organ-on-chips are expected to cause a paradigm shift in healthcare by offering novel approaches to elucidate disease mechanisms, as well as for the prevention and treatment by identifying safe and effective drugs.

The research trend has recently shifted toward the creation of interconnection of multiple organ-on-a-chip systems into a single microfluidic platform in order to emulate inter-organ relationships that mimic a human body-like microphysiological environment. Hence, the "human-on-a-chip" concept has been proposed to replicate organ-to-organ interactions on a chip in order to simulate human body in metabolizing drugs, immune response, transport and clearance, muscle contractility and electrical conduction, bacteria or viral infection, absorption of substances into the lung, skin, and intestines, as well as excretion of metabolized compounds (Fig. 5.8). This opens new perspectives for targeted biomedical and healthcare applications, in particular, in the fields of personalized medicine, drug development, and toxicology. It is anticipated that the efficiency of a particular pharmaceutical compound or therapeutic approach can be effectively evaluated at the whole body level, yet the differential toxicity effects of a chemical can be assessed at each organ level, in respect to the route of exposure or administration. Liver is the first organ in contact with the administrated chemical, which subsequently metabolizes the chemical and releases various metabolites that will be distributed to all organs, particularly for drug that relies on hepatic metabolism to turn into an active compound before it reacts on the targeted organ (e.g., brain). For instance, Skardal A et al. have reported on the fabrication of three organs (liver, heart, and lung) organ-on-a-chip platforms, which are integrated in a closed circulatory perfusion system. This system has facilitated inter-organ responses to drug administration and allowed the assessment of efficacy and side effects associated with the candidate drugs (Skardal et al. 2017).

Apparently, there are a number of challenges need to be tackled before "human-on-a-chip" could become a reality. First, the scientists need

Table 5.9 Examples of "organ-on-a-chip" technologies and their potential field of applications

Types and Aims	Potential Field of Applications	
1. *Brain-on-a-chip* (a) simulates brain tissues and details barrier functions in both healthy and pathological conditions	• *Neurological disorders and neurodegeneration diseases* (e.g., Alzheimer's, Parkinson's, ALS)—study into mutations/malfunction of neuron, microglia, and glia cells, and the pathogenesis of neuronal cell death or brain damage (Park et al. 2015; Kilic et al. 2016; Dauth et al. 2017) • *Blood–brain–barrier dysfunctions* (e.g., due to neuroinflammation, infections, abnormal protein deposition, toxins, or genetic disorders) (Brown et al. 2016; Herland et al. 2016; Feng et al. 2018; Brown et al. 2015)	*Blood–brain–barrier permeability:* 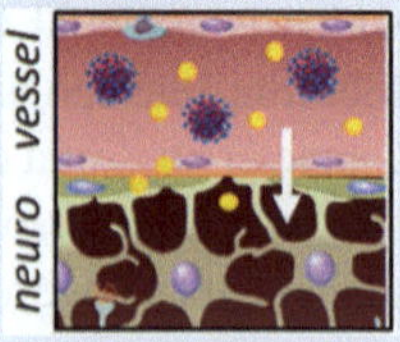
2. *Lung-on-a-chip* (a) Mimics pulmonary gas-exchange microenvironment and alveolar interface	• *Pulmonary failures* due to thinning of alveolar barrier, lung fibrosis, environmental pollutants/toxicants, daily habits (e.g., smoking), infections, etc. (Punde et al. 2015; Fukumoto and Kolliputi 2012; Felder et al. 2019) • *Lung diseases* (e.g., asthma, immunologic disorders, cancer, genetic mutations) (Yang et al. 2018; Gkatzis et al. 2018) • *Drug screening and toxicity* (Huh et al. 2012; Kizilkurtlu et al. 2018; Konar et al. 2016)	*Airway absorption:* 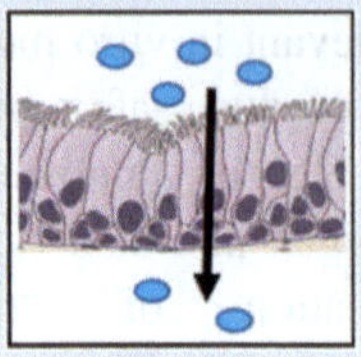
3. *Heart-on-a-chip* (a) Establishes synchronized contractility and electrical conduction properties	• *Acute and chronic heart and cardiovascular failures* (e.g., ischemia, cancer metastasis, infections, hypertension, arrhythmia, and atherosclerosis) (Ugolini et al. 2018; Qian et al. 2017) • *Inherited heart diseases* (e.g., Duchenne and Becker's muscular dystrophy, myotubular myopathy, and metabolism disorders of cardiomyocytes such as Pompe's disease) • *Drug screening and toxicity* (Conant et al. 2017)	*Muscle contractility and conduction:* 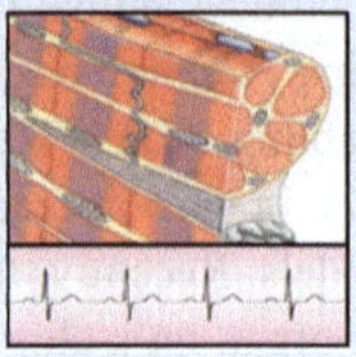
4. Liver-on-a-chip (a) Creates hepatic cultures with clinically relevant liver-specific sinusoidal structures, metabolism, and function profiles	• *Liver diseases* (e.g., cirrhosis, fibrosis, fatty liver, and bacteria/viral infections (hepatitis) (Wang et al. 2018b; Grix et al. 2018; Gori et al. 2016) • *Liver genetic disorders* (e.g., hemophilia, hepatocarcinoma, liver metabolism disorders) (Kong et al. 2016) • *Pharmaceutical compounds metabolism, bioactivation, toxication, and detoxication* (e.g., pharmacokinetics, pharmacodynamics, bioavailability, drug metabolites activation) (Lee et al. 2019; Zakhariants et al. 2016; Prodanov et al. 2016; Ma et al. 2016)	*Drug metabolism, transport and clearance:* 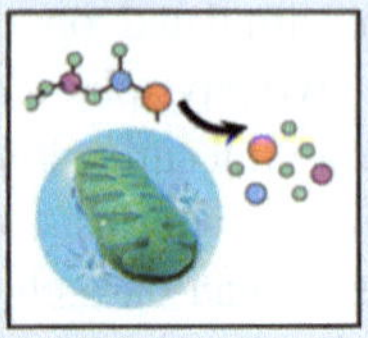
5. Kidney-on-a-chip (a) Culture of kidney epithelial cells with anatomical structures and physiology functions closely mimicking the in vivo excretion system	• *Acute and chronic renal failures* (Zhou et al. 2016; Ashammakhi et al. 2017) • *Pharmaceutical compounds filtration, reabsorption, and excretion* in the urine (Weber et al. 2018; Lee and Kim 2018)	*Kidney absorption, reabsorption and excretion:*

(continued)

Table 5.9 (continued)

Types and Aims	Potential Field of Applications	
	• *Study of transporter function* (Chang et al. 2016; Vriend et al. 2018) • *Nephrotoxicity study* (Kim et al. 2016; Li et al. 2017)	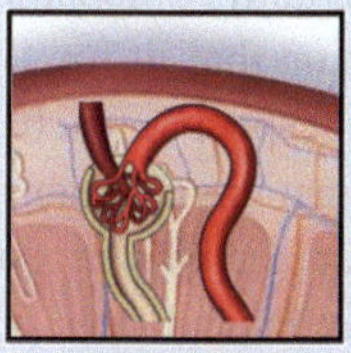
6. Pancreas-on-a-chip (a) Culture of pancreatic islet cells for insulin secretion	• *Diabetes mellitus* (Bandak et al. 2018; Zhang et al. 2010) • *Pancreatic cancer* (Beer et al. 2017)	*Pancreatic endocrine:* 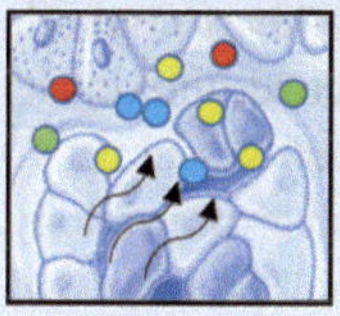
7. Intestine-on-a-chip (a) Establishes gut models that exhibit molecule adsorption and transport mechanisms	• *Gastrointestinal diseases* (e.g., Crohn's disease, ulcerative colitis, irritable bowel syndrome) (Mertz 2016). • *Colon cancer* (Bein et al. 2018) • *Drug metabolism* and transport (Guo et al. 2018; Tan et al. 2018; Kimura et al. 2015) • *Virus infection* (Villenave et al. 2017)	*Absorption:* 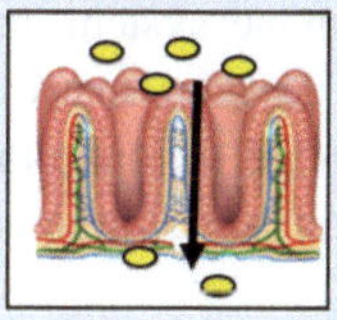

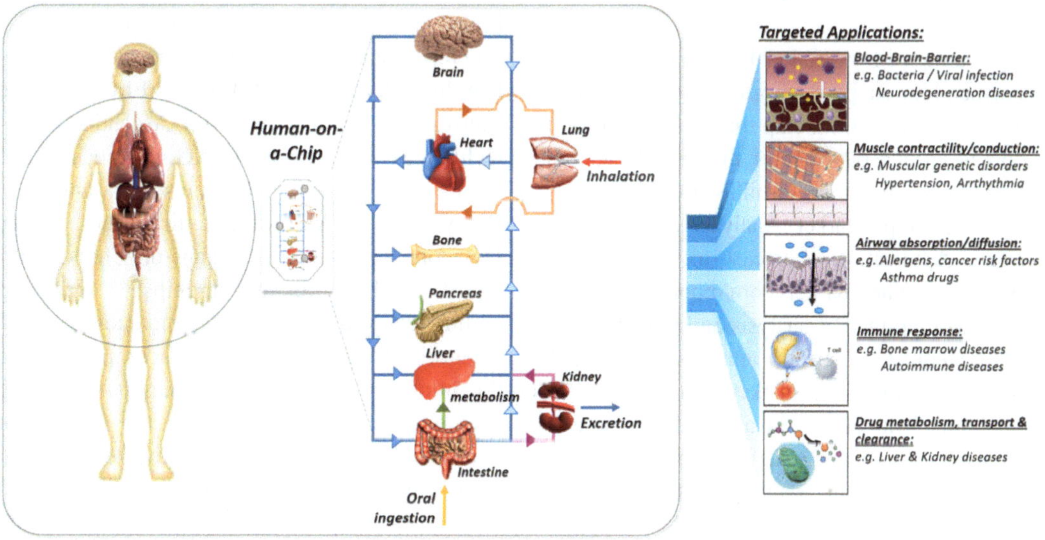

Fig. 5.8 Human-on-a-chip: futuristic or reality? A visionary concept of a human body-like microfluidic chip comprising different engineered "mini-organs" which are integrated into a fluidic system mimicking the human circulatory system in a physiologically relevant manner. It represents an alternative human cognate model to model a complex, dynamic process of drug absorption, distribution, metabolism, and excretion toward reliable evaluation of drug efficacy and toxicity by mimicking the different route of exposure. A number of important scientific and technical challenges need to be addressed before this chip technology can be put in full swing for research, testing, and diagnostic purposes

to overcome the heterogeneity due to different organ culture conditions. A universal culture medium composition and an optimal culture condition that fit for all organs need to be developed in respect to each of their specific and required organ environment both spatially and temporally (e.g., sensitivity to fluid flow and shear stress, mechanical cues on cell differentiation and lineage specification, and growth factors and cytokines controlled release in time and space). Second, the organs need to be interconnected by blood vessel networks as close as possible to the human physiology, including the mimicking of blood pressure and pulsatile rhythms due to contractile action of the heart. Integrating functional immune system and self-generating blood cells by the bone marrow within the microfluidic system are, among others, the most challenging scientific mission to accomplish.

5.12 Microfluidic Tools Toward Industrial Biotechnology: A Brief Market Perspective

Engineering of functional organs is highly demanded to support the growth of the market. The global organ-on-a-chip market is estimated to grow at around 69.4% over the next decade and reach approximately $6.13 billion by 2025 (source: Organ-on-Chip Market Analysis and Trends—Organ, Application—Forecast to 2025; https://www.researchandmarkets.com/reports/3951918/organ-on-chip-market-analysis-and-trends-organ). It is beneficial to perform analysis depicting the global organ-on-a-chip market with current trends and future estimations, in a way to prepare for imminent investment road map including better understanding on the profitable trends in order to gain a stronger foothold. The growth of the global organ-on-a-chip market is mainly driven by the increase in its applications in biomedical research and rise in demand by pharmaceutical companies for drug development. Key market determinants include key drivers, restraints, opportunities, and financial competency of the market (e.g., potency of the buyers and suppliers in the industry) followed by detailed impact quantitative analysis including high cost and nascent stage in research and development. Nonetheless, increase in R&D activities will offer ample opportunities to the stakeholders. The global organ-on-a-chip market is predicted to be impacted largely in the future by the following factors:

- Increase in demand of organ-on-a-chip in drug screening for higher therapeutic efficacy and safety
- Surge in applications of organ-on-a-chip devices in the healthcare sector
- Growth in demand for lung- and kidney-based organ culture devices
- Increase in research activities on organ-on-chips devices
- Preclinical stage of R&D pertaining to organ-on-chips

Current key segments of organ-on-a-chip market include heart, intestine, kidney, liver, lung, brain, as well as multi-organs or "human"--on-chip. Some key market players include Emulate Inc., AxoSim Technologies LLC, CN Bio Innovations, Hurel Corporation, Ascendance Biotechnology Inc., Insphero AG, Mimetas B.V., Nortis Inc., Organovo holdings Inc., and Tara Biosystems. According to the Yole Développement (Yole) "Organs-on-Chips" 2017 report, the organs-on-chips platforms commercialized by these companies can be categorized based on the model complexity and the number of compound throughput for drug testing (Fig. 5.9). Interestingly, the number of compound throughput is in an inverted relationship with the model complexity, whereby an "organ-on-a-chip" platform with simple fluid management system could give rise to up to ~50 times higher compound throughput to drug testing. This intriguing finding suggests that by keeping a system simple may be of higher interest when higher system performance is needed.

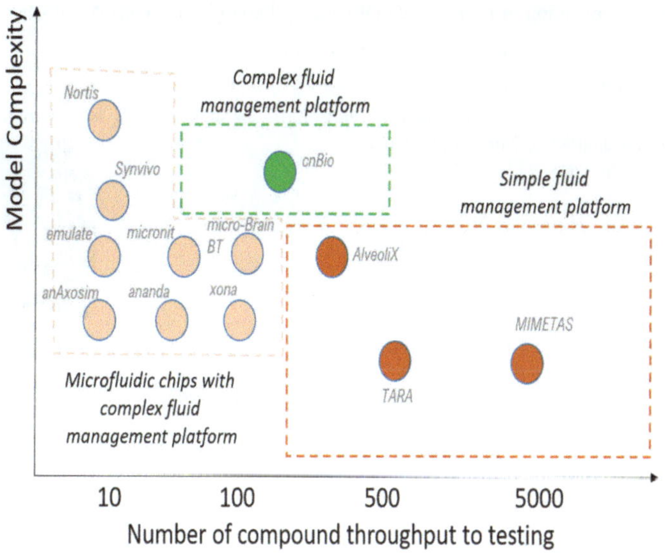

Fig. 5.9 Keep it simple? The model complexity and number of compound throughput to drug testing for the different types of commercialized organs-on-chips platforms (Adapted from: http://www.yole.fr/iso_album/illus_organchip_diffstages_yole_apr2017.jpg)

5.13 Opportunities for Veterinary Biotechnology

As part of the multiple applications of various types of the biotechnological sciences, we would like to point out that the branch of veterinary biotechnologies represents an important opportunity for biotechnologists who wish to approach on the private application of job development. In this regard, it is interesting to understand how, in a similar way as in the veterinary sciences, there are many fields of practical application for the biotechnological investigations.

The veterinary biotechnologist is already directly contemplated as an important specialist in different European biotechnology universities and contemplates experts with multidisciplinary scientific and professional skills, acquiring theoretical knowledge and operating in various specialized diagnostic and research fields of veterinary medicine.

The most relevant fields of application are represented in different veterinary disciplines, which may include microbiology, infectious diseases, pathology, obstetrics and gynecology, animal husbandry and genetic improvement of breeds, pharmacology and toxicology, anatomy, hematology, parasitology, biochemistry and physiology, reconstructive and regenerative surgery, food safety and food control, and so on.

There are important opportunities for a graduate in biotechnology who intends to undertake a business project in the biotechnology sector, with the activation of services aimed at both the public and private sectors (Table 5.10).

In this chapter, we will try to list a series of opportunities and potential aspects that could affect the activation of business projects in the big world of veterinary biotechnology, always keeping in mind the fact that the figure of the biotechnologist, and in this case of the veterinary one, will represent one of the major leaders for the panorama of research, innovation, diagnostics, and general services.

The figure of the veterinary biotechnologist will represent an important point of reference for the development of new diagnostic, prognostic, and therapeutic protocols of many acute and chronic diseases and will be a fundamental

Table 5.10 Examples of application of new biotechnological technologies in veterinary medicine

Types and aims	Potential field of applications	
1. *Serological and parasitologic investigations* (a) To detect antibodies against parasites in the serum of animals of different species	• *Many parasitological diseases, especially the asymptomatic ones* (e.g., Leishmaniasis) (Miro et al. 2014).	*Indirect immunofluorescent antibody test (IFAT)*: 
2. *Cytological, histochemical, ultrastructural, and immunohistochemical investigations* (a) Morphological and immunological diagnostic tools for different diseases	• *Infectious and parasitological diseases* (Leonardi et al. 2012). • *Degenerative, inflammatory, and neoplastic diseases* (Chirullo et al. 2015) • *Toxicological diseases* (Adachi et al. 2016)	*Histochemistry* 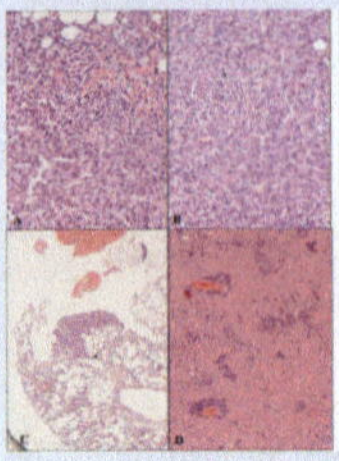
3. Liquid biopsy (a) Detection of circulating markers for different diseases	• *Acute and chronic degenerative diseases, neoplastic diseases* (e.g., cancer and cancer metastasis, muscular dystrophies, metabolism disorders of cardiomyocytes, tumors). (https://www.zomedica.com/home/wp-content/uploads/2018/12/Zomedica_WhitePaper_ZM-017.pdf 2018).	*Detection of circulating tumor cells in liquid biopsy:* 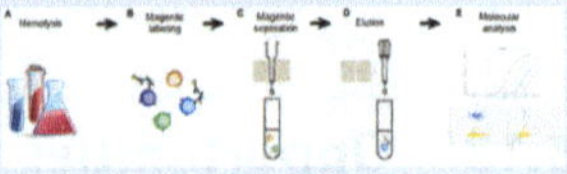
4. Additive manufacturing (a) Produce viable issues and bioscaffolds	• *Tissues damages:* realization of a customized metallic prosthesis developed starting from a canine pelvic computed tomography imaging (Leonardi et al. 2015)	*3D printing:* 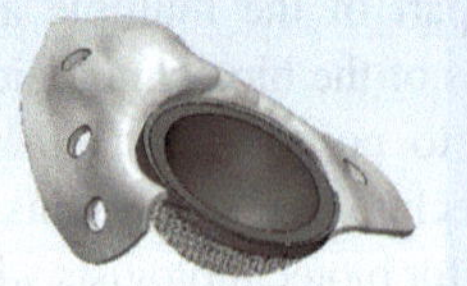

support for the study, prevention, and therapy of many animal diseases, also for zoonotic ones, responsible for serious damages in economic, social, and territorial terms. Degenerative, inflammatory, and neoplastic diseases can be prevented, diagnosed, and appropriately treated now thanks to the application of biomolecular and genetic investigation protocols, which can be applied and developed not only in public research centers but also in structures and centers of private biotechnical diagnostics and researches.

There are already some of these realities, but there are still little private offers that can compete with what is present in most of the EU countries.

The activation of these centers could also represent a reference of excellence in the field of international scientific research, through which they could also open up new panoramas of elaboration of international scientific collaboration projects, in which the energy of diversified and specialized professional figures can be coagulated: biotechnologists, veterinarians, physicians, etc.

The complex chapter of laboratory diagnostics for the study of many diseases, not least those of a neoplastic nature, could represent an optimal substrate for the activation of multidisciplinary diagnostic structures on which to converge also part of sophisticated studies aimed at the biomolecular, phenotypic comparative characterization, genotypic, of many spontaneous, primitive, and metastatic neoplasms.

At this time, pure veterinary diagnostic services are still not widespread and the implementation of these, with a high veterinary and biotechnological sophistication, could represent an important amplification of services at the community level with greater supply and competitiveness (and reduction of prices of services) for all users.

There are many areas of application of these services: veterinary hematology, veterinary parasitology, veterinary microbiology and infectious diseases of animals, general pathology and veterinary pathological anatomy, normal veterinary anatomy, veterinary physiology, animal husbandry, food hygiene and the inspection of food of animal origin, the veterinary medical clinic, the veterinary obstetric clinic, the veterinary surgical clinic, and the anaesthesiology.

The veterinary biotechnologist can be involved in the application of hematological evaluation protocols with professional involvements related to the numerous activities in the centralized clinical laboratories, in performing cytological and serological investigations, and in the search for early markers indicators of disease states of various origins. The serological investigations, together with coprological tests on fresh and preserved samples, may be carried out in the context of the applications of parasitological investigations for the vast parasitic diseases affecting animals of different species, both domestic and non-domestic and tropical.

An important business opportunity could be identified in the field of veterinary microbiology and infectious diseases, in which not only investigative diagnostics can represent an important center for the control of many diseases, both bacterial and viral, but where also the application of genetic and immunological engineering can help to provide a fundamental support for the "construction" of new vaccines and immunological protocols that could protect man and animals from many contagious diseases preserving the health of many and limiting large and serious economic losses.

Being directly involved in diagnostic and research activities of various kinds, I would like to underline the important role of the veterinary biotechnologist in the complex field of veterinary pathology. Private diagnostic services centers in the field of pathological cytology and veterinary histopathology could be the only reference supports for many free veterinarians who daily work in various private sectors of veterinary medicine.

The importance of the biomolecular study in the pathogenetic mechanisms of many diseases, from chemical, physical, or biological causes, will be able to provide important support and cornerstones for the approach and the fight against diseases and for the interpretation of their pathogenesis. The study of diagnostic and early biomarkers represents biomolecular devices on which future diagnostic and therapeutic applications will be based in the complex field of human and veterinary medicine.

The immune modifications and the improvement of therapeutic protocols, more and more sophisticated and precise, will contribute to fight and eradicate many diseases which are currently still sources of huge losses in terms of lives and management costs.

To this end, even for the animals we want to remember the importance of new diagnostic protocols that represent completely innovative processes in the world of veterinary medicine, such as that of liquid biopsy, a methodology of the future that thanks to the application of modern biomolecular diagnostics will allow to diagnose diseases of various kinds early, starting from easy sampling samples, such as saliva, urine, or animal hairs.

Finally, we would like to mention, even if in synthetic form, a last biotechnological reference of marked innovation in the world of veterinary medicine represented by additive manufacturing through the use of 3D printers.

A world of technologies applied to the completely new and futuristic veterinary medicine will be aimed at reparative and regenerative processes toward organs and tissues that have been lost or damaged by diseases from different causes, as well as fractures or tumors.

Data processing centers and production of prostheses and stem cells appropriately built for each individual case may represent references of fundamental importance for laboratories, clinics, or private and public veterinary clinics in which they will perform activities of regenerative and reconstructive medicine by animal species like horses, dogs, birds, and many others.

Finally, and in conclusion we do not want to forget all those application protocols of artificial insemination and genetic control of births, as well as those of genetic selection and improvement of the breeds.

Acknowledgements This work was supported by Erasmus plus consortium, SupBioEnt: Supporting biotechnology students oriented toward an entrepreneurial path (# 2017-1-RO01-KA203-037304). MS is supported by CARIPLO 2015_0634, FWO (G088715 N, G060612 N, G0A8813N) and C1-KUL 3DMUSYC (#C14/17/111) grants. R.D. is a recipient of Rondoufonds voor Duchenne Onderzoek grant (EQQ-FODUCH-O2010). Y.C.C. is a recipient of C1-KUL 3DMUSYC (#C14/17/111) grant.

References

Adachi M, Hoshino Y, Izumi Y, Takagi S (2016) Immunohistochemical detection of a potential molecular therapeutic target for canine hemangiosarcoma. J Vet Med Sci 78:649–656

An R, Hoffman MD, Donoghue MA, Hunt AJ, Jacobson SC (2008) Water-assisted femtosecond laser machining of electrospray nozzles on glass microfluidic devices. Opt Express 16:15206–15211

Andersen P, Tampakakis E, Jimenez DV, Kannan S, Miyamoto M, Shin HK, Saberi A, Murphy S, Sulistio E, Chelko SP, Kwon C (2018) Precardiac organoids form two heart fields via Bmp/Wnt signaling. Nat Commun 9:3140

Ashammakhi N, Elkhammas EA, Hasan A (2017) Glomerulus-on-a-chip life up. Transplantation 101:E343–E344

Bandak B, Yi L, Roper MG (2018) Microfluidic-enabled quantitative measurements of insulin release dynamics from single islets of Langerhans in response to 5-palmitic acid hydroxy stearic acid. Lab Chip 18:2873–2882

Barkauskas CE, Chung MI, Fioret B, Gao X, Katsura H, Hogan BL (2017) Lung organoids: current uses and future promise. Development 144:986–997

Barnard JB, Richardson S, Sheldon S, Fildes J, Pravica V, Hutchinson IV, Leonard CT, Yonan N (2006) The Mdr1/Abcb1 gene, a high-impact risk factor for cardiac transplant rejection. Transplantation 82:1677–1682

Baron MG, Mintram KS, Owen SF, Hetheridge MJ, Moody AJ, Purcell WM, Jackson SK, Jha AN (2017) Pharmaceutical metabolism in fish: using A 3-D hepatic in vitro model to assess clearance. PLoS One 12:E0168837

Barry R, Diggle T, Terrett J, Soloviev M (2003) Competitive assay formats for high-throughput affinity arrays. J Biomol Screen 8:257–263

Beebe D, Wheeler M, Zeringue H, Walters E, Raty S (2002) Microfluidic technology for assisted reproduction. Theriogenology 57:125–135

Beer M, Kuppalu N, Stefanini M, Becker H, Schulz I, Manoli S, Schuette J, Schmees C, Casazza A, Stelzle M, Arcangeli A (2017) A novel microfluidic 3D platform for culturing pancreatic ductal adenocarcinoma cells: comparison with in vitro cultures and in vivo xenografts. Sci Rep 7:1325

Bein A, Shin W, Jalili-Firoozinezhad S, Park MH, Sontheimer-Phelps A, Tovaglieri A, Chalkiadaki A, Kim HJ, Ingber DE (2018) Microfluidic organ-on-a-chip models of human intestine. Cell Mol Gastroenterol Hepatol 5:659–668

Ben-Shaul S, Landau S, Merdler U, Levenberg S (2019) Mature vessel networks in engineered tissue promote graft-host anastomosis and prevent graft thrombosis. Proc Natl Acad Sci U S A 116:2955–2960

Bian S, Zhang L, Duan L, Wang X, Min Y, Yu H (2014) Extracellular vesicles derived from human bone marrow mesenchymal stem cells promote angiogenesis in a rat myocardial infarction model. J Mol Med 92:387–397

Bian S, Repic M, Guo Z, Kavirayani A, Burkard T, Bagley JA, Krauditsch C, Knoblich JA (2018) Genetically engineered cerebral organoids model brain tumor formation. Nat Methods 15:631–639

Birgersdotter A, Sandberg R, Ernberg I (2005) Gene expression perturbation in vitro – a growing case for three-dimensional (3D) culture systems. Semin Cancer Biol 15:405–412

Bollard CM, Heslop HE (2016) T cells for viral infections after allogeneic hematopoietic stem cell transplant. Blood 127:3331–3340

Brown JA, Pensabene V, Markov DA, Allwardt V, Neely MD, Shi M, Britt CM, Hoilett OS, Yang Q, Brewer BM, Samson PC, Mccawley LJ, May JM, Webb DJ, Li D, Bowman AB, Reiserer RS, Wikswo JP (2015) Recreating blood-brain barrier physiology and structure on chip: a novel neurovascular microfluidic bioreactor. Biomicrofluidics 9:054124

Brown JA, Codreanu SG, Shi M, Sherrod SD, Markov DA, Neely MD, Britt CM, Hoilett OS, Reiserer RS, Samson PC, Mccawley LJ, Webb DJ, Bowman AB, Mclean JA, Wikswo JP (2016) Metabolic consequences of inflammatory disruption of the blood-brain barrier in an organ-on-chip model of the human neurovascular unit. J Neuroinflammation 13:306

Buchan NS, Rajpal DK, Webster Y, Alatorre C, Gudivada RC, Zheng C, Sanseau P, Koehler J (2011) The role of translational bioinformatics in drug discovery. Drug Discov Today 16:426–434

Cao Y, Vacanti JP, Paige KT, Upton J, Vacanti CA (1997) Transplantation of chondrocytes utilizing a polymer-cell construct to produce tissue-engineered cartilage in the shape of a human ear. Plast Reconstr Surg 100:297–302. Discussion 303–4

Carpenter S (2007) Science careers. Carving a career in translational research. Science 317:966–967

Chang CM, Chang WH, Wang CH, Wang JH, Mai JD, Lee GB (2013) Nucleic acid amplification using microfluidic systems. Lab Chip 13:1225–1242

Chang SY, Weber EJ, Ness KV, Eaton DL, Kelly EJ (2016) Liver and kidney on chips: microphysiological models to understand transporter function. Clin Pharmacol Ther 100:464–478

Chen YW, Huang SX, De Carvalho A, Ho SH, Islam MN, Volpi S, Notarangelo LD, Ciancanelli M, Casanova JL, Bhattacharya J, Liang AF, Palermo LM, Porotto M, Moscona A, Snoeck HW (2017) A three-dimensional model of human lung development and disease from pluripotent stem cells. Nat Cell Biol 19:542–549

Chin MH, Mason MJ, Xie W, Volinia S, Singer M, Peterson C, Ambartsumyan G, Aimiuwu O, Richter L, Zhang J, Khvorostov I, Ott V, Grunstein M, Lavon N, Benvenisty N, Croce CM, Clark AT, Baxter T, Pyle AD, Teitell MA, Pelegrini M, Plath K, Lowry WE (2009) Induced pluripotent stem cells and embryonic stem cells are distinguished by gene expression signatures. Cell Stem Cell 5:111–123

Chirullo B, Ammendola S, Leonardi L, Falcini R, Petrucci P, Pistoia C, Vendetti S, Battistoni A, Pasquali P (2015) Attenuated mutant strain of Salmonella typhimurium lacking the znuabc transporter contrasts tumor growth promoting anti-cancer immune response. Oncotarget 6:17648–17660

Cho M, Chung S, Jung JH, Rhie GE, Jeon JH, Seo TS (2014) Combination of biobarcode assay with on-chip capillary electrophoresis for ultrasensitive and multiplex biological agent detection. Biosens Bioelectron 61:172–176

Chong JJ, Yang X, Don CW, Minami E, Liu YW, Weyers JJ, Mahoney WM, Van Biber B, Cook SM, Palpant NJ, Gantz JA, Fugate JA, Muskheli V, Gough GM, Vogel KW, Astley CA, Hotchkiss CE, Baldessari A, Pabon L, Reinecke H, Gill EA, Nelson V, Kiem HP, Laflamme MA, Murry CE (2014) Human embryonic-stem-cell-derived cardiomyocytes regenerate non-human primate hearts. Nature 510:273–277

Conant G, Lai BFL, Lu RXZ, Korolj A, Wang EY, Radisic M (2017) High-content assessment of cardiac function using heart-on-a-chip devices as drug screening model. Stem Cell Rev 13:335–346

Crapo PM, Gilbert TW, Badylak SF (2011) An overview of tissue and whole organ decellularization processes. Biomaterials 32:3233–3243

Crespo M, Vilar E, Tsai SY, Chang K, Amin S, Srinivasan T, Zhang T, Pipalia NH, Chen HJ, Witherspoon M, Gordillo M, Xiang JZ, Maxfield FR, Lipkin S, Evans T, Chen S (2017) Colonic organoids derived from human induced pluripotent stem cells for modeling colorectal cancer and drug testing. Nat Med 23:878–884

Cugola FR, Fernandes IR, Russo FB, Freitas BC, Dias JL, Guimaraes KP, Benazzato C, Almeida N, Pignatari GC, Romero S, Polonio CM, Cunha I, Freitas CL, Brandao WN, Rossato C, Andrade DG, Faria Dde P, Garcez AT, Buchpigel CA, Braconi CT, Mendes E, Sall AA, Zanotto PM, Peron JP, Muotri AR, Beltrao-Braga PC (2016) The Brazilian Zika virus strain causes birth defects in experimental models. Nature 534:267–271

Dauth S, Maoz BM, Sheehy SP, Hemphill MA, Murty T, Macedonia MK, Greer AM, Budnik B, Parker KK (2017) Neurons derived from different brain regions are inherently different in vitro: a novel multiregional brain-on-a-chip. J Neurophysiol 117:1320–1341

Dendukuri D, Gu SS, Pregibon DC, Hatton TA, Doyle PS (2007) Stop-flow lithography in a microfluidic device. Lab Chip 7:818–828

Duelen R, Sampaolesi M (2017) Stem cell technology in cardiac regeneration: a pluripotent stem cell promise. EBioMedicine 16:30–40

Duffy DC, Mcdonald JC, Schueller OJ, Whitesides GM (1998) Rapid prototyping of microfluidic systems in poly(dimethylsiloxane). Anal Chem 70:4974–4984

Dutta D, Heo I, Clevers H (2017) Disease modeling in stem cell-derived 3D organoid systems. Trends Mol Med 23:393–410

Dye BR, Dedhia PH, Miller AJ, Nagy MS, White ES, Shea LD, Spence JR (2016) A bioengineered niche promotes in vivo engraftment and maturation of pluripotent stem cell derived human lung organoids. Elife:5

Ebert AD, Yu J, Rose FF Jr, Mattis VB, Lorson CL, Thomson JA, Svendsen CN (2009) Induced pluripotent stem cells from a spinal muscular atrophy patient. Nature 457:277–280

Edmondson R, Broglie JJ, Adcock AF, Yang L (2014) Three-dimensional cell culture systems and their applications in drug discovery and cell-based biosensors. Assay Drug Dev Technol 12:207–218

Eisenstein M (2018) Organoids: the body builders. Nat Methods 15:19

Felder M, Trueeb B, Stucki AO, Borcard S, Stucki JD, Schnyder B, Geiser T, Guenat OT (2019) Impaired wound healing of alveolar lung epithelial cells in a breathing lung-on-a-chip. Front Bioeng Biotechnol 7:3

Feng Y, Chen CS, Ho J, Pearce D, Hu S, Wang B, Desai P, Kim KS, Zhu H (2018) High-throughput chip assay for investigating Escherichia Coli interaction with the blood-brain barrier using microbial and human proteome microarrays (dual-microarray technology). Anal Chem 90:10958–10966

Ferrebee JW, Thomas ED (1960) Transplantation of marrow in man. Arch Intern Med 106:523–531

Figeys D, Aebersold R (1998) High sensitivity analysis of proteins and peptides by capillary electrophoresis-tandem mass spectrometry: recent developments in technology and applications. Electrophoresis 19:885–892

Figeys D, Gygi SP, Mckinnon G, Aebersold R (1998) An integrated microfluidics-tandem mass spectrometry system for automated protein analysis. Anal Chem 70:3728–3734

Fiorotto R, Amenduni M, Mariotti V, Fabris L, Spirli C, Strazzabosco M (2019) Liver diseases in the dish: ipsc and organoids as a new approach to modeling liver diseases. Biochim Biophys Acta Mol Basis Dis 1865 (5):920–928

Folch A, Toner M (2000) Microengineering of cellular interactions. Annu Rev Biomed Eng 2:227–256

Fontanarosa PB, Deangelis CD (2002) Basic science and translational research in JAMA. JAMA 287:1728

Foquet M, Korlach J, Zipfel W, Webb WW, Craighead HG (2002) DNA fragment sizing by single molecule detection in submicrometer-sized closed fluidic channels. Anal Chem 74:1415–1422

Forbester JL, Goulding D, Vallier L, Hannan N, Hale C, Pickard D, Mukhopadhyay S, Dougan G (2015) Interaction of Salmonella Enterica serovar typhimurium with intestinal organoids derived from human induced pluripotent stem cells. Infect Immun 83:2926–2934

Forsythe SD, Devarasetty M, Shupe T, Bishop C, Atala A, Soker S, Skardal A (2018) Environmental toxin screening using human-derived 3D bioengineered liver and cardiac organoids. Front Public Health 6:103

Freedman BS, Brooks CR, Lam AQ, Fu H, Morizane R, Agrawal V, Saad AF, Li MK, Hughes MR, Werff RV, Peters DT, Lu J, Baccei A, Siedlecki AM, Valerius MT, Musunuru K, Mcnagny KM, Steinman TI, Zhou J, Lerou PH, Bonventre JV (2015) Modelling kidney disease with crispr-mutant kidney organoids derived from human pluripotent epiblast spheroids. Nat Commun 6:8715

Fukumoto J, Kolliputi N (2012) Human lung on a chip: innovative approach for understanding disease processes and effective drug testing. Front Pharmacol 3:205

Gabriel E, Ramani A, Karow U, Gottardo M, Natarajan K, Gooi LM, Goranci-Buzhala G, Krut O, Peters F, Nikolic M, Kuivanen S, Korhonen E, Smura T, Vapalahti O, Papantonis A, Schmidt-Chanasit J, Riparbelli M, Callaini G, Kronke M, Utermohlen O, Gopalakrishnan J (2017) Recent Zika virus isolates induce premature differentiation of neural progenitors in human brain organoids. Cell Stem Cell 20:397–406 E5

Gao J, Xu J, Locascio LE, Lee CS (2001) Integrated microfluidic system enabling protein digestion, peptide separation, and protein identification. Anal Chem 73:2648–2655

Garcez PP, Loiola EC, Madeiro Da Costa R, Higa LM, Trindade P, Delvecchio R, Nascimento JM, Brindeiro R, Tanuri A, Rehen SK (2016) Zika Virus impairs growth in human neurospheres and brain organoids. Science 352:816–818

Gherghiceanu M, Barad L, Novak A, Reiter I, Itskovitz-Eldor J, Binah O, Popescu LM (2011) Cardiomyocytes derived from human embryonic and induced pluripotent stem cells: comparative ultrastructure. J Cell Mol Med 15:2539–2551

Gilbert TW, Sellaro TL, Badylak SF (2006) Decellularization of tissues and organs. Biomaterials 27:3675–3683

Gkatzis K, Taghizadeh S, Huh D, Stainier DYR, Bellusci S (2018) Use of three-dimensional organoids and lung-on-a-chip methods to study lung development, regeneration and disease. Eur Respir J 52

Glasgow IK, Zeringue HC, Beebe DJ, Choi SJ, Lyman JT, Chan NG, Wheeler MB (2001) Handling individual mammalian embryos using microfluidics. IEEE Trans Biomed Eng 48:570–578

Gori M, Simonelli MC, Giannitelli SM, Businaro L, Trombetta M, Rainer A (2016) Investigating nonalcoholic fatty liver disease in a liver-on-a-chip microfluidic device. PLoS One 11:E0159729

Grix T, Ruppelt A, Thomas A, Amler AK, Noichl BP, Lauster R, Kloke L (2018) Bioprinting perfusion-enabled liver equivalents for advanced organ-on-a-chip applications. Genes (Basel) 9

Guo Y, Li Z, Su W, Wang L, Zhu Y, Qin J (2018) A biomimetic human gut-on-a-chip for modeling drug metabolism in intestine. Artif Organs 42:1196–1205

Han C, Sun X, Liu L, Jiang H, Shen Y, Xu X, Li J, Zhang G, Huang J, Lin Z, Xiong N, Wang T (2016) Exosomes and their therapeutic potentials of stem cells. Stem Cells Int 2016:7653489

Hardy RE, Ikpeazu EV (1989) Bone marrow transplantation: a review. J Natl Med Assoc 81:518–523

He Y, Zhang YH, Yeung ES (2001) Capillary-based fully integrated and automated system for nanoliter polymerase chain reaction analysis directly from cheek cells. J Chromatogr A 924:271–284

Heo I, Dutta D, Schaefer DA, Iakobachvili N, Artegiani B, Sachs N, Boonekamp KE, Bowden G, Hendrickx APA, Willems RJL, Peters PJ, Riggs MW, O'connor R, Clevers H (2018) Modelling cryptosporidium infection in human small intestinal and lung organoids. Nat Microbiol 3:814–823

Herland A, Van Der Meer AD, Fitzgerald EA, Park TE, Sleeboom JJ, Ingber DE (2016) Distinct contributions of astrocytes and pericytes to neuroinflammation identified in a 3D human blood-brain barrier on a chip. PLoS One 11:E0150360

Hopwood AJ, Hurth C, Yang J, Cai Z, Moran N, Lee-Edghill JG, Nordquist A, Lenigk R, Estes MD, Haley JP, Mcalister CR, Chen X, Brooks C, Smith S, Elliott K, Koumi P, Zenhausern F, Tully G (2010) Integrated microfluidic system for rapid forensic DNA analysis: sample collection to DNA profile. Anal Chem 82:6991–6999

Hu H, Gehart H, Artegiani B, LÖpez-Iglesias C, Dekkers F, Basak O, Van Es J, Chuva De Sousa Lopes SM, Begthel H, Korving J, Van Den Born M, Zou C, Quirk C, Chiriboga L, Rice CM, Ma S, Rios A, Peters PJ, De Jong YP, Clevers H (2018) Long-term expansion of functional mouse and human hepatocytes as 3D organoids. Cell 175(1591–1606):E19

Huh D, Hamilton GA, Ingber DE (2011) From 3D cell culture to organs-on-chips. Trends Cell Biol 21:745–754

Huh D, Leslie DC, Matthews BD, Fraser JP, Jurek S, Hamilton GA, Thorneloe KS, Mcalexander MA, Ingber DE (2012) A human disease model of drug toxicity-induced pulmonary edema in a lung-on-a-chip microdevice. Sci Transl Med 4:159ra147

Jacob A, Morley M, Hawkins F, Mccauley KB, Jean JC, Heins H, Na CL, Weaver TE, Vedaie M, Hurley K, Hinds A, Russo SJ, Kook S, Zacharias W, Ochs M, Traber K, Quinton LJ, Crane A, Davis BR, White FV, Wambach J, Whitsett JA, Cole FS, Morrisey EE, Guttentag SH, Beers MF, Kotton DN (2017) Differentiation of human pluripotent stem cells into functional lung alveolar epithelial cells. Cell Stem Cell 21 (472–488):E10

Joachim J, Tooze SA (2018) Control of gabarap-mediated autophagy by the golgi complex, centrosome and centriolar satellites. Biol Cell 110:1–5

Justice BA, Badr NA, Felder RA (2009) 3D cell culture opens new dimensions in cell-based assays. Drug Discov Today 14:102–107

Karakikes I, Ameen M, Termglinchan V, Wu JC (2015) Human induced pluripotent stem cell-derived cardiomyocytes: insights into molecular, cellular, and functional phenotypes. Circ Res 117:80–88

Keatch RP, Armoogum K, Schor SL, Pridham MS, Banks K, Khor TY, Matthew C (2002) Microengineering as a tool to study substratum modulation and cell behaviour. Biomed Tech (Berl) 47 (Suppl 1 Pt 1):386–389

Kehat I, Khimovich L, Caspi O, Gepstein A, Shofti R, Arbel G, Huber I, Satin J, Itskovitz-Eldor J, Gepstein L (2004) Electromechanical integration of cardiomyocytes derived from human embryonic stem cells. Nat Biotechnol 22:1282–1289

Khanlarkhani N, Baazm M, Mohammadzadeh F, Najafi A, Mehdinejadiani S, Sobhani A (2016) Multipotent stem cell and reproduction. J Stem Cells 11:219–229

Kilic O, Pamies D, Lavell E, Schiapparelli P, Feng Y, Hartung T, Bal-Price A, Hogberg HT, Quinones-Hinojosa A, Guerrero-Cazares H, Levchenko A (2016) Brain-on-a-chip model enables analysis of human neuronal differentiation and chemotaxis. Lab Chip 16:4152–4162

Kim S, Lesherperez SC, Kim BC, Yamanishi C, Labuz JM, Leung B, Takayama S (2016) Pharmacokinetic profile that reduces nephrotoxicity of gentamicin in a perfused kidney-on-a-chip. Biofabrication 8:015021

Kimura H, Ikeda T, Nakayama H, Sakai Y, Fujii T (2015) An on-chip small intestine-liver model for pharmacokinetic studies. J Lab Autom 20:265–273

Kizilkurtlu AA, Polat T, Aydin GB, Akpek A (2018) Lung on a chip for drug screening and design. Curr Pharm Des 24:5386–5396

Konar D, Devarasetty M, Yildiz DV, Atala A, Murphy SV (2016) Lung-on-a-chip technologies for disease modeling and drug development. Biomed Eng Comput Biol 7:17–27

Kong J, Luo Y, Jin D, An F, Zhang W, Liu L, Li J, Fang S, Li X, Yang X, Lin B, Liu T (2016) A novel microfluidic model can mimic organ-specific metastasis of circulating tumor cells. Oncotarget 7:78421–78432

Lancaster MA, Renner M, Martin CA, Wenzel D, Bicknell LS, Hurles ME, Homfray T, Penninger JM, Jackson AP, Knoblich JA (2013) Cerebral organoids model human brain development and microcephaly. Nature 501:373–379

Lee J, Kim S (2018) Kidney-on-a-chip: a new technology for predicting drug efficacy, interactions, and drug-induced nephrotoxicity. Curr Drug Metab 19:577–583

Lee JH, Lee JB, Shapovalova Z, Fiebig-Comyn A, Mitchell RR, Laronde S, Szabo E, Benoit YD, Bhatia M (2014) Somatic transcriptome priming gates lineage-specific differentiation potential of human-induced pluripotent stem cell states. Nat Commun 5:5605

Lee H, Chae S, Kim JY, Han W, Kim J, Choi Y, Cho DW (2019) Cell-printed 3D liver-on-a-chip possessing a liver microenvironment and biliary system. Biofabrication 11:025001

Lenigk R, Liu RH, Athavale M, Chen Z, Ganser D, Yang J, Rauch C, Liu Y, Chan B, Yu H, Ray M, Marrero R, Grodzinski P (2002) Plastic biochannel hybridization devices: a new concept for microfluidic DNA arrays. Anal Biochem 311:40–49

Leonardi L, Mechelli L, Lepri E, Manuali E, And Vitellozzi G (2012) PMWS (Postweaning multisystemic wasting syndrome) in central italy: histopathological, immunohistochemical and ultrastructural findings. Int J Vet Sci 1(2):49–54

Leonardi L, Piro F, Pannaioli V, Bellezza E, Marsili R, Angeli G, Bucci A, Moretti M, Rossi G (2015) Additive manufacturing and advanced testing techniques applied to bone reconstructive sciences in animals and humans. Congresso Aip Vet, Perugia, 15–17 June 2015

Levenberg S, Rouwkema J, Macdonald M, Garfein ES, Kohane DS, Darland DC, Marini R, Van Blitterswijk CA, Mulligan RC, D'amore PA, Langer R (2005) Engineering vascularized skeletal muscle tissue. Nat Biotechnol 23:879–884

Li X, Ballerini DR, Shen W (2012) A perspective on paper-based microfluidics: current status and future trends. Biomicrofluidics 6:11301–1130113

Li Z, Jiang L, Tao T, Su W, Guo Y, Yu H, Qin J (2017) Assessment of cadmium-induced nephrotoxicity using a kidney-on-a-chip device. Toxicol Res (Camb) 6:372 380

Li Y, Hermanson DL, Moriarity BS, Kaufman DS (2018) Human iPSC-derived natural killer cells engineered with chimeric antigen receptors enhance anti-tumor activity. Cell Stem Cell 23:181–192 E5

Lin JH (1995) Species similarities and differences in pharmacokinetics. Drug Metab Dispos 23:1008–1021

Lion N, Gellon JO, Jensen H, Girault HH (2003) On-chip protein sample desalting and preparation for direct coupling with electrospray ionization mass spectrometry. J Chromatogr A 1003:11–19

Liu RH, Lenigk R, Druyor-Sanchez RL, Yang J, Grodzinski P (2003) Hybridization enhancement using cavitation microstreaming. Anal Chem 75:1911–1917

Liu C, Oikonomopoulos A, Sayed N, Wu JC (2018) Modeling human diseases with induced pluripotent stem cells: from 2D to 3D and beyond. Development 145(5), pii: dev 156166

Ma C, Zhao L, Zhou EM, Xu J, Shen S, Wang J (2016) On-chip construction of liver lobule-like microtissue and its application for adverse drug reaction assay. Anal Chem 88:1719–1727

Maitre JL (2017) Mechanics of blastocyst morphogenesis. Biol Cell 109:323–338

Mak IW, Evaniew N, Ghert M (2014) Lost in translation: animal models and clinical trials in cancer treatment. Am J Transl Res 6:114–118

Mandai M, Watanabe A, Kurimoto Y, Hirami Y, Morinaga C, Daimon T, Fujihara M, Akimaru H, Sakai N, Shibata Y, Terada M, Nomiya Y, Tanishima S, Nakamura M, Kamao H, Sugita S, Onishi A, Ito T, Fujita K, Kawamata S, Go MJ, Shinohara C, Hata KI, Sawada M, Yamamoto M, Ohta S, Ohara Y, Yoshida K, Kuwahara J, Kitano Y, Amano N, Umekage M, Kitaoka F, Tanaka A, Okada C, Takasu N, Ogawa S, Yamanaka S, Takahashi M (2017) Autologous induced stem-cell-derived retinal cells for macular degeneration. N Engl J Med 376:1038–1046

Marban E (2018) The secret life of exosomes: what bees can teach us about next-generation therapeutics. Journal of the american college of cardiology 71:193–200

Mariani J, Coppola G, Zhang P, Abyzov A, Provini L, Tomasini L, Amenduni M, Szekely A, Palejev D, Wilson M, Gerstein M, Grigorenko EL, Chawarska K, Pelphrey KA, Howe JR, Vaccarino FM (2015) Foxg1-dependent dysregulation of gaba/glutamate neuron differentiation in autism spectrum disorders. Cell 162:375–390

Maude SL, Laetsch TW, Buechner J, Rives S, Boyer M, Bittencourt H, Bader P, Verneris MR, Stefanski HE, Myers GD, Qayed M, De Moerloose B, Hiramatsu H, Schlis K, Davis KL, Martin PL, Nemecek ER, Yanik GA, Peters C, Baruchel A, Boissel N, Mechinaud F, Balduzzi A, Krueger J, June CH, Levine BL, Wood P, Taran T, Leung M, Mueller KT, Zhang Y, Sen K, Lebwohl D, Pulsipher MA, Grupp SA (2018) Tisagenlecleucel in children and young adults with B-cell lymphoblastic leukemia. N Engl J Med 378:439–448

Mccauley KB, Hawkins F, Serra M, Thomas DC, Jacob A, Kotton DN (2017) Efficient derivation of functional human airway epithelium from pluripotent stem cells via temporal regulation of WNT signaling. Cell Stem Cell 20:844–857 E6

Mccracken KW, Cata EM, Crawford CM, Sinagoga KL, Schumacher M, Rockich BE, Tsai YH, Mayhew CN, Spence JR, Zavros Y, Wells JM (2014) Modelling human development and disease in pluripotent stem-cell-derived gastric organoids. Nature 516:400–404

Mcgonigle P, Ruggeri B (2014) Animal models of human disease: challenges in enabling translation. Biochem Pharmacol 87:162–171

Mertz L (2016) Omics tech, gut-on-a-chip, and bacterial engineering: new approaches for treating inflammatory bowel diseases. IEEE Pulse 7:9–12

Miller AJ, Dye BR, Ferrer-Torres D, Hill DR, Overeem AW, Shea LD, Spence JR (2019) Generation of lung organoids from human pluripotent stem cells in vitro. Nat Protoc 14:518–540

Miro G, Ruperez C, Checa R, Galvez R, Hernandez L, Garcia M, Canorea I, Marino V, Montoya A (2014) Current status of L. infantum infection in stray cats in the madrid region (Spain): implications for the recent outbreak of human leishmaniosis? Parasites and Vectors 7:112

Moffat JG, Vincent F, Lee JA, Eder J, Prunotto M (2017) Opportunities and challenges in phenotypic drug discovery: an industry perspective. Nat Rev Drug Discov 16:531–543

Moretti A, Bellin M, Welling A, Jung CB, Lam JT, Bott-Flugel L, Dorn T, Goedel A, Hohnke C, Hofmann F, Seyfarth M, Sinnecker D, Schomig A, Laugwitz KL (2010) Patient-specific induced pluripotent stem-cell models for long-QT syndrome. N Engl J Med 363:1397–1409

Morrison C (2018) Boom: 2018's Biotech Ipos. Nat Rev Drug Discov 18:3–6

Morrison JI, Loof S, He P, Simon A (2006) Salamander limb regeneration involves the activation of a multipotent skeletal muscle satellite cell population. J Cell Biol 172:433–440

Moslem M, Olive J, Falk A (2018) Stem cell models of schizophrenia, what have we learned and what is the potential? Schizophr Res pii: S0920-9964(18)30719-9

Murphy SV, Atala A (2014) 3D bioprinting of tissues and organs. Nat Biotechnol 32:773–785

Muscari C, Giordano E, Bonafe F, Govoni M, Guarnieri C (2014) Strategies affording prevascularized cell-based constructs for myocardial tissue engineering. Stem Cells Int 2014:434169

Nakayama KH, Batchelder CA, Lee CI, Tarantal AF (2010) Decellularized rhesus monkey kidney as a three-dimensional scaffold for renal tissue engineering. Tissue Eng Part A 16:2207–2216

Neuville A, Renaud L, Luu TT, Minde MW, Jettestuen E, Vinningland JL, Hiorth A, Dysthe DK (2017) Xurography for microfluidics on a reactive solid. Lab Chip 17:293–303

Ni Z, Knorr DA, Clouser CL, Hexum MK, Southern P, Mansky LM, Park IH, Kaufman DS (2011) Human pluripotent stem cells produce natural killer cells that mediate anti-HIV-1 activity by utilizing diverse cellular mechanisms. J Virol 85:43–50

Ni Z, Knorr DA, Bendzick L, Allred J, Kaufman DS (2014) Expression of chimeric receptor Cd4zeta by natural killer cells derived from human pluripotent stem cells improves in vitro activity but does not enhance suppression of HIV infection in vivo. Stem Cells 32:1021–1031

Nie YZ, Zheng YW, Miyakawa K, Murata S, Zhang RR, Sekine K, Ueno Y, Takebe T, Wakita T, Ryo A, Taniguchi H (2018a) Recapitulation of hepatitis B virus-host interactions in liver organoids from human induced pluripotent stem cells. EBioMedicine 35:114–123

Nie YZ, Zheng YW, Ogawa M, Miyagi E, Taniguchi H (2018b) Human liver organoids generated with single donor-derived multiple cells rescue mice from acute liver failure. Stem Cell Res Ther 9:5

Nowakowski TJ, Pollen AA, Di Lullo E, Sandoval-Espinosa C, Bershteyn M, Kriegstein AR (2016) Expression analysis highlights AXL as a candidate Zika virus entry receptor in neural stem cells. Cell Stem Cell 18:591–596

Nugraha B, Buono MF, Von Boehmer L, Hoerstrup SP, Emmert MY (2019) Human cardiac organoids for disease modeling. Clin Pharmacol Ther 105:79–85

Ogawa M, Ogawa S, Bear CE, Ahmadi S, Chin S, Li B, Grompe M, Keller G, Kamath BM, Ghanekar A (2015) Directed differentiation of cholangiocytes from human pluripotent stem cells. Nat Biotechnol 33:853–861

Ott HC, Matthiesen TS, Goh SK, Black LD, Kren SM, Netoff TI, Taylor DA (2008) Perfusion-decellularized matrix: using nature's platform to engineer a bioartificial heart. Nat Med 14:213–221

Pampaloni F, Reynaud EG, Stelzer EH (2007) The third dimension bridges the gap between cell culture and live tissue. Nat Rev Mol Cell Biol 8:839–845

Park J, Lee BK, Jeong GS, Hyun JK, Lee CJ, Lee SH (2015) Three-dimensional brain-on-a-chip with an interstitial level of flow and its application as an in vitro model of Alzheimer's disease. Lab Chip 15:141–150

Paul SM, Mytelka DS, Dunwiddie CT, Persinger CC, Munos BH, Lindborg SR, Schacht AL (2010) How to improve R&D productivity: the pharmaceutical industry's grand challenge. Nat Rev Drug Discov 9:203–214

Pawlak M, Schick E, Bopp MA, Schneider MJ, Oroszlan P, Ehrat M (2002) Zeptosens' protein microarrays: a novel high performance microarray platform for low abundance protein analysis. Proteomics 2:383–393

Peela N, Truong D, Saini H, Chu H, Mashaghi S, Ham SL, Singh S, Tavana H, Mosadegh B, Nikkhah M (2017) Advanced biomaterials and microengineering technologies to recapitulate the stepwise process of cancer metastasis. Biomaterials 133:176–207

Perestrelo AR, Aguas AC, Rainer A, Forte G (2015) Microfluidic organ/body-on-a-chip devices at the convergence of biology and microengineering. Sensors (Basel) 15:31142–31170

Petersen TH, Calle EA, Zhao L, Lee EJ, Gui L, Raredon MB, Gavrilov K, Yi T, Zhuang ZW, Breuer C, Herzog E, Niklason LE (2010) Tissue-engineered lungs for in vivo implantation. Science 329:538–541

Pillekamp F, Haustein M, Khalil M, Emmelheinz M, Nazzal R, Adelmann R, Nguemo F, Rubenchyk O, Pfannkuche K, Matzkies M, Reppel M, Bloch W, Brockmeier K, Hescheler J (2012) Contractile properties of early human embryonic stem cell-derived cardiomyocytes: beta-adrenergic stimulation induces positive chronotropy and lusitropy but not inotropy. Stem Cells Dev 21:2111–2121

Pjescic I, Tranter CA, Haywood JC, Paidipalli M, Ganveer A, Haywood SE, Tham J, Crews ND (2011) Real-time damage monitoring of irradiated DNA. Integr Biol (Camb) 3:937–947

Pollen AA, Bhaduri A, Andrews MG, Nowakowski TJ, Meyerson OS, Mostajo-Radji MA, Di Lullo E, Alvarado B, Bedolli M, Dougherty ML, Fiddes IT, Kronenberg ZN, Shuga J, Leyrat AA, West JA, Bershteyn M, Lowe CB, Pavlovic BJ, Salama SR, Haussler D, Eichler EE, Kriegstein AR (2019) Establishing cerebral organoids as models of human-specific brain evolution. Cell 176:743–756 E17

Prodanov L, Jindal R, Bale SS, Hegde M, Mccarty WJ, Golberg I, Bhushan A, Yarmush ML, Usta OB (2016) Long-term maintenance of a microfluidic 3D human liver sinusoid. Biotechnol Bioeng 113:241–246

Punde TH, Wu WH, Lien PC, Chang YL, Kuo PH, Chang MD, Lee KY, Huang CD, Kuo HP, Chan YF, Shih PC, Liu CH (2015) A biologically inspired lung-on-a-chip device for the study of protein-induced lung inflammation. Integr Biol (Camb) 7:162–169

Qian F, Huang C, Lin YD, Ivanovskaya AN, O'hara TJ, Booth RH, Creek CJ, Enright HA, Soscia DA, Belle AM, Liao R, Lightstone FC, Kulp KS, Wheeler EK (2017) Simultaneous electrical recording of cardiac electrophysiology and contraction on chip. Lab Chip 17:1732–1739

Raies AB, Bajic VB (2016) In silico toxicology: computational methods for the prediction of chemical toxicity. Wiley Interdiscip Rev Comput Mol Sci 6:147–172

Raja WK, Mungenast AE, Lin YT, Ko T, Abdurrob F, Seo J, Tsai LH (2016) Self-organizing 3D human neural tissue derived from induced pluripotent stem cells recapitulate Alzheimer's disease phenotypes. PLoS One 11:E0161969

Reininger-Mack A, Thielecke H, Robitzki AA (2002) 3D-biohybrid systems: applications in drug screening. Trends Biotechnol 20:56–61

Robertson JA (2001) Human embryonic stem cell research: ethical and legal issues. Nat Rev Genet 2:74–78

Robinton DA, Daley GQ (2012) The promise of induced pluripotent stem cells in research and therapy. Nature 481:295–305

Rossant J (2007) Stem cells and lineage development in the mammalian blastocyst. Reprod Fertil Dev 19:111–118

Rowe RG, Daley GQ (2019) Induced pluripotent stem cells in disease modelling and drug discovery. Nat Rev Genet 20:377–388

Saeidnia S, Gohari AR, Manayi A (2016) Reverse pharmacognosy and reverse pharmacology; two closely related approaches for drug discovery development. Curr Pharm Biotechnol 17:1016–1022

Saheli M, Sepantafar M, Pournasr B, Farzaneh Z, Vosough M, Piryaei A, Baharvand H (2018) Three-dimensional liver-derived extracellular matrix hydrogel promotes liver organoids function. J Cell Biochem 119:4320–4333

Sampaolesi M, Blot S, D'antona G, Granger N, Tonlorenzi R, Innocenzi A, Mognol P, Thiibaud JL, Galvez BG, Barthelemy I, Perani L, Mantero S, Guttinger M, Pansarasa O, Rinaldi C, Cusella De Angelis MG, Torrente Y, Bordignon C, Bottinelli R, Cossu G (2006) Mesoangioblast stem cells ameliorate muscle function in dystrophic dogs. Nature 444:574–579

Samuel R. 40 Cycle PCR using human genomic DNA in less than 1 minute. Proceedings of microtas, 2016, Dublin, Theatr Irel

Sato T, Clevers H (2013) Growing self-organizing mini-guts from a single intestinal stem cell: mechanism and applications. Sciencei 340:1190–1194

Sato T, Vries RG, Snippert HJ, Van De Wetering M, Barker N, Stange DE, Van Es JH, Abo A, Kujala P, Peters PJ, Clevers H (2009) Single Lgr5 stem cells build crypt-villus structures in vitro without a mesenchymal niche. Nature 459:262–265

Schuster TG, Cho B, Keller LM, Takayama S, Smith GD (2003) Isolation of motile spermatozoa from semen samples using microfluidics. Reprod Biomed Online 7:75–81

Schwach V, Passier R (2016) Generation and purification of human stem cell-derived cardiomyocytes. Differentiation 91:126–138

Schwamborn JC (2018) Is Parkinson's disease a neurodevelopmental disorder and will brain organoids help us to understand it? Stem Cells Dev 27:968–975

Scrivener E, Barry R, Platt A, Calvert R, Masih G, Hextall P, Soloviev M, Terrett J (2003) Peptidomics: a new approach to affinity protein microarrays. Proteomics 3:122–128

Sendi H, Mead I, Wan M, Mehrab-Mohseni M, Koch K, Atala A, Bonkovsky HL, Bishop CE (2018) Mir-122 inhibition in a human liver organoid model leads to liver inflammation, necrosis, steatofibrosis and dysregulated insulin signaling. PLoS One 13: E0200847

Skardal A, Murphy SV, Devarasetty M, Mead I, Kang HW, Seol YJ, Shrike Zhang Y, Shin SR, Zhao L, Aleman J, Hall AR, Shupe TD, Kleensang A, Dokmeci MR, Jin Lee S, Jackson JD, Yoo JJ, Hartung T, Khademhosseini A, Soker S, Bishop CE, Atala A (2017) Multi-tissue interactions in an integrated three-tissue organ-on-a-chip platform. Sci Rep 7:8837

Soloviev M, Barry R, Scrivener E, Terrett J (2003) Combinatorial peptidomics: a generic approach for protein expression profiling. J Nanobiotechnol 1:4

Sonnen KF, Merten CA (2019) Microfluidics as an emerging precision tool in developmental biology. Dev Cell 48:293–311

Sumi T, Tsuneyoshi N, Nakatsuji N, Suemori H (2008) Defining early lineage specification of human embryonic stem cells by the orchestrated balance of canonical WNT/beta-catenin, Activin/Nodal and BMP signaling. Development 135:2969–2979

Sung NS, Crowley WF Jr, Genel M, Salber P, Sandy L, Sherwood LM, Johnson SB, Catanese V, Tilson H, Getz K, Larson EL, Scheinberg D, Reece EA, Slavkin H, Dobs A, Grebb J, Martinez RA, Korn A, Rimoin D (2003) Central challenges facing the national clinical research enterprise. JAMA 289:1278–1287

Taapken SM, Nisler BS, Newton MA, Sampsell-Barron TL, Leonhard KA, Mcintire EM, Montgomery KD (2011) Karotypic abnormalities in human induced pluripotent stem cells and embryonic stem cells. Nat Biotechnol 29:313–314

Taguchi A, Nishinakamura R (2017) Higher-order kidney organogenesis from pluripotent stem cells. Cell Stem Cell 21:730–746 E6

Takahashi K, Yamanaka S (2006) Induction of pluripotent stem cells from mouse embryonic and adult fibroblast cultures by defined factors. Cell 126:663–676

Takahashi K, Tanabe K, Ohnuki M, Narita M, Ichisaka T, Tomoda K, Yamanaka S (2007) Induction of pluripotent stem cells from adult human fibroblasts by defined factors. Cell 131:861–872

Tan HY, Trier S, Rahbek UL, Dufva M, Kutter JP, Andresen TL (2018) A multi-chamber microfluidic intestinal barrier model using Caco-2 cells for drug transport studies. PLoS One 13:E0197101

Teh SY, Lin R, Hung LH, Lee AP (2008) Droplet microfluidics. Lab Chip 8:198–220

Themeli M, Kloss CC, Ciriello G, Fedorov VD, Perna F, Gonen M, Sadelain M (2013) Generation of tumor-targeted human t lymphocytes from induced pluripotent stem cells for cancer therapy. Nat Biotechnol 31:928–933

Thomson JA, Itskovitz-Eldor J, Shapiro SS, Waknitz MA, Swiergiel JJ, Marshall VS, Jones JM (1998) Embryonic stem cell lines derived from human blastocysts. Science 282:1145–1147

Tulloch NL, Muskheli V, Razumova MV, Korte FS, Regnier M, Hauch KD, Pabon L, Reinecke H, Murry CE (2011) Growth of engineered human myocardium with mechanical loading and vascular coculture. Circ Res 109:47–59

Turco MY, Gardner L, Kay RG, Hamilton RS, Prater M, Hollinshead MS, Mcwhinnie A, Esposito L, Fernando R, Skelton H, Reimann F, Gribble FM, Sharkey A, Marsh SGE, O'rahilly S, Hemberger M, Burton GJ, Moffett A (2018) Trophoblast organoids as a model for maternal-fetal interactions during human placentation. Nature 564:263–267

Ugolini GS, Visone R, Cruz-Moreira D, Mainardi A, Rasponi M (2018) Generation of functional cardiac microtissues in a beating heart-on-a-chip. Methods Cell Biol 146:69–84

Unger MA, Chou HP, Thorsen T, Scherer A, Quake SR (2000) Monolithic microfabricated valves and pumps by multilayer soft lithography. Science 288:113–116

Uygun BE, Soto-Gutierrez A, Yagi H, Izamis ML, Guzzardi MA, Shulman C, Milwid J, Kobayashi N, Tilles A, Berthiaume F, Hertl M, Nahmias Y, Yarmush ML, Uygun K (2010) Organ reengineering through development of a transplantable recellularized liver graft using decellularized liver matrix. Nat Med 16:814–820

Villenave R, Wales SQ, Hamkins-Indik T, Papafragkou E, Weaver JC, Ferrante TC, Bahinski A, Elkins CA, Kulka M, Ingber DE (2017) Human gut-on-a-chip supports polarized infection of coxsackie B1 virus in vitro. PLoS One 12:E0169412

Vriend J, Nieskens TTG, Vormann MK, Van Den Berge BT, Van Den Heuvel A, Russel FGM, Suter-Dick L, Lanz HL, Vulto P, Masereeuw R, Wilmer MJ (2018) Screening of drug-transporter interactions in A 3D microfluidic renal proximal tubule on a chip. AAPS J 20:87

Vyas D, Baptista PM, Brovold M, Moran E, Gaston B, Booth C, Samuel M, Atala A, Soker S (2017) Self-assembled liver organoids recapitulate hepatobiliary organogenesis in vitro. Hepatology 67(2):750–761

Waheed S, Cabot JM, Macdonald NP, Lewis T, Guijt RM, Paull B, Breadmore MC (2016) 3D printed microfluidic devices: enablers and barriers. Lab Chip 16:1993–2013

Wang Y, Vaidya B, Farquar HD, Stryjewski W, Hammer RP, Mccarley RL, Soper SA, Cheng YW, Barany F (2003) Microarrays assembled in microfluidic chips fabricated from poly(methyl methacrylate) for the detection of low-abundant DNA mutations. Anal Chem 75:1130–1140

Wang W, Wang Y, Qu C, Wang S, Zhou J, Cao W, Xu L, Ma B, Hakim MS, Yin Y, Li T, Peppelenbosch MP, Zhao J, Pan Q (2018a) The RNA genome of hepatitis E virus robustly triggers an antiviral interferon response. Hepatology 67:2096–2112

Wang Y, Wang H, Deng P, Chen W, Guo Y, Tao T, Qin J (2018b) In situ differentiation and generation of functional liver organoids from human iPSCS in a 3D perfusable chip system. Lab Chip 18:3606–3616

Weber EJ, Lidberg KA, Wang L, Bammler TK, Macdonald JW, Li MJ, Redhair M, Atkins WM, Tran C, Hines KM, Herron J, Xu L, Monteiro MB, Ramm S, Vaidya V, Vaara M, Vaara T, Himmelfarb J, Kelly EJ (2018) Human kidney on a chip assessment of polymyxin antibiotic nephrotoxicity. JCI Insight 3, pii: 123673

Wells MF, Salick MR, Wiskow O, Ho DJ, Worringer KA, Ihry RJ, Kommineni S, Bilican B, Klim JR, Hill EJ, Kane LT, Ye C, Kaykas A, Eggan K (2016) Genetic ablation of AXL does not protect human neural progenitor cells and cerebral organoids from Zika virus infection. Cell Stem Cell 19:703–708

Whitesides GM (2006) The origins and the future of microfluidics. Nature 442:368–373

Wimmer RA, Leopoldi A, Aichinger M, Wick N, Hantusch B, Novatchkova M, Taubenschmid J, Hammerle M, Esk C, Bagley JA, Lindenhofer D, Chen G, Boehm M, Agu CA, Yang F, Fu B, Zuber J, Knoblich JA, Kerjaschki D, Penninger JM (2019) Human blood vessel organoids as a model of diabetic vasculopathy. Nature 565:505–510

Workman MJ, Mahe MM, Trisno S, Poling HM, Watson CL, Sundaram N, Chang CF, Schiesser J, Aubert P, Stanley EG, Elefanty AG, Miyaoka Y, Mandegar MA, Conklin BR, Neunlist M, Brugmann SA, Helmrath MA, Wells JM (2017) Engineered human pluripotent-stem-cell-derived intestinal tissues with a functional enteric nervous system. Nat Med 23:49–59

Yan Y, Song L, Bejoy J, Zhao J, Kanekiyo T, Bu G, Zhou Y, Li Y (2018) Modeling neurodegenerative microenvironment using cortical organoids derived from human stem cells. Tissue Eng Part A 24:1125–1137

Yang X, Pabon L, Murry CE (2014) Engineering adolescence: maturation of human pluripotent stem cell-derived cardiomyocytes. Circ Res 114:511–523

Yang X, Li K, Zhang X, Liu C, Guo B, Wen W, Gao X (2018) Nanofiber membrane supported lung-on-a-chip microdevice for anti-cancer drug testing. Lab Chip 18:486–495

Ye L, Wang J, Beyer AI, Teque F, Cradick TJ, Qi Z, Chang JC, Bao G, Muench MO, Yu J, Levy JA, Kan YW (2014) Seamless modification of wild-type induced pluripotent stem cells to the natural CCR5delta32 mutation confers resistance to HIV infection. Proc Natl Acad Sci U S A 111:9591–9596

Yeager A (2018) As brain organoids mature, ethical questions arise. The Scientist. https://www.the-scientist.com/features/brain-organoids-mature%2D%2Draise-ethical-questions-64533

Yu W, Mackerell AD Jr (2017) Computer-aided drug design methods. Methods Mol Biol 1520:85–106

Yu J, Vodyanik MA, Smuga-Otto K, Antosiewicz-Bourget J, Frane JL, Tian S, Nie J, Jonsdottir GA, Ruotti V, Stewart R, Slukvin I, Thomson JA (2007) Induced pluripotent stem cell lines derived from human somatic cells. Science 318:1917–1920

Zakhariants AA, Burmistrova OA, Shkurnikov MY, Poloznikov AA, Sakharov DA (2016) Development of a specific substrate-inhibitor panel (liver-on-a-chip) for evaluation of cytochrome P450 activity. Bull Exp Biol Med 162:170–174

Zhang N, Tan H, Yeung ES (1999) Automated and integrated system for high-throughput DNA genotyping directly from blood. Anal Chem 71:1138–1145

Zhang X, Grimley A, Bertram R, Roper MG (2010) Microfluidic system for generation of sinusoidal glucose waveforms for entrainment of islets of Langerhans. Anal Chem 82:6704–6711

Zhou M, Zhang X, Wen X, Wu T, Wang W, Yang M, Wang J, Fang M, Lin B, Lin H (2016) Development of a functional glomerulus at the organ level on a chip to mimic hypertensive nephropathy. Sci Rep 6:31771

Zhu Z, Zhang W, Leng X, Zhang M, Guan Z, Lu J, Yang CJ (2012) Highly sensitive and quantitative detection of rare pathogens through agarose droplet microfluidic emulsion PCR at the single-cell level. Lab Chip 12:3907–3913

Zhu W, Qu X, Zhu J, Ma X, Patel S, Liu J, Wang P, Lai CS, Gou M, Xu Y, Zhang K, Chen S (2017) Direct 3D bioprinting of prevascularized tissue constructs with complex microarchitecture. Biomaterials 124:106–115

Creating Products and Services in Industrial Biotechnology

6

Eleonora Calzoni, Alessio Cesaretti, and Carla Emiliani

Abstract

Biotechnologies represent a set of enabling technologies that find application in various industrial and economic sectors. There are more and more companies that, while operating in "traditional" fields, integrate biotechnological products and technologies into their production processes, in order to improve their quality and yield and reduce their environmental impact. Not surprisingly, no production process is less invasive for the environment than natural processes from which biotechnology originates.

Biocatalysis has now fully entered into all industrial sectors and is crucial for the development of sustainable chemistry. It is based on the use of bioprocesses mediated by microorganisms or enzymes capable of accelerating the speed of the reaction without the production of toxic substances and on the use of mild temperature conditions. Biocatalysis has therefore become a valuable tool, not only for its lower environmental impact, but also as it is increasingly becoming a cost-effective alternative to classical chemical processes. Biocatalysis is now used in a variety of fields, from the pharmaceutical and food sectors to the production of biofuels to the restoration and conservation of cultural heritage.

Bioeconomy is the challenge that Europe is taking on for the establishment of a new model of sustainable development, capable of generating value and employment. This new type of industry will therefore implement traditional processes with biotechnological processes with the aim being both to enhance its production yields and standards and to lower its impact on the environment.

Keywords

Bioeconomy · Biocatalysis · Bio-based products · Enzymes · Microorganisms

6.1 Industrial Biotechnology Is the Driving Force of Sustainable Development

Industrial biotechnology is one of the most promising new approaches to pollution prevention, resource conservation, and cost reduction. In fact, industrial biotechnology is defined as any application of biochemical, molecular biology, and microbiology techniques aimed at facilitating industrial processes, producing bio-products and

E. Calzoni · A. Cesaretti · C. Emiliani (✉)
Department of Chemistry, Biology and Biotechnology, University of Perugia, Perugia, Italy

Centro di Eccellenza su Materiali Innovativi Nanostrutturati (CEMIN), University of Perugia, Perugia, Italy
e-mail: carla.emiliani@unipg.it

F. Matei, D. Zirra (eds.), *Introduction to Biotech Entrepreneurship: From Idea to Business*,
https://doi.org/10.1007/978-3-030-22141-6_6

bioenergy and reclaiming environmentally compromised areas.

Biotechnologies have been defined by the European Commission as key enabling technologies (KET), as they represent a resource that can increase the productivity of a system, through the improvement of the efficiency of existing processes (Preparing for our future: Developing a common strategy for key enabling technologies in the EU, 2009).

The main target of industrial biotechnology is to replace the current chemical synthesis processes of useful compounds with low environmental impact processes through the use of enzymes and microorganisms or plant/animal cells to produce products in a wide range of industrial sectors including chemical products, pharmaceuticals, food and feed, detergents, textiles, energy, materials, and polymers (Fig. 6.1).

Men have already benefited from biotechnology for a long time, but with the progress of new technologies and a much deeper understanding of cellular metabolism and material sciences, many new opportunities have been identified and others continue to emerge. Moreover, a very interesting aspect associated with industrial biotechnology is the reduced consumption of energy, greenhouse gas emissions, and waste production, thanks to the reuse of biomasses of various origins (Singh 2014). The bio-waste from the agri-food industry, for instance, has a high potential to such an extent that the EU has set itself the goal to derive the production of 2% of renewable energy from this source. The theme of the reuse of waste and by-products is central to those industrial sectors that traditionally use biological resources as their main source of supply (forest sector, starches, sugar, biofuels/bio-energy, biotechnology) and in others for which biomass is among the raw materials used (chemical, plastic, and consumer goods) (Le imprese di biotecnologie in Italia. Facts and Figures, 2018).

For example, residual biomass deriving from agriculture and forestry is used to produce biofuels, one of the most promising and profitable applications of industrial biotechnology. This, in fact, not only guarantees the reuse of waste but

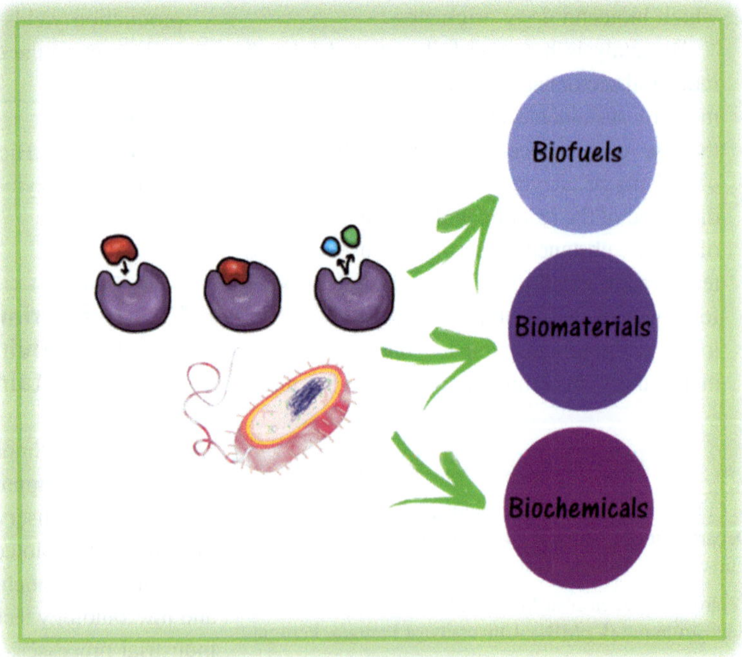

Fig. 6.1 Enzyme and cell factories generate white biotechnology products

also the abatement of emissions into the atmosphere. The European RED directive has required that 10% of the energy used in transport will come by 2020 from biofuels, electricity, and hydrogen produced from renewable sources (Bianchi 2013).

Industrial biotechnology promotes a virtuous circle between research, technological development, and new applications that will enable the implementation of a sustainable development model. In fact, this model aims at creating bio-renewable resources, in order to reduce CO_2 emissions and conserve fossil resources; using processes based on bioconversions, thus featuring low environmental and economic impact and being more selective; and generating bio-based products. To date, there are many applications of industrial biotechnology that find fertile ground in the food, pharmaceutical, textile, biofuels, and bioplastics industries and in the conservation and restoration of archaeological and cultural heritage. In the pharmaceutical industry, industrial biotechnology has enormous potential, as many of its products do not require the long overhaul times that pharmaceutical products must undergo, thus creating a faster and easier way into their market. Today, new industrial processes can be taken from laboratory study to commercial application in two to 5 years, compared to a decade needed for drugs. The application of biotechnology to industrial processes is not only to transform the way to synthetize products but also to provide us with new products that could not even be imagined a few years ago. Since industrial biotechnology is a new-born area, its benefits are not yet well known or understood by industry, policy makers, or consumers.

Overall, industrial biotechnology involves the microbial production of enzymes, which are specialized proteins. These enzymes have evolved in nature to be super-performing biocatalysts that facilitate and accelerate complex biochemical reactions. They were also used in the prevention of pollution, as is the case with phosphate water pollution in the 1970s caused by the use of phosphates in laundry detergents. Biotechnology companies have developed enzymes that remove stains from clothing better than phosphates, thus allowing the replacement of a polluting material with a non-polluting bio-based additive while improving the performance of the final product. This innovation has dramatically reduced algal proliferations related to phosphate in surface waters around the world and simultaneously allowed consumers to get clean garments with lower wash water temperatures and concomitant energy savings. Biotechnological applications have advanced progressively with the cultural evolution of mankind; as early as 6000 BC, the grape fermentation process was known to produce wine and the use of yeasts to make beer. In the early 1900s, Alexander Fleming extracted penicillin from the mold and later large-scale fermentation techniques were developed to produce industrial quantities of this drug. Only after the Second World War, however, the biotechnological revolution began, giving rise to modern industrial biotechnology. Since then, industrial biotechnology has produced enzymes for our daily lives and for the manufacturing sector. In the industrial field, enzymes have always been the subject of great interest and a source of undisputed utility. Thanks to their role as catalysts, enzymes are not only fundamental for life, but they also find many applications in different industrial fields, since they are more efficient and advantageous than traditional organic synthetic products. In fact, their use as biocatalysts has greatly increased, as they are economical and more environmentally friendly, being biocompatible, biodegradable, and derived from renewable resources (Sheldon and van Pelt 2013). Industrial biotechnology companies use many specialized techniques to find and improve natural enzymes. Information from genomic studies on microorganisms is helping researchers to capitalize on the richness of genetic diversity in microbial populations. Researchers first look for microorganisms that produce enzymes in their natural environment and then use DNA probes to search for genes that produce enzymes with specific biocatalytic abilities at the molecular level. Once isolated, these enzymes can be identified and characterized for their ability to function in specific industrial processes. If necessary, they can also be improved through biotechnological techniques.

Although many biocatalytic tools are rapidly becoming available for industrial applications, biocatalysts or whole cell processes are often not yet available for implementation, thus generating a "technology gap" where there is a delay between availability and widespread use of new technology. This gap must be overcome in order to accelerate progress in the development of cheaper and more sustainable production processes. Industrial biotechnology is therefore the cornerstone of the bioeconomy, an economy based on the sustainable use of renewable natural resources and their transformation into final or intermediate goods and services (European Commission 2012).

The definition of bioeconomy includes the sectors of agriculture, food, fishing, forestry, the wood and paper industry, and that of bio-based industries. An OECD study (OECD Annual Report, 2009) estimated that in 2030 in developed countries biotechnology will represent 35% of chemical and industrial products, 80% of pharmaceuticals and diagnostics, and 50% of agricultural products. Industrial biotechnologies have therefore enormous innovative potential and may be the answer to many problems that in the coming years will have to be addressed as environmental remediation, the problems of climate change, and the invention of new medicines. However, this new approach also requires a change in the political setup and in the field of research. In the first case, the transition from sectoral governance mechanisms to an integrated strategy is necessary (EuropaBio 2011). At the same time, the system of research and university education must be reoriented. This has already been happening at different levels, as shown, for example, by the launch of the Horizon 2020 program. Hence, industrial biotechnology has great potential not only at the economic level but also for development and employment, representing the tool for achieving totally unimaginable goals up to a few years ago.

6.2 Biocatalysis and Industrial Processes

Biocatalysis, or catalysis operated by enzymes, is nowadays used in a significant number of industrial production processes and represents a crucial strategy in what is called "sustainable chemistry" or green chemistry (Fig. 6.2).

Biocatalysis is the basis of some of the oldest chemical transformations known to human beings such as the fermentation of grapes and beer and the production of dairy products. In the context of biocatalysis, enzymes are fundamental proteins because they have considerable catalytic power, very often higher than synthetic or inorganic catalysts. They have a high degree of specificity for their substrates, greatly accelerate chemical reactions, and act in aqueous solution under very mild temperature and pH conditions. Enzymes catalyze various reactions, such as those implied in the degradation of molecules, in the conservation and transformation of chemical energy, and in the synthesis of macromolecules from simpler precursors. They manage to have such a wide range of applications because they exist in very large numbers. Although there are already many enzymatic processes in use, many more would be desirable. Unfortunately, known and available biocatalysts are sometimes inadequate, or not competitive, with respect to traditional chemical processes, hence the need for the identification of new enzymes with promising structural and functional characteristics for biotechnological applications. Biocatalysis has thus become a valuable tool, not only because of its lower environmental impact, but also as it is an economically advantageous alternative to classical chemical processes. Enzymatic transformations effected by partially purified enzymes or whole cells as catalysts are increasingly used for the production of a multitude of chemicals. Biocatalysis is now used in various fields, and there are numerous examples that can be reported, such as the production of acrylamide by the nitrile hydratase of *Rhodococcus rhodochrous* and the production of lactose-free milk through the use of β-galactosidase, which splits lactose into glucose and galactose; similarly, fructose is produced by different companies

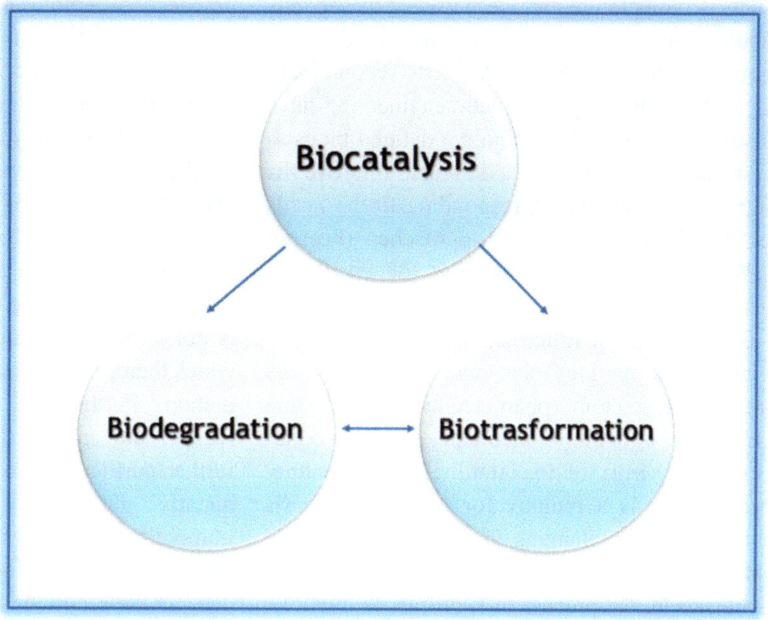

Fig. 6.2 Biocatalysis is used in many industrial processes of biodegradation and biotransformation

starting from glucose through the use of glucose isomerase. In the textile industry, enzymes such as proteases and lipases are used instead of chemical additives and allow us to wash at low temperatures with considerable energy savings and reduced environmental impact. The diffusion of biotech in this market reaches now 95% in Europe and Japan, 70% in North America, and 50% in Latin America and Asia (Rapporto sulle Biotecnologie in Italia, 2012).

Enzymes are widely used today in many fields, so that in 2011 the global market reached 2.7 billion euros (The Novozyme Report, 2011). The textile, food, cosmetics, and waste treatment sector are just some of the sectors in which these molecules are used. In the food sector, for example, biotechnologies aim to act on improving both the quantity and quality of animal and plant products and on improving the use, processing, and conservation of the raw material.

In recent years, there has been a significant increase in the use of industrial biocatalysts for the production of high value-added molecules, as an industrial production process conducted by biocatalysis has the advantage of being environmentally friendly and free of costly solvent or unwanted product disposal, but above all characterized by the chemoselectivity, regioselectivity, and especially stereoselectivity that these enzymes often possess, necessary for the production of enantiomerically pure products. In fact, chemoselectivity allows the specific interaction with particular functional groups; regioselectivity is guaranteed by the formation or break of a bond that allows the synthesis of a specific compound compared to all the other possible; and finally, stereoselectivity allows us to produce a specific stereoisomer. Generally speaking, hydrolyses involving ester or amide bonds are the easiest to carry out using enzymes such as lipases, esterases, and proteases. One of the best-known transformations, in the preparation of pharmaceutical products, is represented by the selective hydrolysis of penicillin G, by means of penicilline amidase, to give the fundamental precursor for the preparation of a broad category of antibiotics, the 6-aminopennicillanic acid.

In recent years, thanks to metabolic and protein engineering techniques, it has been possible to improve the biocatalysis processes allowing the limitations of enzyme stability under different conditions to be overcome, where parameters such as temperature and the presence of solvents and/or reactive products may cause the denaturation of the biocatalyst. Modern recombinant DNA technologies and genetic engineering have also made it feasible to obtain a defined biocatalyst for any kind of reaction and improve its intrinsic characteristics, responding to the needs of the different synthetic approaches (Fessner 1998). The non-negligible role of enzymatic catalysis in synthesis, thanks to contributions from various scientific disciplines, constitutes a veritable "technology" that can be used in various application sectors (pharmaceutical, food, fine chemicals, agriculture, medicine, energy production). By virtue of its interdisciplinary nature, biocatalysis is a toolbox for organic synthesis that enables the planning of a synthetic rationale taking into account the molecule to be synthesized, the process of synthesis as a whole, and the biological nature of the catalyst (Bommarius and Riebel-Bommarius 2004).

To date, the application of biocatalysis on an industrial scale is possible through the optimization of the biocatalyst as a function of both the substrate and the reaction to be catalyzed and the reaction conditions required by industrial technology. Commonly, the use of biocatalysts can have disadvantages such as thermal instability, protease susceptibility, activity inhibition, high sensitivity to different denaturing agents, and the difficulty of separating or reusing the free catalyst at the end of the reaction from the reaction mixture (Khan and Alzohairy 2010). All these problems have been solved by using a more innovative biocatalysis system, i.e., the use of immobilized enzymes. As of today, the immobilization approach is indeed one of the most promising strategy, widely used in various fields and in continuous development, because it is full of advantages from various points of view. Enzyme immobilization is an extremely interesting technique, mainly due to its potential applicability in various industrial fields. It consists in the binding of free enzymes on diverse types of supports, with the aim of making them lose mobility and improve their stability and recyclability with respect to the corresponding free enzyme. The use of immobilized enzymes creates practical, economic, and ecological advantages. Their use grants, first of all, a simpler and more effective manipulation, as they are found in a solid formulation rather than the liquid one characteristic of free enzymes. Immobilized enzymes are therefore easier to use than other technologies and other methods of synthesis. This technique also has the advantage of completely avoiding contamination of the final product with enzyme residues. In fact, since the latter is immobilized, it is not found within the product of the reaction, which therefore does not undergo protein contamination. Furthermore, it is not necessary to recover the enzyme from the product through further purifications and the procedure is consequently faster. Biocatalysis mediated by immobilized enzymes is now used in various industrial fields starting from the pharmaceutical one, for the production of drugs such as β-lactam or anti-thrombotic antibiotics; in the food industry where they can be used both as biosensors and as catalysts of reactions of production, processing, and degradation; and in the biofuels synthesis industry, where biodiesel is produced, as well as through the classic chemical way, also through reactions based on the use of immobilized enzymes. This guarantees greater selectivity and specificity, the occurrence of reactions under mild conditions of temperature, pH, and pressure, and, in addition, the absence of by-products which should otherwise be removed. In most cases, lipases from different sources such as *Thermomyces lanuginosus*, *Candida antarctica* and *Candida rugosa*, *Pseudomonas fluorescens*, *Pseudomonas cepacia*, and *Saccharomyces cerevisiae* are used as enzymes to be immobilized for the production of biodiesel (Khan and Alzohairy 2010). Lipases are the chosen enzymes because they are able to conserve their activity even with low water content, such as organic solvents, and because, in addition to catalyzing the hydrolysis of triglycerides, they also catalyze esterification and transesterification. To date, biocatalysis has therefore more applications than conventional chemical methods both in laboratories and in industries with attributes such as efficiency, faster performance, and low environmental impact. Furthermore, research, both scientific and industrial, is investing to overcome the problems still encountered and to improve its performance. At present, efforts are being made to develop biocatalysis processes with greater control and fewer limitations and to achieve further progress in the various fields in which they are used.

6.3 Bio-Based Products

Bio-based products are commodities entirely or partially derived by biomasses such as plants, trees, or animals (such biomasses can also be subject to physical, chemical, or biological treatments, Fig. 6.3) (United States Secretary of Agriculture in the Farm Security and Rural Investment Act, 2002).

Some examples of agricultural resources that make up many bio-based products include soybeans, corn, kenaf, flax, jute, and numerous other types of crops that are harvested. Current applications of these agricultural resources create products such as ethanol (corn-based), soy candles, soy-based lubricants, kenaf office paper, and bioplastics to name but a few.

Bio-based products can make economy more sustainable by reducing its dependence on fossil fuels. In this light, EU stated that biological products represent a sector of upmost importance in that they have a great potential of future growth, re-industrialization, and answer to the challenges that society is going to face. Being derived from renewable raw materials such as plants, bio-based products can help in reducing the CO_2 emissions and offer other advantages such as minor toxicity and new features for the product itself (i.e., biodegradable plastic materials). From a technical point of view, most industrial products obtained from fossil feedstocks could indeed be substituted by their bio-based counterparts. However, as of today, the costs associated with the production of bio-based materials often exceed those related to the more common petrochemical industry.

Moreover, it is required to prove that on the one hand, these brand-new materials can deliver performances at least as good as the industrial products to be substituted, and on the other hand, that they exert a lower environmental impact. By doing so, bio-based products can make a positive contribution to savings in greenhouse gas emissions, helping in meeting the EU's climate goal.

The industries that produce bio-based products are constantly growing and represent a capital of more than 2000 billion euros a year (European Commission Recommendation 2004).

6.4 Restoration and Conservation of Artworks

Another application field of biotechnology concerns the restoration and conservation of cultural heritage, where scientific skills are increasingly combined to historical and artistic skills in

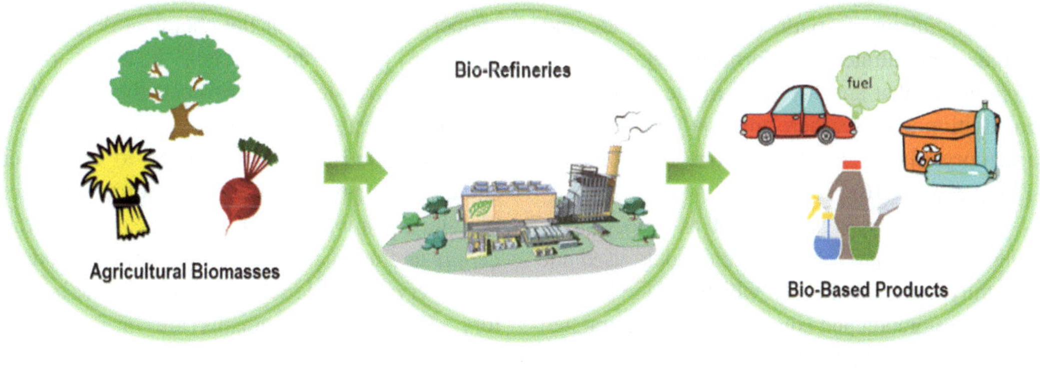

Fig. 6.3 Biomass or agri-food waste products are transformed by biorefineries into bio-based products with high added value

Fig. 6.4 Biocleaning and biorestoration mediated by microorganisms

order to identify innovative and effective approaches for the restoration and conservation of works of art. In fact, biotechnological research allows the development of innovative restoration methods, mainly based on the use of microorganisms called "biorestorers," but also allows the use of advanced diagnostic techniques to identify and characterize the origin of the agents that induce deterioration. As a rule, non-pathogenic microorganisms or particular enzymes such as hydrolases are used to remove unwanted organic matter or crusts of sulfates and nitrates from the surface of the works, which alter the appearance and health of the cultural good. They are innovative approaches based mainly on the enzymatic hydrolysis performed by microorganisms or parts of them that degrade sulfates and nitrates, transforming them into non-toxic gases that are dispersed in the atmosphere; in addition to these, there are also biocalcificating microorganisms, used as stone consolidators (Palla et al. 2016).

These processes are defined as "biocleaning" (Fig. 6.4) and have already been widely used successfully, especially in Italy, which is at the forefront for the restoration and conservation of cultural heritage, being the cradle of an immense historical and artistic heritage. In 2004, the first biorestoration was carried out on some frescoes of the Camposanto Monumentale of Pisa that was damaged due to the use of a glue of animal origin, which was removed thanks to the use of a specific microorganism, *Pseudomonas stutzeri* (Ranalli et al. 2005).

However, there are many other works and monuments that have undergone bioremediation techniques, such as Milan Cathedral, Florence Cathedral, the church of Santos Juanes in Valencia, the cathedral of Riga, as well as the "Pietà Rondanini" of Michelangelo.

Bacteria are not the only organisms used in the conservation of cultural heritage: even micro-fungi and certain yeasts can do a great job. The *Beauveria bassiana* microform, for example, is able to transform the corrosion compounds of copper, responsible for the chemical degradation of some metal objects, into a stable patina (of copper oxalate) which protects the metal itself from deterioration, as well.

Microorganisms are also used to strengthen stone works, giving rise to a phenomenon called biomineralization, as in the case of *Bacillus cereus* and *Myxococcus xanthus* able to produce calcite crystals (calcium carbonate) in the laboratory. These microorganisms are in fact called calcinogenic bacteria, and they are used to consolidate the works in calcareous carbon stone, inasmuch as the calcite (called biocalcite) they produce is more resistant and better integrates with the substrate than that produced with more traditional methods. In addition, some light-emitting proteins naturally produced by bacteria, fireflies, and jellyfish have been used to build biosensors capable of detecting contamination even in works of art (Ramírez et al. 2005).

Cultural heritage and works of art are an important asset for mankind, and as such, it is extremely important to be able to guarantee their conservation for future generations. The use of biotechnologies together with other scientific disciplines such as physics or chemistry is therefore an extremely promising and effective approach to safeguarding human inheritance.

6.5 The Commitment of Companies in the Sector of the Biotechnological Drug

Companies operating in the field of biotechnology applied to human health, using modern biotechnological methods for research, development, and production of drugs, new therapies, vaccines, diagnostics systems, molecular farming, etc., represent an important percentage of European biotech companies (Evens and Kaitin 2015).

In this case, the term biotechnology refers to the use of living systems or molecular engineering to create and manufacture drugs, biologic therapies, and products for patient care such as antibodies, proteins, peptides, cells, tissues, but also gene therapies, molecularly engineered proteins, and cells.

Traditionally, drugs are chemically produced molecules manufactured by pharmaceutical companies. Since the 1980s, small biotechnology companies, whose research, development, and marketing activities have focused on biotechnological drugs, have started to develop. These companies were often launched after the identification of new molecular pathways involved in the pathophysiology of a disease or the discovery of new molecules implicated in the treatment of a particular pathology.

Currently, biopharmaceuticals are the main focus of pharmaceutical and biotechnology companies. Almost all therapeutic areas are involved, ranging from oncology to diabetes and infectious diseases. The large pharmaceutical companies have acquired over the years small biotech companies or converted part of their production into biotechnological processes.

The first commercialization of biotech products dates back to the 1980s when recombinant DNA proteins, monoclonal antibodies, and vaccines were discovered and developed.

As for recombinant proteins, their commercialization requires some essential steps: firstly, protein isolation, identification, sequencing, and cloning need to be performed in nonhuman host cells; then, production techniques have to be scaled up in large-scale procedures which meet quantity and quality standards adequate for commercialization. Monoclonal antibody technology began when a hybridoma cell (made up of a plasma cell merged with a multiple myeloma cell) was developed. In fact, hybridoma cells exhibit the ability to produce large quantities of the targeted antigen-specific monoclonal antibodies.

As is the case with the antigen vaccine for hepatitis, vaccines were developed using recombinant DNA technology, as well. In this case, once identified, a protein antigen is generated and then incorporated into a vaccine formulation.

In the 1990s, biotech companies were increasingly interested in the manipulation of proteins and their use; for example, thanks to genetic engineering, it was possible to modify a fibrinolytic molecule, tenecteplase, used in patients affected by myocardial infarction (Evens 2013). Modification of its amino-acid chain allowed the oral administration of the drug, thus replacing the intravenous route. More recently, during the 2000s, the focus has shifted toward fusion proteins, an important molecular engineering technique by which seven new molecules were developed, and on the use of peptide molecules which are slowly replacing the recombinant and synthetic proteins used in the treatment of diseases such as diabetes. (Therapeutic peptides in clinical study in 2000–07 nearly doubled 1990s rate. Medford (MA): The Center, 2009.) All these developments in the biotechnological field have led to the acquisition by large pharmaceutical companies of biotech companies, which in this way were asked to share patent rights related to the development of new molecules.

Hence, during the past 30 years, health care has been profoundly affected by biotechnology, as the latter has progressively allowed the pathophysiology of many previously untreatable diseases to be better understood. This represents a turning point in the identification and development of new drugs and treatments for groundbreaking innovations in patient care.

6.6 Conclusions

Industrial biotechnology represents the driving force of bioeconomy, which is intended as an economy based on biological resources used as inputs for energy, industrial, food, and animal production. A report presented by the Organisation for Economic Co-operation and Development OECD, entitled "The Bioeconomy to 2030: designing a policy agenda," recognized the possibility for bioeconomy to promote an actual and practical impulse toward a new "industrial revolution" (Oborne 2010). Such revolution would be grounded on the research into the field of renewable raw materials, allowing constant innovation in a plethora of industrial sectors and thus ensuring environmental, economic, and social sustainability of the world economic system in the long term.

The growth model suggested by the European Union is increasingly focusing on the management, production, and sustainable use of renewable biological resources, through the help of science and biotechnology, integrated with other technologies.

In this light, the European Union believes in the possibility for a strong acceleration, seeing as how bioeconomy possesses all the potentialities required to face many of the major challenges for the future of Europeans: from food security to energy needs and the reduction of the environmental impact coming from agriculture and industry; from the provision of healthy food at accessible costs to the encouragement of coastal and rural development; and from the fight against climate changes to the achievement of zero waste in landfills.

References

Bianchi D (2013) La filiera delle biomasse e dei rifiuti. Lab4energy: il futuro dell'energia

Bommarius AS, Riebel-Bommarius BR (2004) Biocatalysis, fundamentals and application. Wiley-VCH, Hoboken, NJ

European Association for Bioindustries (EuropaBio) (2011)

European Commission (2012) Innovating for a sustainable growth: a bioeconomy for Europe

European Commission Recommendation (2004)

Evens RP (2013) Biotechnology and biological preparations. In: Swarbrick J (ed) Encyclopedia of pharmaceutical science and technology, vol 1, 4th edn. Taylor and Francis, Boca Raton, FL, pp 239–266

Evens R, Kaitin K (2015) The evolution of biotechnology and its impact on health care. Health Aff 34(2): 210–219

Fessner WD (1998) Enzyme mediated C-C bond formation. Curr Opin Chem Biol 2(1):85–97

Khan AA, Alzohairy MA (2010) Recent advances and applications of immobilized enzyme technologies: a review. Res J Biol Sci 5(8):565–575

Le imprese di biotecnologie in Italia. Facts and Figures, BioInItaly, Report 2018

Oborne M (2010) The bioeconomy to 2030: designing a policy agenda. OECD Obs (278):35–38

Organisation for Economic Co-operation and Development- OECD Annual Report 2009

Palla F, Barresi G, Giordano A, Trapani M, Rotolo V, Parisi M, Cammarata M (2016) Cold-active molecules for a sustainable preservation and restoration of historic-artistic manufacts. Int J Conserv Sci 7(Special Issue 1):239–246

Preparing for our future: Developing a common strategy for key enabling technologies in the EU, 2009

Ramírez JL, Santana MA, Galindo-Castro I, Gonzalez A (2005) The role of biotechnology in art preservation. Trends Biotechnol 23(12):584–588

Ranalli G, Alfano G, Belli C, Lustrato G, Colombini MP, Bonaduce I, Zanardini E, Abbruscato P, Cappitelli F, Sorlini C (2005) Biotechnology applied to cultural heritage: biorestoration of frescoes using viable bacterial cells and enzymes. J Appl Microbiol 98(1):73–83

Rapporto sulle Biotecnologie in Italia, BioInItaly, 2012

Sheldon RA, van Pelt S (2013) Enzyme immobilisation in biocatalysis: why, what and how. Chem Soc Rev 42(15):6223–6235

Singh RS (2014) Industrial biotechnology: an overview. International Publishing House Pvt. Ltd, New Delhi
The Novozyme Report, 2011
Tufts Center for Study of Drug Development. Therapeutic peptides in clinical study in 2000–07 nearly doubled 1990s rate. Medford (MA): The Center; 2009. (Impact report)
United States Secretary of Agriculture in the Farm Security and Rural Investment Act of 2002

Creating Products and Services in Food Biotechnology

7

Mona Elena Popa, Amalia Carmen Mitelut, Elisabeta Elena Popa, and Florentina Matei

Abstract

This chapter includes new material on applications of food biotechnology products in industry, e.g., in food fermentations, enzymes, probiotics, prebiotics, synbiotics, microbial food cultures, or improving food nutrition using bioactive compounds—functional and nutraceutical food through development of functional food products with good examples of this kind of food. At the same time, the chapter offers information regarding new trends on the market, biopreservation of food using bacteriocins, bacteriophages, and natural preservatives from plants. Considering the sustainability issues which are the latest trends in bioeconomy, the chapter also presents aspects regarding bio-valorization of food waste through biofuel and bioenergy production, biomaterials production, and food ingredients recovery.

Written by professors in food biotechnology, food microbiology, food conditioning and preservation, food safety, and food sustainability and very good researchers, the chapter includes up-to-date information which could be a useful tool for knowledge improvement in this field and a starting point for future entrepreneurs.

Keywords

Food biotechnology · Fermentations · Enzymes · Starter cultures · Functional food · Bio-preservatives · Bio-valorization · Food waste

7.1 Introduction

One of the most demanding challenges of the food industry in the twenty-first century is the increasing demand for food and specifically for healthy and nutritionally rich food products. Biotechnology is one of the solutions, its intention being to improve food, food ingredients, and functional food at the processing stage, beyond agricultural production. In this respect, food biotechnology deals with current developments and applications of modern genetics, enzymatic, metabolic, and systems-based biochemical processes in food, and food-related biological systems. The economic importance of food biotechnology becomes particularly evident when we look at few food products produced in very large quantities worldwide using biotechnological methods: wine, beer, and cheese.

The targeted use of biotechnological methods can, among other things, help reduce the quantity and number of unhealthy ingredients in foods as

M. E. Popa (✉) · A. C. Mitelut · E. E. Popa · F. Matei
Faculty of Biotechnologies, University of Agronomic Sciences and Veterinary Medicine of Bucharest, Bucharest, Romania

F. Matei, D. Zirra (eds.), *Introduction to Biotech Entrepreneurship: From Idea to Business*,
https://doi.org/10.1007/978-3-030-22141-6_7

well as degrade allergenic substances. Genomic research and targeted breeding also greatly facilitate progress in agriculture. Functional food, a concept originated in the late 1980s in Japan, refers to a food product "designed" to have physiological benefits and/or reduce the risk of chronic disease beyond basic nutritional functions and may be similar in appearance to conventional food and consumed as part of a regular diet. In order to produce functional and nutraceutical food, several techniques are available which include microbial and fermentation-based metabolic processing. Fermentation has many applications not only in improving food for health but also in food waste remediation. Other examples of food biotechnology uses are the enzymatic degradation of lactose in dairy products or acrylamide precursors in bread and potato crisps, which results in a significant reduction of potentially carcinogenic acrylamide in the final product.

Biotechnology also offers new sustainable ways to preserve the existing resources. On the one hand, the addition of specific enzymes that stop biological degradation processes can prolong the shelf life of food. This leads to lowering sales losses, as well as the quantity of food waste. On the other hand, biotechnological methods can break up nutrient-rich compounds such as woody plant constituents to make such compounds part of human and animal diets.

7.2 Applications of Food Biotechnology in Industry

7.2.1 Fermentations

Fermented foods are usually defined as "foods or beverages made through controlled microbial growth and enzymatic conversions of major and minor food components" (Marco et al. 2017).

Fermented foods, such as cheese and wine which are obviously food products produced by fermentation and some others that are not as obvious, are an important part of our diet. The transformation carried out by the microorganisms in fermented foods such as tea and bread can be delicate thus less obvious to laypeople. In addition to fermented foods themselves, a large number of ingredients are produced using microorganisms' fermentation.

Fermentation is a technology that utilizes the growth and metabolic activities of microorganisms for the preservation and transformation of food materials. During food fermentation, the growth of spoilage and pathogenic organisms is inhibited by the metabolites generated by the fermenting organisms, thereby extending the shelf life of perishable produce. For instance, during lactic acid fermentation, lactic acid bacteria synthesize metabolites such as lactic acid, acetic acid, carbon dioxide, ethanol, hydrogen peroxide, bacteriocins, and antimicrobial peptides (Di Cagno et al. 2013), which through synergy suppress the survival and growth of pathogenic and spoilage microorganisms.

The fermentation processes are wide, and the responsible microorganisms include bacteria, yeasts, and fungi. Bacteriophages also play a role in modulating the microbial flora. Fermentation processes are sometimes simple involving one substrate component (e.g., milk) and one microorganism (e.g., Lactococcus lactis), but sometimes can involve a complex mixture of substrates and several microorganisms. In some cases, the microbial flora responsible for a certain fermented food has not been clearly defined by science. A well-documented history is available to produce beer with perhaps the earliest documented cases occurring in Egypt nearly four centuries ago (Delween 1996). Anyway, the Egyptians did not know that microorganisms were the substantiating factor in the transformation of the liquid bread mixture into beer. It would take many centuries till the work of Pasteur and van Leeuwenhoek led to the realization that microorganisms were the agents that turned food substrates into fermented food products.

A great number of foods are transformed by the result of microbial action. Many of transformations represent spontaneous fermentations which occur when endogenous microorganisms can grow and metabolize the substrate under the influence of the extrinsic conditions. Food products such as sauerkraut and tea are transformed as a result of harvesting an agricultural commodity such as cabbage and tea, respectively, and then man-made alteration of the environment to promote the action of these endogenous microorganisms. The cascade of seemingly innocuous events is now being dissected using high-resolution metagenomic analyses which reveal the succession of microbial populations including those previously unable to be cultured (Jung et al. 2011). The complex micro-flora that is present in the raw ingredients and the dynamic nature of the population during fermentation contribute to the product (Wolfe and Dutton 2015). The deliberate introduction of microorganisms into food substrates was first reduced to practice by the work of scientists such as Christian Hansen. Starter cultures represent individual and collections of microorganisms that are deliberately added to induce a change in the food substrate. Some of these microorganisms, for example, lactic acid bacteria, are extraordinary in their capacity to convert substrates into products reaching efficiencies of over 98%. Fermentation end products include simple alcohols and acids, but the changes induced by these starter cultures range from taste to texture and nutritional content to safety. Besides preservation action, fermentation can improve and sometimes give characteristic aroma, flavor, texture, and nutritional profile into food.

Thus, although ancient civilizations developed fermentation primarily as a way of preserving perishable agricultural produce, nowadays the technology has become more and more a real tool for developing desirable organoleptic profiles in foods and improving their palatability. Bread is a good example, where the primary function of dough fermentation is to create the characteristic structure, texture, and organoleptic profile of bread after the baking process. Fermentation also may help to remove antinutritional factors and toxins in food materials and improve their nutritional profile. For instance, fermentation of soybean into products such as tempeh (fermented dehulled soybean with meat like flavor and texture), natto (a fermented soybean dish from Japan with strong smell and flavor and a slimy texture), and soy sauce (a dark brown condiment made from fermentation of soybean, wheat, and salt) leads to reduction of antinutritional factors such as phytic acid and trypsin inhibitors and results in the hydrolysis of complex soy proteins into more digestible and bioavailable peptides and amino acids (Chen et al. 2013; Soni and Dey 2014).

Traditional food fermentation processes can be broadly classified into lactic acid fermentation, fungal fermentation, and alkaline fermentation. Examples of lactic acid fermented products, i.e., products primarily fermented by lactic acid bacteria, include yogurt, sausages, cheese, sauerkraut (fermented cabbage from eastern and central Europe), and kimchi (fermented and spiced Napa cabbage from Korea). Yeast spp. are also involved in the fermentation of many of the lactic acid-fermented products, including kefir (a slightly alcoholic dairy beverage from the Caucasus) and kombucha (a fermented sweetened tea from China). Most of the well-known soy-based fermented foods from Asia such as tempeh and soy sauce are produced by fungal fermentation, except for natto, which is produced by alkaline fermentation (Dirar 1994).

Industrial fermentation processes use either submerged or solid-state bioreactors that are operated in batch, semi-batch, or continuous mode. Most food fermentation processes from sauerkraut and kimchi to miso and tempeh use solid-state fermentation processes operated in batch mode, where microorganisms are cultivated on the surface of a water-insoluble substrate (Paulova et al. 2013). Submerged fermentation processes are used in the production of yogurt and other dairy-based beverages, alcoholic beverages, and food condiments such as vinegar.

In the last decade, fermented food products make up a significant part of the diet in developing nations and the Far East. In the West, with the exceptions of bread, cheese, and sausages, fermented foods have largely faded to the sidelines with the advent of modern technologies such as refrigeration. Nevertheless, there is a renewed interest in traditional fermented foods in recent times, mainly driven by the purported health benefits of fermented foods as vehicles of both probiotic organisms and health-promoting metabolites.

Fermented foods are currently being promoted to prevent or cure a range of diseases from obesity to cancer. For instance, kimchi is claimed to have anticancer, antiobesity, antiaging, and anticonstipation effects (Kim et al. 2011), whereas kefir is claimed to reduce lactose intolerance symptoms, stimulate the immune system, and lower cholesterol and to have antimutagenic and anticarcinogenic properties (Guzel-Seydim et al. 2011). Although most of the health claims around fermented foods are based on folk beliefs with no scientific substantiation, findings from recent in vitro and animal models, as well as human intervention studies, support some of these claims (Terefe 2016). For example, a recent study by a Korean group reported that consumption of kimchi for 2 weeks by obese and overweight individuals led to a decrease in body mass index, body fat, waist–hip ratio, blood pressure, fasting blood sugar, and total cholesterol (Kim et al. 2011).

Fermented foods are one of the top 10 food trends in 2016 (Terefe 2016), continuing the trend over the last few years. Food companies are responding to this growing trend either by commercializing traditional fermented foods (e.g., kefir and kombucha, whose market value in North America alone was $130 million and $480 million, respectively, in 2014) or developing novel fermented foods based on the traditional ones (e.g., Bionade, flavored malt-based beverages fermented using the starter culture of kombucha, and Rythem, coconut milk-based and fruit juice-based beverages fermented using kefir grains). Several soy- and cereal-based probiotic products are also in the market in response to the growing prevalence of allergies to dairy proteins, lactose and gluten intolerances, and lifestyle choices such as veganism (Gupta and Abu-Ghannam 2012).

Beyond foods, food ingredients are also produced as a result of fermentation. Examples include amino acids, vitamins, and flavoring agents. Amino acids, notably glutamic acid, are produced by fermentation; the current processes are the result of intense research and development since the discovery in the 1950s that *Corynebacterium* could produce it (Hermann 2003). Increases in its production were driven using this amino acid as a flavor enhancer initially in Japan, but eventually worldwide. Through a combination of classical mutagenesis and selection, molecular biology, and fermentation, the yields were currently optimized. New efforts are under way to elucidate fermentation processes using metagenomics and to develop novel fermented foods as part of the evolving culinary arts. Several world-class culinary institutions and their innovative chefs are bringing fermentation to a new level of fine dining (Felder et al. 2012).

7.2.2 Enzymes

Enzymes are biological catalysts (also known as biocatalysts) that speed up biochemical reactions in living organisms. They can also be extracted from cells and then used to catalyze a wide range of commercially important processes. For example, they have important roles in the production of sweetening agents and the modification of antibiotics, they are used in washing powders and various cleaning products, and they play a key role in analytical devices and assays that have clinical, forensic, and environmental applications (Robinson 2015).

Enzymes are biological catalysts, responsible for accelerating six types of biochemical reactions which form the basis for their systematic classification. Like all other catalysts, enzymes are characterized by two fundamental properties. First, they increase the rate of chemical reactions without themselves being consumed or permanently altered by the reaction. Second, they increase reaction rates without altering the chemical equilibrium between reactants and products (Cooper 2000). All the enzymes are macromolecular compounds of protein nature, with simple or complex structure, found in every vegetal and animal tissue and fluid. Historically enzymes are nontoxic and not of safety concern for consumers since they are naturally present in ingredients used to make food. However, food enzymes produced industrially by extraction from plant and animal tissues, or by fermentation of microorganisms, are assessed for safety.

Enzymes typically have common names (often called "trivial names") which refer to the reactions that they catalyze, with the suffix *-ase* (e.g., oxidase, dehydrogenase, carboxylase), although individual proteolytic enzymes generally have the suffix *-in* (e.g., trypsin, chymotrypsin, papain). Often the trivial name also indicates the substrate on which the enzyme acts (e.g., glucose oxidase, alcohol dehydrogenase, pyruvate decarboxylase). However, some trivial names (e.g., invertase, diastase, catalase) provide little information about the substrate, the product, or the reaction involved.

Due to the growing complexity of and inconsistency in the naming of enzymes, the International Union of Biochemistry set up the Enzyme Commission to address this issue. The first Enzyme Commission Report was published in 1961 and provided a systematic approach to the naming of enzymes. The sixth edition, published in 1992, contained details of nearly 3200 different enzymes, and supplements published annually have now extended this number to over 5000.

Within this system, all enzymes are described by a four-part Enzyme Commission (EC) number. For example, the enzyme with the trivial name lactate dehydrogenase has the EC number 1.1.1.27 and is more correctly called l-lactate: NAD^+ oxidoreductase (Robinson 2015).

An enzyme can be succinctly described as having two characteristics: (1) its *structure* almost always consists of a protein (sometimes nucleic acid) tertiary or quaternary structure that is configured in a precise three-dimensional arrangement and (2) its *function* arises from that structure's tendency to exclusively bind one or more particular organic or inorganic molecules in a configuration that causes a reaction to take place under thermodynamic conditions and at a rate that would not otherwise be conducive to the required physicochemical changes.

Enzyme–substrate complex—a substrate is a molecule upon which an enzyme acts. Enzymes catalyze chemical reactions involving the substrate. In the case of a single substrate, the substrate bonds with the enzyme active site, and an enzyme–substrate complex is formed such as in Fig. 7.1.

In addition to binding their substrates, the active sites of many enzymes bind other small molecules that participate in catalysis. These are called coenzymes because they work together with enzymes to enhance reaction rates. Prosthetic groups are small molecules bound to proteins in which they play critical functional roles. For example, the oxygen carried by myoglobin and hemoglobin is bound to heme, a prosthetic group of these proteins. In many cases, metal ions (such as zinc or iron) are bound to enzymes and play central roles in the catalytic process. In contrast to substrates, coenzymes are not irreversibly altered by the reactions in which they are involved. Rather, they are recycled and can participate in multiple enzymatic reactions (Cooper 2000).

Processing strategies were historically focused on eliminating enzymes to preserve the quality of agricultural commodities and now more and more enzymes are used as processing improvers through adding exogenous enzymes.

Enzymes have been used unknowingly in food production, e.g., dough making, for centuries. They can be obtained by extraction from plants or animals or from microorganisms. They are usually purified but may contain varying traces

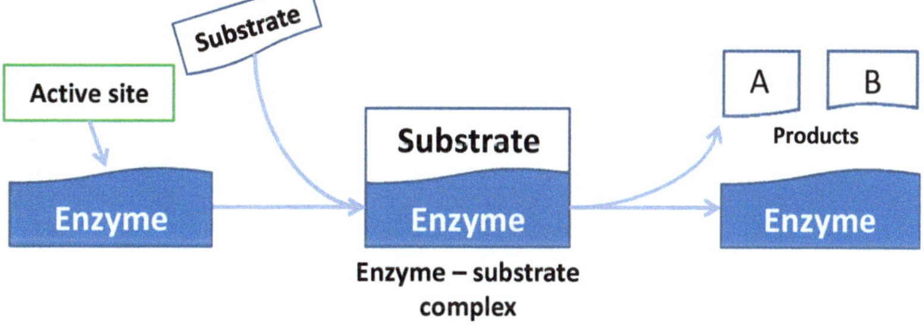

Fig. 7.1 Mechanism of enzyme–substrate formation

of the other naturally occurring constituents of these three sources.

The advancements in food enzyme technologies appeared when people tried to see how process parameters affect thermodynamics and kinetics of enzymatic reaction in relation to their molecular structures. The word enzyme was first used by the German physiologist Wilhelm Kühne and is derived from the Greek term meaning "in yeast." Although the use of enzymes in the production of food and drink, such as bread and wine, had been practiced for thousands of years, the underlying process (and indeed many food processes) which used enzymes from microbial sources was not fully understood until the late nineteenth century.

The industrial processing of many animal and plant origin raw materials is based nowadays on the understanding of enzyme technologies. They are normally added in food to perform a technological function in the manufacture, processing, preparation, and treatment of foods. Examples include enzymes used to break down the structure of fruit so that manufacturers can extract more juice, or to convert starch into sugar in alcohol production.

Immobilization of enzymes expanded a lot using exogenous enzyme in food processing.

The harsh conditions of industrial processes, however, increase propensity of enzyme destabilization, shortening their industrial lifespan. Consequently, the technology of enzyme immobilization provides an effective means to circumvent these concerns by enhancing enzyme catalytic properties and simplify downstream processing and improve operational stability. There are several techniques used to immobilize the enzymes onto supports which range from reversible physical adsorption and ionic linkages to the irreversible stable covalent bonds. Such techniques produce immobilized enzymes of varying stability due to changes in the surface microenvironment and degree of multipoint attachment. Few of them are presented in the scheme from Fig. 7.2.

Utilization of immobilized enzymes as biocatalysts will continue to attract significant

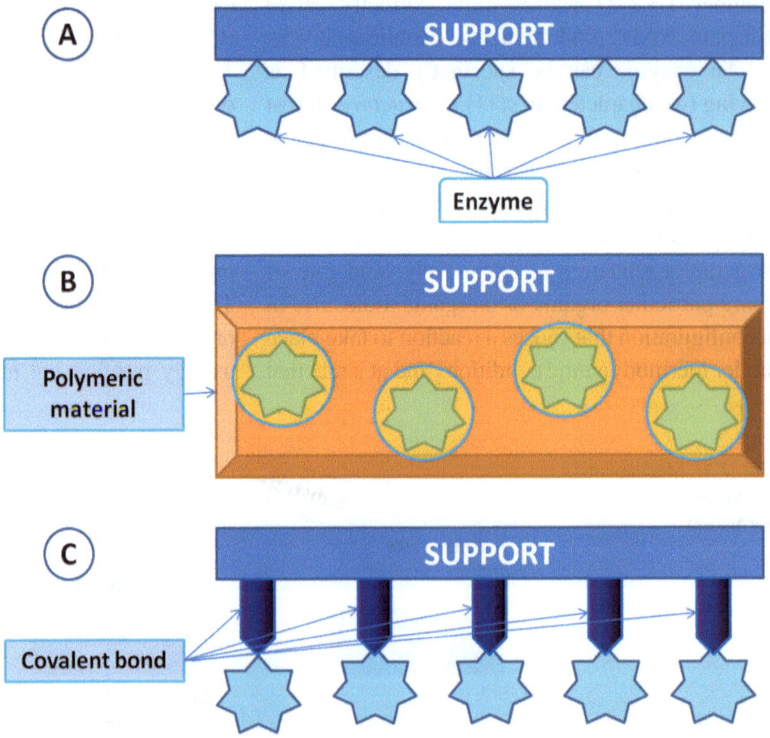

Fig. 7.2 Schematics of the three most common enzyme immobilization techniques: (**a**) physical adsorption, (**b**) entrapment, and (**c**) covalent attachment/cross-linking (Adapted after Spahn and Minteer 2008)

attention from industries as the technique is highly efficient, environmentally friendly, and potentially cost saving when further research is done to seek out or manufacture new matrices that are cheaper and more robust, which can be used as supports for enzyme immobilization.

Food industry enzymes can now be precisely engineered to improve their rate of reaction and the specificity of their action to improve process efficiency, to improve product quality, and to decrease various food safety risks.

The precise uses of enzymes are many, but here are some common examples: removal of oxygen in bottled soft drinks to reduce browning due to oxidation, breakdown of plant cells to help improve efficiency of vegetable oil extraction, removal of pectin from fruit juice in order to produce clear fruit juices, conversion of starch to sugar in alcohol production, curdling of milk in order to produce cheese, hydrolysis of lactose from milk in order to give lactose-free milk, tenderization of meat products, reduction of acrylamide in cooked potato products such as frozen chips, etc.

Previously, food enzymes other than those used as food additives were not regulated at EU level or were regulated as processing aids under the legislation of member states. Only France and Denmark have required safety evaluations for enzymes used as processing aids before they could be used in food production.

Due to differences between national rules on the assessment and authorization of food enzymes, new EU framework legislation on food enzymes was adopted in 2008. This legislation has the aim eventually to establish an EU list of enzymes. Until such a list is established national rules on the marketing and use of food enzymes and food produced with food enzymes will continue to apply in EU countries.

Regulation EC 1331/2008 introduced a common approval procedure for additives, enzymes, and flavorings used in food. Regulation EC 1332/2008 harmonizes rules on enzymes used in foods in the EU and requires the submission of applications for authorization of all existing and new enzymes prior to their inclusion in an official EU list of approved food enzymes at a future date. This legislation requires EFSA to evaluate the safety of all these food enzymes before they can be authorized in the EU and added to the EU list. A food enzyme will be included in the EU list if it does not pose a health concern to the consumer; there is a technological need for its use; and its use does not mislead consumers. Labeling rules related to food enzymes are set down in Regulation (EU) No. 1169/2011 on the provision of food information to consumers and in Articles 10 to 13 of Regulation (EC) No. 1332/2008 on food enzymes.

7.2.3 Probiotics, Prebiotics, Synbiotics

Looking at the food market over the past years, it can be easily noticed the surge of health-conscious food choices making their way into the consumers' eyes. There is an increased demand for functional foods having positive effect beyond their basic nutritional value. Probiotic-rich drinks are nowadays popular beverages on the supermarket shelves. In the recent past, the beverage companies have been innovating around the probiotic products as fermented teas and drinks which have gained in popularity, like kefir; kombucha, one of the most popular beverages in the probiotic space, reported 25% annual market growth (Conik 2016). Probi otic niche products include ice cream, cheese, candy, and chewing gum, although they do not play a major role in the European marketplace (Saxelin 2008).

The FAO/WHO issued in 2001 a definition of **probiotics**, namely, that they are live microorganisms which when consumed in adequate amounts confer a health effect on the host. It is important to notice that this statement is not covering the use of probiotics in feed, as pharmaceuticals, as cosmetics, or as food additives. Naturally, we can find probiotics in fermented foods like yogurt, kefir, cider, vinegar, pickles, some cheeses, or fermented teas.

Compared with fermented foods, probiotics are relatively new products used for health maintenance which may not have a therapeutic claim.

Probiotics may increase the resistance of the gut to invasion by pathogens, prevent the growth of pathogenic bacteria, enhance epithelial barrier function, or ameliorate disease processes by inducing the secretion of soluble factors (Cremon et al. 2018). However, in people with a critical immune status, probiotics should only be used after careful consideration (SKLM 2010).

Probiotic foods are **manufactured** by adding the probiotic strains directly in the non-fermented food or simultaneously with the standard cultures in the case of fermented food, after careful selection and studies proving beneficial effect to consumers. Some of the popularly used probiotic microorganisms are lactic acid bacteria, like *Lactobacillus rhamnosus*, *Lactobacillus reuteri*, and certain strains of *Lactobacillus casei*, *Lactobacillus acidophilus* group, bifidobacteria, *Bacillus coagulans*, *Escherichia coli* strain Nissle 1917, certain enterococci, especially *Enterococcus faecium* SF68, and the yeast *Saccharomyces boulardii*. In an alternative process, fermentation takes place separately, and probiotic cultures (e.g., *Lactobacillus acidophilus* and *Bifidobacterium* species) are combined to form the final product (Saxelin 2008).

According to WHO/FAO guidelines, probiotic manufacturers should register their strains with an international depository (FAO 2006). SKLM's report from 2010 emphasizes the fact that only lactobacilli and bifidobacteria were present in probiotic foods in the European market at that time, and species like *Bifidobacterium*, *Enterococcus*, *Lactobacillus*, *Lactococcus*, and *Streptococcus* were listed especially in yogurt-like products. Still, the probiotic dietary supplements contain a wide variety of organism groups, e.g., *Bacillaceae* (e.g., *B. coagulans*, *B. subtilis*), enterococci (*E. faecium*), *Propionibacterium freudenreichii*, and yeast *(S. boulardii)*. On the market there are also available as probiotic foods other products, like breakfast cereals, muesli, ice cream, cheese, various beverages, and uncooked sausages.

Nowadays, "probiotics" are big **business** all over the world. According to Casseli et al. (2013), the global functional food market has been estimated at up to $50 billion annual share, while the world probiotic market was estimated at $15 billion. Since then, the demand for probiotics has heavily increased due to increase in functional foods' consumption. Another factor may be the ban on the usage of antibiotics in animal feed in several European and North American countries, as well as the increasing product innovations in the form of chocolates or biscuits. In this context, according to Allied Market Research (2016) report, probiotics market is expected to reach USD 57.4 billion by 2022 and to register an annual growth rate of 7.7% during the forecast period 2016–2022. The same source indicates that key players operating in this market include Danone, Chr. Hansen Holding A/S, BioGaia AB, Yakult Honsha Co. Ltd, Probi AB, Lifeway Foods, Inc, Nestle S.A., E. I. du Pont de Nemours and Company, Ganeden, Inc, and Protexin. A more recent report, from Global Market Insight (2018), evidences Europe, led by France, Spain, Italy, and Germany, its market size which exceeded USD 630 million in 2017; the region is characterized by a rise in demand for animal nutrition probiotic products.

Several companies have developed innovative products to enhance their product portfolio. Product launch is considered the key strategy adopted by the manufacturers, followed by collaborations and agreement, expansion, and acquisition. Based on end use, the market addresses human probiotics and animal probiotics; the first is the major revenue contributor in the overall probiotic market. According to Allied Market Research (2016) consumers' option leans toward probiotic dairy products such as yogurt, ice cream, and cheese; the probiotic yogurt is the most common probiotic product preferred by consumers. Any business strategy should take into account several factors: the market niche; access to distribution channels; and pricing, profit margins, and proprietary insulation in the marketplace. Hence, many companies underestimate the full impact of regulatory requirements on both business objectives and plans for product development (Hoffmann 2008).

In terms of probiotics, **regulation** varies between regions. Generally, probiotics are regulated as food supplements and regulation is focused on the legitimacy of any claims, rather than efficacy, safety, and quality. Ingredients, manufacturing processes, and conditions are important determinants of product characteristics and changes of these factors may give rise to a product not identical to the "original" in efficacy and safety if proper measures are not taken. The lack of stringent regulation of probiotic manufacturing means that the manufacturer (trademark owner) can commercialize any formulation under the same brand, even if significantly different from the original (De Simone 2019). In this respect, an important issue is to keep in the final product the ratio live/dead microorganisms; the whole technological process (harvesting, centrifugation, lyophilization, etc.) leads to an increase in dead microorganisms. The current regulations for labeling of probiotic products require that the consumer is informed about the number of live bacteria expressed as CFU per dose.

Talking about regulation in Europe, probiotics will be a major focus during the Probiotics Summit (7–9th May 2019, Geneva, Switzerland), developed in conjunction with the International Probiotics Association (IPA); the event, part of the Vitafoods Education Programme, will explore market and regulatory issues in the sector, as well as the latest developments in related R&D. IPA Europe is engaged to develop a European framework giving appropriate information to consumers on the presence of probiotics in food, helping to differentiate nutrition information from evidence-based health claims.

Generally, the **prebiotic** indicates dietary substances with the ability to modify host microbiota inducing benefit to the host; the most recent definition of a prebiotic, proposed by the International Scientific Association for Probiotics and Prebiotics, is "a substrate that is selectively utilized by host microorganisms conferring a health benefit" (Cremon et al. 2018). Generally, prebiotics are derived from different fiber-rich product that humans can't easily digest (Jerusalem artichoke, garlic, and leeks). These prebiotic fibers pass through the human/animal digestive system with no immediate benefits, while the "good" bacteria in the gut actually consume this fiber which can lead to better digestive health. The most characterized prebiotics are galactans, fructans (fructo-oligo-saccharides and inulin), oligo-fructose, galacto-oligosaccharides, lactulose, and breast milk oligosaccharides. These substances are used as food ingredients mostly consisting of non-starch polysaccharides and oligosaccharides and are able to stimulate lactobacilli, bifidobacteria, and other beneficial microorganisms.

Despite the huge success of the probiotic, prebiotics haven't seen as much publicity until recently. The lack of consumers' demand might be the fact that they haven't really been aware of prebiotics, but recent reports show that the 2020 market is estimated to hit 424 million USD by 2020 (Daniells 2016). The addition of chicory inulin on the prebiotic fibers list is estimated to impact positively the market.

In the last few years, there has been an increase in the market demand for food supplements containing both probiotics and prebiotics, especially in non-traditional beverages. The research and innovation teams are now looking for novel methods of inclusion to preserve the integrity of probiotics in such products.

The concept of **symbiotic** represents a combination of probiotic and prebiotics that affects the host beneficially by improving the survival and implantation of selected live microbial strains in gastrointestinal tract; their combination can cause the release of antibacterial substances such as bacteriocin, which can retard the growth of pathogenic bacteria (Khurana and Kanawjia 2007). Actually, synbiotics were developed to overcome possible survival difficulties for probiotics. Among the commonly used probiotic strains for synbiotic product formulations are *Bifidobacteria* spp., *Lactobacilli*, *Saccharomyces boulardii* and *Bacillus coagulans*, whereas the major prebiotics used include oligosaccharides such as fructo-oligosaccharide (FOS), galacto-oligosaccharides (GOS), xyloseoligosaccharides (XOS) or prebiotic from natural sources like inulin (Fazilah et al. 2018).

Nowadays, probiotic, prebiotic, and synbiotic products are marketed toward use in gastroenterology, immunology, gynecology, cardiology, urology, anti-aging, skin care, dietetics, and oral care. The broad applications of this limited group of organisms suggest that there is a need for more extensive clinical and epidemiological evaluation of these products and their efficacies. Generally, for probiotics to be marketed as pharmaceutical products, the burden of proof for efficacy will be much greater than for similar formulations marketed as functional food products or supplements. Most probably, in the coming years, the research and regulations may shift the market share of probiotics toward pharmaceutical companies, which have infrastructure and revenue models to accommodate clinical trials (Kearney and Gibbons 2017).

Tri-biotic products represent a new trend in dietary supplement area. At first glance, the term implies the presence of at least three antibiotics in the same product, but actually, from the manufacturer point of view in such products are mixed prebiotics, probiotics/synbiotics and postbiotics. An example of such product is the Kombucell line developed by Laboratoarele Medica S.A., Otopeni, Romania. They have developed an innovative pollen fermentation technology, based on the cultivation and use in the pollen fermentation process of yeasts and bacteria symbiotic colonies (SCOBY), commonly named kombucha, to use it to produce a new food supplement. The prebiotic effect is conferred by the bioactive carbohydrates and fibers (especially insoluble fibers) content, the probiotic effect comes from the live beneficial microorganisms from SCOBY consortium (bacteria and yeast) and the postbiotic effect that is determined by the probiotic microorganisms' metabolism products, like short chain fatty acids. In the conditioned product the level of probiotics reaches 10^9 CFU/mg (Utoiu et al. 2018). On economic side, after scaling-up the process from laboratory (2 l) to industrial level (100 l) the price for a 60 capsules tube arrived to 3 €; the product is already on the market and the estimated revenue is 50,000 €/year.

7.2.4 Microbial Food Cultures

Fermented food is known since centuries considered as a feasible solution in preserving different perishable products. The need for an inoculum it was obvious and this need was covered by keeping a sample from the previous production. The deep scientific knowledge related to the microorganisms used for food fermentation came later on, around 1858 when Louis Pasteur demonstrated that the yeast are responsible for the alcoholic fermentation in wine and beer making. Since then, thousands of scientists and practitioners have been focused on the isolation and selection of microbial food cultures, known as "starter cultures" to be used in food manufacture.

Starting with the nineteenth century the use of starter cultures became a norm for beer production, wine, vinegar and bakery yeast. Later on, after almost another century, the dairy and the meat industry started to use well characterized and defined starter cultures (Hansen 2002). **Yeasts, mouds, and bacteria** are microorganisms used to make food products such as beer, bread, wine, vinegar, yogurt, cheese, other dairy products, fermented meat and vegetables. The main food products obtained with the use of microbial starter cultures are presented in Table 7.1.; conducting well the fermentation will favor useful flora, to the detriment of undesirable flora and consequently preventing spoilage and promote taste and texture. Still, should be mentioned that in the last years has been noticed a trend in the production of "traditional" fermented food which keeps the autochthonous flavors and, in this case, the inoculation is not necessary as naturally occurring

Table 7.1 The use of starter cultures in fermented food products

Food product	Microorganisms used as starter culture
Bread	Yeast/lactic acid bacteria
Wine	Yeast/malolactic bacteria
Beer	Yeast
Dairy products	Lactic bacteria/propionic bacteria Bifidobacteria/yeast/molds
Pickled vegetables	Lactic bacteria
Fermented sausages/ salami	Lactic bacteria/yeast/molds
Soy sauce	Lactic bacteria/moulds (*Aspergillus*)

microorganisms in the raw materials could be a reliable source of the microbial flora, under proper conditions.

Basically, Microbial Food Cultures (MFC) are concentrates of one or more microbial species and/or strains including unavoidable media components carried over from the fermentation and components, which are necessary for their survival, storage, standardization and to facilitate their application in the food processing. These products may be classified as following: single-strain cultures, which contain one strain of a species; multi-strain cultures, which contain more than one strain of a single species; multi-strain mixed cultures, containing different strains from different species.

Relatively recently, practitioners have started to explore the so-called functional starter cultures. Compared to classical starter cultures, functional starter cultures offer an additional functionality and represent a way of improving and optimizing the food fermentation process and achieving tastier, safer, and healthier products (Leroy et al. 2006). These starters include microorganisms that generate health-promoting molecules, antimicrobials, including bacteriocins, aroma compounds or contribute to cured meat color for example; very important, they possess probiotic qualities, or lack negative properties such as the production of biogenic amines and toxic compounds.

The production of starter cultures follows a simple and basic technology conducted in fermenters under strict hygienic conditions. Usually, the original microbial strains are stored in a microbiology laboratory as fresh or conditioned cultures; from the stock small quantity of microorganism, representing the inoculation material, is prepared to start the production process. This inoculation material is transferred in different growth media placed in fermenters for liquid fermentation. Media are tailored to the specific requirements of the microbial species and typically contain carbohydrates, proteins, vitamins and minerals. The culture is allowed to multiply and grow under carefully defined and monitored conditions. After the microbial growth, the cultured cells are harvested, usually by centrifugation, and the biomass is conditioned in liquid, frozen or powder form. This form may undergo a final formulation, which involve blending of multiple cultures, prior to shipment to the food manufacturer. As a general rule, the microbial load of the starter culture should be higher than 10^8 CFU/ml or mg for each microbial specie. The manufacturer should know that is compulsory to verify the viable cells in the final matrix and to insert the information on the product label.

An important issue for the manufacturer is to maintain the genetic stability of the employed strains. In this sense here are some good practice to maintain this stability and to avid genetic drifting: the presence of an "in house" microbial bank to provide the original strain to each batch and to each production site; storage the "mother" culture below −80 °C; setting a clear internal system for comprehensive documentation and traceability for each strain; a long-term storage plan that minimizes the number of generation times should be implemented; DNA-fingerprinting of new batches and comparison to previous batches should be routine; basic phenotypic characteristics of new batches should be verified.

Case Study. Scaling Up a Freeze-Dried Kefir Culture Starter (Kourkoutas et al. 2007)

The authors have published an economic study for the industrial scale production of freeze-dried kefir starter culture based on results obtained on a laboratory scale. The industrial scale-up was based on a 3-step process using 3 bioreactors of 100, 3000, and 30,000 L for 300 kg of freeze-dried culture/d of plant capacity. The major cost component of the total investment was the freeze-drying machinery, which consisted of 57% of the total investment. Production cost was reduced from 15.4 €/kg to 2.9 €/kg when the production capacity was increased from 30 to 900 kg/d. An economic analysis revealed a 3.5-fold increase in production cost compared with the corresponding production cost of the wet product, with an added value of up to 10.8 € × 10^9.

Microbial food cultures used directly in food production are regarded as food ingredients in the EU like other food ingredients, and consequently must fulfill the requirements set out in the General Food Law, EU Regulation No. 178/2002 Article 14 ("Food shall not be placed on the market if it is unsafe and it is the food business operator's responsibility for ensuring food safety"). According to DFG Senate Commission on Food Safety from Germany, on the level of 2010 in the EU, it was no specific legal regulations regarding microbial food cultures. Since that report, no regulation has been approved on European level, except the Regulation (EU) 1169/2011 related to labeling. This regulation provides the information which shall be included in the labeling of all foodstuffs; the label should include the exact nature and characteristics of the product enabling the consumer to choose in full knowledge of the facts. However, there is a need for regulations and these regulations differ from country to country, all aimed at assuring the safe use of food cultures which has to be guaranteed by the food culture supplier (Laulund et al. 2017). Like the EU, the USA has no specific regulation for food culture. Some species are regarded as "safe and suitable" for human consumption while others have a status as Generally Recognized As Safe (GRAS) and are notified to the Food and Drug Administration (FDA) and published. In this context, the food starter suppliers have the obligation to reach the safety target and their main tools are the one used by other food ingredient manufacturers: HACCP and good manufacturing practices.

The **microbial starter culture market** is one of the most competitive and comprises of large number of players. However, the market is dominated by few players such as Chr. Hansen A/S, Danisco A/S, and Lasaffre group. Other key players in this market are C.S.K. Food enrichment B.V., D.S.M. Food Specialities, Biena, Lallemand Inc., Wyeast laboratories and others. As all over the world, the types of starter cultures market in Europe are yeast, bacteria, and mold, while their applications in the market are alcoholic beverages and non-alcoholic beverages. The main operational companies involved in the starter cultures market in Europe are presented in Table 7.2.

Table 7.2 Players on microbial culture starter production on European market

Company	Country	Food industry application
Chr. Hansen	Denmark	Cultures, enzymes and probiotics for the food industry, including wine and meat, and in particular for the dairy industry
CSK Food Enrichment	The Netherland	Bacterial cultures for dairy products
Danisco (Du Pont)	Denmark	Dairy cultures and probiotics
Lesaffre	France	Yeast starters for bakery
Bioprox	France	Freeze-dried lactic starter cultures for dairy products (yogurt, fermented milk, cheese); liquid lactic acid concentrate (NIZO method) of specific lactic bacteria for butter production
THT	Belgium	Probiotics for human and animal application Lactic starter and bioprotective cultures
Frutarom	Austria	Lactic bacteria, catalase-positive cocci, molds for meat industry
Alce	Italy	Lactic bacteria, yeast and molds for dairy industry
CSL (Centro Sperimentale del Latte)	Italy	Frozen and lyophilized lactic bacteria, yeast, and molds for dairy industry, bakery, meat industry
Lactina by Magnis International Trading	Bulgaria	Dairy cultures
BDF Ingredients	Spain	Dairy cultures including probiotics
Lallemand	Canada/ International	*Meat, dairy, bakery starters*

According to a report of Global View Research (GVR) from 2014, the increasing demand for alcoholic and non-alcoholic beverages is expected to positively influence the global starter culture market growth. While alcoholic beverages made with yeast, were dominating the market in the near past, it is expected that non-alcoholic beverages, made by the use of bacteria starters, will offer the highest growth rate in the next years (GVR 2014). Meanwhile, the fluctuant raw material prices are anticipated to restrain the market growth, while increasing functional drinks demand offer the requisite opportunities for the market. The same source makes the statement that Europe dominated the regional starter culture market based on high alcoholic beverage demand, while the forecast point on Asia Pacific and Latin America, anticipated to have the highest market growth owing to high alcoholic and non-alcoholic beverage demand in emerging economies such as India, China and Brazil. Regarding North America, the starter culture market is expected to be positively affected due to the high nutritional beverages demand.

In the latest market report published by Credence Research, Inc. (2016) the starter culture market was valued at USD 814.4 million by value and 1107.5 kilo ton by volume in 2015, and is expected to reach USD 1426.9 million by value and 1896.1 kilo ton by volume by 2024, expanding at an annual growth rate of 6.1% by value and 5.8% by volume from 2016 to 2024. The same report emphasizes the fact that lactic acid starter culture held largest market share in the very past years, being a key ingredient of various probiotics and beverages, as well as baked products; meanwhile, the mix strain starters is recognized currently as major revenue generating segment due to its consistency and predictability widely accepted in cheese manufacturing industries. Generally, non-alcoholic starter is considered as major revenue contributor because it plays important role in manufacturing of cheese, baked and dairy products. The expected key market movements are the following: change in consumers' dietary pattern, rising demand for fermented, baked and dairy based products, substantial increase in the demand for alcoholic beverages; a preventive and better healthcare, the increased use of probiotic drinks, as well as the discovery of novel probiotics (Credence Research report 2016).

7.3 Improving Food Nutrition Using Bioactive Compounds: Functional and Nutraceutical Food

7.3.1 Introduction

In the last decade, consumer demand was focused on healthier food products. This is the reason for why the food products developers looking for a variety of ingredients to develop new category of foods who deliver more than basic nutrition, in other words functional foods or functional and nutraceutical foods.

But what should understand, as product developer or consumer, through functional and nutraceutical foods? Into a large acceptance the functional foods products are defined as products that may provide benefits beyond basic nutrition and consist of a variety of food components/ingredients that potentially may reduce disease and promote health. The term "nutraceuticals" is often interchanged with "functional foods" even though there are subtle differences between the two. Nutraceuticals are "naturally derived bioactive compounds that are found in foods, dietary supplements and herbal products, and have health promoting, disease preventing, or medicinal products". "Nutraceuticals" appears to be the favored term used by industry which includes functional foods and supplements. As yet the term has no legal status (Ref: Nutraceuticals Institute). So, through functional foods products consuming, the consumers are looking for an added health benefit over and above the food product's traditional nutritional value (Khan et al. 2013).

7.3.2 Definition

The term "functional food" itself was first used in Japan, in the 1980s, for food products fortified with special constituents that possess advantageous physiological effects and the concept of functional food was first promoted in 1984 by Japanese scientists who studied the relationships between nutrition, sensory satisfaction, fortification and modulation of physiological systems (Siro et al. 2008; Bigliardi and Galati 2013; Vincentini et al. 2016). In 1991, have been approved a specific health-related food category called FOSHU (FOod for Specified Health Uses—often considered as first generation functional foods), having a specific health claims (Siro et al. 2008). In this way, a novel term-functional food-was defined and introduced, referring to prevention and/or curing effects of food beyond its nutritional value (Homayouni et al. 2012).

The study of many literature reviews on functional food highlighted that, although the term "functional food" has already been defined in numerous ways, so far there is no unitary accepted definition for this group of food. In most countries there is no legislative definition of the term and drawing a border line between conventional and functional foods is challenging even for nutritionists and food experts. Moreover, the European legislation does not consider functional food as specific food categories, but rather as a concept. To date, different authorized actors from food chains have proposed different definitions for functional food, from simply definitions which suggest that any food able to improve the health, is a functional food, to others, more complex or more specifically, maintain that only fortified, enriched, or enhanced food with a component having a health benefit beyond basic nutrition can be considered functional foods (Rincon-Leon 2003; Siro et al. 2008; Bigliardi and Galati 2013). After reviewing over 100 definitions, Bigliardi and Galati (2013) selected 39 on the basis of their representation of functional foods and the selection was classified in three conceptual categories:

1. The concept of health benefits: almost all definitions (35 out of 39) mention the health benefits that a food have to bring to its consumer in order to be labeled as functional food products;
2. The technological process for functional food products obtaining: some definitions (18 out of 39) outlined the fact that the food must have been fortified, enriched or had an ingredient added, while others mention the removal of allergens or of components considered detrimental to the health if over consumed (e.g., salt, sugar);
3. The nutritional function: all food to be functional must have some nutritional functions, as pointed out by 25 definitions out of the total.

Despite of these hard work, there is no official definition of functional foods common to all States, but the EU project "Functional Food Science in Europe" (FUFOSE) gives an appropriate working definition: "A food can be regarded as 'functional' if it is satisfactorily demonstrated to affect beneficially one or more target functions in the body, beyond adequate nutritional effects, in a way that is relevant to either an improved state of health and well-being and/or reduction of risk of disease". Functional foods are similar in appearance to conventional foods: must remain foods (they are not pills or capsules, tablet, or powder) and they must demonstrate their effects, in amounts that can normally be expected to be consumed in the normal diet, such as: enhancement of biological defense mechanisms, prevention of specific diseases, recovery from specific diseases, control of physical and mental disorders, maintenance of gut health, slowing of the aging process (Vincentini et al. 2016; Cencic and Chingwaru 2010).

7.3.3 Classification of Functional Foods

There are some functional foods able to improve healthy by intrinsic composition as (a) conventional food with high quantities of functional ingredients (e.g., β-glucaninoatbran), or could became functional products through adding some functional ingredients or remove some antinutritive compounds, as (b) foods modified through the enrichment with or the removal of functional ingredients (e.g., margarine that contains added phytosterol), and (c) food in which the nature of functional ingredients is changed. Recently, a wide range of functional foods were developed and many of them are being produced in all over the world including: (1) probiotic, prebiotic and synbiotic foods as well as (2) foods enriched with antioxidants, isoflavones, phytosterols, anthocyanins and (3) fat-reduced, sugar-reduced or salt-reduced foods (Homayouni et al. 2012).

The extant literature proposes different classification of functional foods depends of the considered criteria.

From a **product point of view**, literature proposed the following classification:

- food fortified with additional nutrients (fortified products), such as: fruit juices fortified with vitamin C, vitamin E, folic acid, zinc and calcium;
- food with additional new nutrients or components not normally found in a particular food (enriched products), like probiotics or prebiotics;
- food from which a deleterious component has been removed, reduced or replaced by another with beneficial effects (altered products), for example fibers as fat releasers in meat or ice cream;
- food in which one of the components have been naturally enhanced (enhanced commodities), e.g., eggs with increased omega-3 content.

According to alternative classification based on the **aim of functional foods**, they can be classified as follows:

- functional foods that add good to life or improve children's life, like prebiotics and probiotics;
- functional foods that reduce an existing health risk problem such as high cholesterol or high blood pressure;
- functional foods which makes life easier, such as lactose-free or gluten-free products;

(Bigliardi and Galati 2013; Polia et al. 2018)

According to the **potential medical benefits** and **properties** of their ingredients, they can be classified as follows: dietary fibers; oligosaccharides; amino acids, peptides, and proteins; glycosides; vitamins; cholines; lactic acid bacteria; polyunsatured fatty acids; phytochemicals and antioxidants (Rincon-Leon 2003).

7.3.4 Functional Foods Products Applications

Functional foods have been developed in different food categories as: dairy-, confectionery, soft-drinks, bakery and baby-food products (Bigliardi and Galati 2013).

7.3.4.1 Probiotic Foods

The current probiotics' definition is formulated in 2002 by FAO (Food and Agriculture Organization of the United Nations) and WHO (World Health Organization) working group experts, as live selected microorganisms which, when administered in adequate amounts confer a health benefit on the host (Markowiak and Slizewska 2017). Probiotics means "let good microbes work for you in different fields get their benefits and take a rest". Such "work" will include, food digestion, production of useful products to destroy the bad microbes, complement the functions of the missed digestive enzymes (due to missed or defective genes), and to maintain the digestive system's pH, and so on. Probiotics will improve the efficiency of consumers' biological fermentors, the digestive system (Amara and Shibl 2015). Probiotics were originally used to influence human health through intestinal microbiota alterations. At present, probiotics and their effects on human health have been demonstrated both within different food matrices and as single or mixed microbial culture preparations. Furthermore, the health-promoting properties of probiotics are now known to be strain dependent. An international expert group of the International Life Sciences Institute (ILSI) has evaluated the categorized and published evidence of functionality of different probiotics in four areas of (human) application, namely, (1) metabolism, (2) chronic intestinal inflammatory and functional disorders, (3) infections, and (4) allergy (Cencic and Chingwaru 2010).

The largest representative category of probiotics on the market is dairy products. Milk has an outstanding position in the development of functional foods because it has Omega-3, phytosterols, isoflavins, conjugated linoleic acid, minerals, and vitamins. Dairy products such as ice cream, cheese, yogurt, Acidophilus-Bifidus-milk, Ayran, Kefir, Kumis, Doogh containing probiotics and dairy beverages (both fermented and non-fermented) have long been considered as important vehicles for the delivery of probiotics. In milk fermentation process, acids such as lactic acid, acetic acid and citric acid are naturally produced. These acids are commonly used as organic acids to enhance sensorial qualities as well as safety of food products. Lactic acid bacteria are found to be more tolerant to acidity and organic acids than most of the pathogens and spoilage microorganisms. These types of probiotic products are already launched on the market and some of most commune are presented in Table 7.3.

Probiotic Ice Cream Produced by incorporation of probiotic bacteria in both of Fermented and unfermented mix. *Lactobacillus* and *Bifidobacterium* are the most common species of lactic acid bacteria used as probiotics for fermented dairy products. The researches have demonstrated that it is possible to select the appropriate probiotic strains for use in probiotic ice cream. *Lactobacillus casei* (Lc01) and *Bifidobacterium lactis* (Bb12) had the highest resistance to simulated acidic, alkaline and ice cream conditions in comparison with other

Table 7.3 Some commercial examples of probiotic products (Siro et al. 2008)

Brand/ Trade name	Description	Producer
Actimel	Probiotic drinking yogurt with *L. casei Imunitass*© cultures	Danone, France
Activia	Creamy yogurt containing *Bifidus ActiRegularis*©	Danone, France
Gefilus	A wide range of LGG products	Valio, Finnland
Hellus	Dairy products containing *Lactobacillus fermentum ME-3*	Tallinna Piimatoostuse AS, Estonia
Jovita Probiotisch	Blend of cereals, fruit and probiotic yogurt	H&J Bruggen, Germany
Pohadka	Yogurt milk with probiotic cultures	Valasske Mezirıcı Dairy, Czech Republic
ProViva	Refreshing natural fruit drink and yogurt in many different flavo–rs containing *Lactobacillus plantarum*	Skane mejerier, Sweden
Rela	Yogurts cultured milks and juices with *L. reuteri*	Ingman Foods, Finland
Revital	Active Yogurt and drink yogurt with probiotics	Olma, Czech Republic
Snack Fibra	Snacks and bars with natural fibers and extra minerals and vitamins	Celigueta, Spain
SOYosa	Range of products based on soy and oats and includes a refreshing drink and a probiotic yogurt-like soy–oat product	Bioferme, Finland
Yosa	Yogurt-like oat product flavored with natural fruits and berries containing probiotic bacteria (*Lactobacillus acidophilus*, *Bifidobacterium lactis*)	Bioferme, Finland
Soytreat	Kefir type product with six probiotics	Lifeway, USA
Yakult	Milk drink containing *Lactobacillus casei* Shirota	Yakult, Japan
Vitality	Yogurt with pre- and probiotics and omega-3	Muller, Germany
Vifit	Drink yogurts with LGG, vitamins and minerals	Campina, the Netherlands

probiotic strains, making them suitable probiotic strains for use in probiotic ice cream.

Probiotic Cheese Provides a valuable vehicle for probiotic delivery, due to creation of a buffer against the high acidic environment in the gastrointestinal tract, and thus creates a more favorable environment for probiotic survival throughout the gastric transit, ought to higher pH. Moreover, the dense matrix and relatively high total solids as well as fat content of cheese may offer additional protection to probiotic bacteria in stomach. The presence of the prebiotics inulin and oligofructose can promote growth rates of *Bifidobacterium* and *Lactobacillus*, besides increased lactate and short chain fatty acids production in petit-suisse cheese.

Probiotic Yogurt Yogurt has been historically recognized to be "a healthy food" with therapeutic effects. Recently has been evidenced a high increasing of the popularity of yogurt especially probiotic yogurt. The conventional yogurt starter bacteria, *L. bulgaricus* and *Streptococcus thermophilus*, do not have ability to survive passage through intestinal tract and consequently so, they are not considered as probiotics. But the addition of *L. acidophilus* and *B. bifidum* into yogurt can add extra nutritional and physiological values.

Probiotic Milk Technology of bifidus milk and acidophilus-bifidus milk manufacturing is similar to acidophilus milk. Milk is standardized to desire protein and fat levels in both products. Then, for manufacture of bifidus milk, milk is heat-treated at 80–120 °C for 5–30 min and rapidly cooled to 37 °C. Heat-treated milk is inoculated with frozen culture of *Bifidobacterium bifidum* and *Bifidobacterium longum* at a level of 10% and left to ferment until pH 4.5. After fermentation, the product is cooled to <10 °C and packaged. Final product has a slightly acidic flavor and the ratio of lactic acid to acetic acid is 2:3. It is also possible to produce probiotic milks by simply adding mix culture of *Lactobacillus acidophilus* and *Bifidobacterium bifidum* to cold pasteurized milk (Homayouni et al. 2012).

Prebiotics Prebiotics are non-digestible food ingredients (a selectively fermented ingredient, or a fiber) that beneficially affect the host by stimulating the growth and/or activity of one or a limited number of bacteria in the colon, thus improving host health. The most representative prebiotic ingredients are: fructo-oligosaccharide (FOS), inulin, isomalto-oligosaccharides (IMO), polydextrose, lactulose (constitutes a significant part of produced oligosaccharides, as much as 40%) and resistant starch.

Primarily oligosaccharides, such as soy oligosaccharides (SOS), galacto-oligosaccharides (GOS) and xylo-oligosaccharides (XOS) play an important role in obesity control through resulting increased satiety and reduced hunger. Fructans, such as inulin and oligofructose, are the most used and effective in relation to many species of probiotics.

Fruit, vegetables, cereals, and other edible plants as: tomatoes, artichokes, bananas, asparagus, berries, garlic, onions, chicory, green vegetables, legumes, as well as oats, linseed, barley, and wheat, are sources of carbohydrates constituting potential prebiotics (Siro et al. 2008; Cencic and Chingwaru 2010; Markowiak and Slizewska 2017).

The main effects of prebiotics consisting in enhance the growth and survival of the probiotic cultures by influencing the growth and metabolites of both the probiotic and the starter. The beneficial effects depend on the nature of probiotics and prebiotics and the food matrixes involved and are shortly presented in Table 7.4.

Synbiotic Synbiotics are food products containing a synergistic combination of pro and prebiotics (Cencic and Chingwaru 2010; Ranadheera et al. 2010). To describe a combination of synergistically action of probiotics and prebiotics, Gibson and Roberfroid introduced in 1995, the term "synbiotic". As the word "synbiotic" implies synergy, the term should be reserved for those products in which a prebiotic component selectively favors a probiotic microorganism, having a healthy effect on subject. The principal purpose of that type of combination is the improvement of survival of probiotic

Table 7.4 Beneficial effects of prebiotics on probiotic bacteria in foods (Ranadheera et al. 2010)

Food matrix	Prebiotic ingredients	Probiotic ingredients	Effect
Yogurt	Hi-maize/resistant starch	*L. acidophilus* *L. casei*	Growth and viability
	Inulin	*L. acidophilus* *L. casei* *L. rhamnosus* *L. reuteri* *Bifidobacterium*	Growth and viability
	Fructooligosaccharides	*L. acidophilus* *L.casei* *L. rhamnosus* *Bifidobacterium* *B. animalis* *B. longum*	Viability and fatty acid production
Fermented milk	Polydextrose Oligofructose	*L. acidophilus* *L. rhamnosus* *B. animalis* subsp. *lactis*	Growth, viability and fatty acid production
Ice cream	Inulin	*L. acidophilus* *B. lactis*	Viability
Cheese and cheese-based products	Oligofructose Inulin	*L. acidophilus* *B. animalis* subsp. *lactis*	Growth, viability, sensory and fatty acid production
	Carboxy methyl cellulose	*P. freudenreichii* subsp. *shermanii*	Growth

microorganisms in the gastrointestinal tract. Synbiotics have both probiotic and prebiotic properties and were created in order to overcome some possible difficulties in the survival of probiotics in the gastrointestinal tract. Therefore, an appropriate combination of both components in a single product should ensure a superior effect, compared to the activity of the probiotic or prebiotic alone. Synbiotics are used not only for the improved survival of beneficial microorganisms added to food or feed, but also for the stimulation of the proliferation of specific native bacterial strains present in the gastrointestinal tract. The effect of synbiotics on metabolic health remains unclear. It should be mentioned that the health effect of synbiotics is probably associated with the individual combination of a probiotic and prebiotic. Considering a huge number of possible combinations, the application of synbiotics for the modulation of intestinal microbiota in humans seems promising.

A combination of *Bifidobacterium* or *Lactobacillus* genus bacteria with fructooligosaccharides in synbiotic products seems to be the most popular, and some example of these cases are presented in Table 7.5.

Markowiak and Slizewska (2017), outlined the following beneficial effects of synbiotics on humans' health:

1. Increased *Lactobacillus* and *Bifidobacterium* genus count and maintenance of balance of the intestinal microbiota;
2. Improved hepatic function in patients suffering from cirrhosis;
3. Improved immunomodulative abilities;
4. Prevention of bacterial translocation and reduced incidence of nosocomial infections in patients' post-surgical procedures and similar interventions.

Functional Drinks Another important product category within the functional food segment is non-alcoholic beverages fortified with vitamins (A, C, E) or other functional ingredients. Although, there is a relatively high number of a product available in this segment, the market is

Table 7.5 Composition of some commonly synbiotics used in human nutrition (Markowiak and Slizewska 2017)

Prebiotics	Synbiotics
FOS	*Lactobacillus* genus bacteria + inulin
GOS	*Lactobacillus, Streptococcus* and
Inulin	*Bifidobacterium* genus bacteria + FOS
XOS	*Lactobacillus, Bifidobacterium, Enterococcus* genus bacteria + FOS
Lactitol	*Lactobacillus* and *Bifidobacterium* genus bacteria + oligofructose
Lactosucrose	*Lactobacillus* and *Bifidobac*terium genus bacteria + inulin
Lactulose	
Soy oligosaccharides	
TOS	

FOS fructooligosaccharides, *GOS* galactooligosaccharides, *TOS* transgalactooligosaccharides, *XOS* xylooligosaccharides

still small and fragmented in most European countries. Germany is the only country in Europe with a sizeable functional drink market, mainly due to the success of ACE drinks in this country. In 1999 these beverages reached a market volume of 89 million US$ up from sales of around 15 million US$ in 1996. In 2000 more than 117 million of vitaminized non-alcoholic beverages were consumed in Germany, which equals to around 1% of the total consumption of these beverages. Other types of functional drinks are those of cholesterol lowering drinks (with combination of omega-3 and soy), "eye health" drinks (with lutein) or "bone health" drinks (with calcium and inulin). In Estonia, for example fortified juices are produced under the trade name of Largo containing inulin, L-carnitine, vitamins, calcium and magnesium as functional ingredients. The European functional drink market was estimated to be around 7% of the total soft drink market in 2004, with a further increase to 8% in 2005. According to the predictions the consumption will reach 5.1 billion l by the year of 2009, which corresponds to 23% increase compared to that of 2005 (Siro et al. 2008).

Functional Cereals Cereals, in particular oat and barley, offer another alternative for the production of functional foods. The multiple beneficial effects of cereals can be exploited in different ways leading to the design of novel cereal foods or cereal ingredients that can target specific populations. Cereals can be used as fermentable substrates for the growth of probiotic microorganisms. Additionally, cereals can be applied as sources of non-digestible carbohydrates that besides promoting several beneficial physiological effects can also selectively stimulate the growth of lactobacilli and bifidobacteria present in the colon and act as prebiotics. Cereals contain water soluble fiber, such as beta-glucan and arabinoxylan, oligosaccharides, such as galacto- and fructo-oligosaccharides and resistant starch, which have been suggested to fulfill the prebiotic concept. Finally, cereal constituents, such as starch, can be used as encapsulation materials for probiotics in order to improve their stability during storage and enhance their viability during their passage through the adverse conditions of the gastrointestinal tract. Some functional cereal components such as beta-glucan, however, applied also in the dairy and bakery industries. Many researches have focused on the use of beta-glucans, in the manufacture of low-fat ice creams and yogurts. Incorporation of beta-glucans with other soluble dietary fiber, into low-fat dairy products can make their mouthfeel, scoopability and sensory properties resemble those of full-fat products (Siro et al. 2008).

Bakery Products While functional foods are rapidly gained in popularity in such sectors as dairy products or confectionery, in bakery it is still relatively underdeveloped. For example, in Germany about 20–21% of the new functional food products launched in 2001 were dairy and confectionary product and only about 13% were from the bakery industry. This difference was found to be even higher in Spain in 2006, where about 45% of the launched functional food products were dairy food compared to about 13% of the functional bakery product. Despite these, bakery products provide ideal matrix by which functionality can be delivered to the consumer in an acceptable food. In late 2003, Unilever introduces a white bread called Blue Band Goede Start, which was the first white bread containing the nutritional elements normally available in brown bread including fibers, vitamins B1, B3, and B6; iron; zinc; inulin, a starch that comes from wheat. In developing functional bakery products (including bread), it is important to realize that achieving functional food quality does not simply involve delivering the active principle at the appropriate level for physiological effectiveness, but also supplying a product which meets the consumer's requirements in terms of appearance, taste, and texture (Siro et al. 2008).

Spreads as Functional Food It can be assumed that cholesterol-lowering spreads will gain increasing relevance in the coming years due to the market introduction of, e.g., a functional variety of margarine of Unilever (named "Becel pro-activ"), containing phytostanol esters which are supposed to lower the cholesterol level of consumers. A product with similar characteristics named "Benecol©" has already been launched by the Finnish company Raisio in some Scandinavian countries in the mid-1990s. More recently the Benecol© brand was completed with spreads containing camelina oil as a source of omega-3 fatty acids. Low-cholesterol butter under the trade name of Balade™ has been produced and marketed in Belgium since 1992. In this case more than 90% of the cholesterol in milk fat has been removed by the addition of crystalline betacyclodextrin to the molten butter. Other low-cholesterol milk products, like cheese, cream, or even low-cholesterol egg, are produced by this technology (Siro et al. 2008).

Functional Meat Meat and its derivatives may also be considered functional foods due to intrinsic composition in the numerous functional compounds. The idea of using food for health purposes rather than for nutrition opens up a whole new field for the meat industry. In addition to traditional presentations, meat industry can explore various possibilities, including the control of the composition of raw and processed materials via reformulation of fatty acid profiles or inclusion of antioxidants, dietary fiber or probiotics, etc. (Siro et al. 2008).

Functional Eggs Eggs are of particular interest from a functionality point of view, because they are relatively rich in fatty acids and the associated fat-soluble compounds. The type and ratio of fatty acids is an important determinant of human health. The idea of eggs enrichment with omega-3 fatty acids simultaneously with antioxidants and other vitamins has recently been used to produce VITA Eggs by Freshlay Foods (Devon, UK). They state that their eggs were enriched with omega-3 fatty acids, Se, vitamins D, E, B12 and folic acid. Eggs enriched in omega-3 and vitamin E produced by Belovo under the trade name of Columbus first appeared in Belgium in 1997, and since then they have been sold in the UK (from 1998), The Netherlands (from 1999), India, Japan and South Africa (from 2000). Currently, production of Columbus egg exceeds 50 millions/year in Europe. Similar eggs are produced by Pilgrim's Pride Company, Gold Circle Farms and OmegaTech in the USA (Siro et al. 2008).

Farmed Fish as a Functional Food Being a rich source of important nutrients, including highly digestible proteins, vitamins (A, D3), trace minerals (iodine, selenium) and $n-3$ long chain polyunsaturated fatty acids ($n-3$ LCPUFA), fish consumption is generally regarded as part of a healthy dietary pattern. Exogenous feeding in aquaculture unlocks the possibility to tailor fish composition with healthy valuable nutrients (Ribeiro et al. 2019).

Mushrooms "As" and "In" Functional Foods Based on their content in bioactive compounds, some authors consider mushrooms as functional foods or a good source for the "mushroom nutraceuticals" which should not be confused with the general nutraceuticals, functional foods and/or pharmaceuticals. A mushroom nutraceutical is considered a refined or partially refined extract or dried biomass from either mycelium or the fruiting body of a mushroom which is consumed in the form of pharmaceutical formulations, capsules or tablets, as a dietary supplement and has potential therapeutic applications. The principal nutraceuticals found in mushrooms include: (1) lipids, especially unsaturated fatty acids; (2) vitamins, such as vitamin E and vitamin C; (3) proteins, peptides and amino acids, including lectins, leucine and valine; (4) carbohydrates, especially polysaccharides, such as lentinan. Due to the existence of a wide variety of bioactive compounds in this matrix, this has great potential to produce new nutraceutical formulations.

These bioactive properties include immunomodulating, antitumour, anti-hypercholesterolemia, antibacterial and antifungal, anti-inflammatory, antiviral, anti-diabetic, and cardiovascular beneficial effects.

Most of the nutraceutical products which already existed on the market are in fact different formulation containing different functional extracts from mushroom species rich in bioactive compounds as: *Ganoderma lucidum*, *Agaricus blazei*, *Cordyceps sinensis*, *Grifola frondosa*, *Lentinula edodes*, *Trametes versicolor*, *Pleurotus ostreatus*, *Armillerla mellea*, *Cordyceps militaris*, *Cordyceps sinensis*, *Ganoderma applanatum* (Reis et al. 2017).

Food Plants and Edible Flowers as Functional Foods Artichokes (*Cynara cardunculus* L.) may be classified as functional food having a broad range of applications within the food industry itself where it is primarily used as a vegetable. The edible flower of artichoke is known to have a health protective potential. In vivo and in vitro studies have demonstrated their hepatoprotective, anticarcinogenic and hypocholesterolemic functions (Gostin and Waisundara 2019).

Sorghum has immense potential as a food ingredient to benefit human health due to its versatility and composition. Its unique phytochemical profile, coupled with the relatively slow digesting nature of its endosperm, have been demonstrated to contribute to superior benefits against various chronic disease targets compared to other cereal grains. The emerging evidence indicates whole grain sorghum and its components can improve glucose metabolism and insulin sensitivity, enhance lipid metabolism, reduce low-grade chronic inflammation and promote satiety through slowed gastric emptying. The new discoveries on unique functional properties of sorghum components, like the proanthocyanidins and the 3-deoxyanthocyanins, will likely lead to isolation and use of such components as high value ingredients to enhance food quality and health attributes, as well as use as therapeutic agents. Application of biotechnology, e.g., through microbial structural modification, appears poised to further enhance the functionality and value of the sorghum components (Girard and Awika 2018).

Plantain flour, after several chemical modifications made with the aim of producing a suitable raw material for the development of functional gluten-free foods, offers better nutritional solution for celiac deficiencies. Some good results were obtained utilizing phosphated plantain flour which represents a promising alternative for the development of mold bread (sliced bread) and cookie with improved nutritional properties due to the cross-linking reaction produced by the sodium trimetaphosphate (Gutiérrez 2018).

Melon by-products (seeds and peels) represent a good source of minerals (potassium, sodium, magnesium, calcium) and phenolic compounds. It also contains carotenoids, namely, lycopene and β-carotene. In turn, melon seeds are a good source of protein (15–36%) and fiber (7–44%). Melon seed oil is a good source of tocopherols and sterols (β-sitosterol, 0.06–289 mg/100 g oil), as well as phenolic compounds. It has an interesting fatty acid profile, very similar to soybean and sunflower oils. Moreover, melon by-products present different biological activities including antioxidant, anti-inflammatory, antidiabetic, antiulcer, antibacterial, and anti angiogenic, fully justified by the presence of bioactive compounds. Therefore, these by-products can be considered good candidates for the development

Table 7.6 Target function of some functional foods

Functional foods	Active food component	Target function
Yogurts	*Probiotics:* Foods with beneficial live cultures as a result of fermentation or that have been added to improve intestinal microbial balance, such as *Lactobacillus* sp. *Bifidobacteria* sp. *Prebiotics:* A non-digestible component that has beneficial affects by stimulating the growth of bacteria in the colon. Examples include inulin and oligofructose.	Optimal intestinal function and intestinal microbial balance
Margarines	Added plant sterols and stanols esters	Decreased LDL-cholesterol (bad cholesterol) Decreased risk of coronary heart disease (CHD)
Omega-3 fatty acids enriched eggs	Omega-3 fatty acids	Control of hypertension, lipids metabolism

of novel functional foods, contributing to promote sustainability across food chain (Silva et al. 2018).

According to EUFIC Report (2006), most of the well-known functional food products are consumed for the potential positive benefit on health, as are presented in Table 7.6.

7.3.4.2 Development of Functional Food Products. New Trends on the Market

Innovation is today's business demand and development of a new functional food is an expensive process and is very important for both food companies and consumers. Regulations should encourage food companies to follow functional food development. Development of new functional food products is very challenging and it has to complete the consumer's expectations for quality (safety, nutritional, sensorial) and healthy products. Key points regarding for a successful functional food product development are consumer demands, technological conditions, and legislative regulatory background. There are several ways to produce a functional food product:

- to eliminating an allergenic protein, lactose, phenylanine and etc. from the natural food product;
- by fortification with a micronutrient;
- by adding antioxidants, probiotics and/or prebiotics;
- by replacing a antinutritive component;
- by increasing bioavailability or stability of a component known to produce a functional effect or to reduce the disease-risk potential of the food

Field of functional probiotic foods requires the cooperation of food technologists, nutritionists, medical doctors, and food chemists in order to obtain innovative products. In this way, these foods may be able to adjust physiological parameters related to health status or disease prevention in human. So, the design and development of functional probiotic foods is a scientific work which is an expensive and multistage process that takes into account many factors, such as sensory acceptance, physical and microbial stability, price, and chemical and other intrinsic functional properties to be successful in the marketplace. Moreover, consumer attitude toward the functional probiotic product also needs to be understood and taken into consideration (Homayouni et al. 2012). The global functional foods market size was USD 129.39 billion in 2015. Growing consciousness among consumers regarding their health and proper diet is expected to aid the overall industry over the next 8 years. Foods are not only intended to satisfy one's hunger but also to eliminate nutrition related diseases. Such a factor is anticipated to affect the global industry demand positively. The production of enhanced levels of functional compounds can be through processes such as microencapsulation, enzyme technology, and nanoencapsulation of the ingredients. As perspective, according to the report on functional food production forecast, the global functional foods market revenue, in Europe will increase by 50% between 2018 and 2024. Key products include: carotenoids, dietary fibers, fatty acids, minerals, prebiotics & probiotics, vitamins, minerals and others. Others segment commonly include phytochemicals, enzymes, and antioxidants. Dietary fibers product segment is expected to grow at a CAGR (Compound Annual Growth Rate) of 8.4% over the next eight years. Fiber intake has exhibited chemo protective effects for cancer proliferation applications. They also help in combating cardiovascular diseases, obesity and diabetes. These factors are expected to cater the overall demand over the forecast period. Applications commonly include bakery and cereals, dairy products, meat, fish and eggs, soy products, fats and oils and other applications. Others segment include beverages, nutritional bars, and various snack types. In 2015, global functional foods market revenue share by application was in decreasing order: Dairy products, Bakery and Cereal products, Meat, Fish and Eggs, Fats and Oils, Soy products, others. Fermented dairy products such as yogurts that contain probiotics in it are considered to benefit the immune system, strengthen the mucosal barrier and suppress intestinal infection of human health. Margarine, which consists of fatty acid esters, is designed to lower the cholesterol level for people. On the other hand, eggs containing omega-3 which are produced by chickens are fed as a microalgal feed ingredient. Such factors are expected to drive the total industry growth in future years (Functional Foods Market Analysis, 2018 to 2024).

The global nutraceutical market should reach $285.0 billion by 2021 from $198.7 billion in 2016 at a compound annual growth rate (CAGR) of 7.5%, from 2016 to 2021. The functional beverages market should reach $105.5 billion by 2021 from $71.5 billion in 2016 at a CAGR of 8.1%, from 2016 to 2021. The functional food market should reach $92.3 billion by 2021 from $64.6 billion in 2016 at a CAGR of 7.4%, from 2016 to 2021 (Global Nutraceuticals Industry Report: 2017–2021).

The functional food ingredients market was valued at USD 64.75 billion in 2017; this is projected to reach USD 94.21 billion by 2023, at a CAGR of 6.6% during the forecast period. The increase in consumption of nutritive convenience food and fortified food and the growth in health awareness among consumers leading to increased consumption of healthier diets are the factors driving this market (Functional Food Ingredients Market, Global Forecast to 2023).

7.3.4.3 Conclusions

The development of functional foods is increasing, as their market increases day by day, and the consumer's information about these foods is increasing without relation to gender, age, and educational or economic levels of the consumers. The therapeutically effect of a functional food may depend on the consumer's characteristics and the type of carrier and enrichment considered. For instance, yogurt is most preferred by its enrichment with calcium and fiber. Ingredients such as vitamins and minerals applied in fortification of functional foods are widely recognized and accepted by consumers, but new functional ingredients such as some probiotics and prebiotics are not common to them. So, there is a need for increasing the consumer knowledge with respect to these new special ingredients. Functional probiotic food industry should communicate with consumer in a clear way and this is one of the most important aspects for success of this sector.

The future success of functional foods products in marketplace depends on consumer acceptance of such kind of products. The consumers must be convinced by its health claims through clear, honest, and definite messages to agree to pay the cost associated with functional foods products (Homayouni et al. 2012). Functional foods need to be promoted with the aim of making them much more visible and recognizable to final consumers, in order to avoid confusion with other generic health foods, such as light or diet products (Annunziata and Vecchio 2011).

The main consumer motivation for purchasing functional foods is the growing desire to use foods either to help prevent chronic illnesses such as cardiovascular disease, Alzheimer's disease and osteoporosis, or to optimize health, for example by increasing energy, boosting the immune system, generation of wellbeing (Khan et al. 2013). In this regard, functional foods play an outstanding role, as demonstrated by their increasing demand derived from the increasing cost of healthcare, the steady increase of life expectancy, and the desire of older people for improved quality of their later years. (Bigliardi and Galati 2013). These needs conduct to rapidly growing and highly dynamic market of functional foods (Khan et al. 2013; Vincentini et al. 2016).

7.4 Bio-Preservation of Food

7.4.1 Bacteriocins

Bacteriocins are ribosomal-synthesized peptides with antimicrobial properties, produced by different groups of bacteria (Gálvez et al. 2007; Des Field et al. 2018). Bacteriocins are generally known as natural compounds that can influence food quality and safety (Settanni and Corsetti 2008). They are toxic to the producing bacteria as well, but using a suite of immunity proteins, they can protect themselves.

The sensitivity of microorganisms to bacteriocins is due to their interaction with bacterial cell surface and cell membrane. Cell permeabilization and pore formation represents a major mechanism by which bacteriocins attack the target bacteria. However, reports of resistance development by several food-spoilage or pathogenic bacteria against bacteriocins like nisin and pediocin have implied that increased resistance may compromise the potential role of these antimicrobial peptides in biopreservation (Kumariya et al. 2019).

7.4.2 Classification

Class I contains bacteriocins termed lantibiotics (from lanthionine-containing antibiotic). They typically comprise 19–50 amino acids (Kumariya et al. 2019), being small (<5 kDa) peptides containing the unusual post-translational modified amino acids such as lanthionine (Lan), α-methyl-lanthionine (MeLn), dehydroalnine, and dehydrobutyrine (Hoover and Chen 2005; Woraprayote et al. 2016). Class I is further subdivided into class Ia (lantibiotics), class Ib (labyrinthopeptins) and class Ic (sanctibiotics).

Class II bacteriocins are the largest group that have been investigated and characterized. They contain small (<10 kDa) heat-stable, nonlanthionine-containing, unmodified peptides (Hoover and Chen 2005). They can be further subdivided into class IIa (pediocin-like bacteriocins), class IIb (two-peptides unmodified bacteriocins), class IIc (circular bacteriocins) and class IId (unmodified, linear, non-pediocin-like bacteriocins) (Kumariya et al. 2019).

Class III bacteriocins are large (>30kDa) heat-labile proteins that are not as well characterized. Class III bacteriocins include helveticin M, helveticin J and enterolysin A produced by *Lactobacillus crispatus*, *Lactobacillus helveticus* and *E. faecalis*, respectively. Helveticin M has been recently characterized to be effective against both Gram-positive and Gram-negative bacteria (Kumariya et al. 2019).

7.4.3 Bacteriocins and Food Systems

Foods can be supplemented with ex situ produced bacteriocin preparations, or by inoculation with the bacteriocin-producing strain under conditions that favor production of the bacteriocin in situ (Gálvez et al. 2007). The most studied bacteriocins in food biopreservation are lactic acid bacteria (LAB) ones, due to the fact that most of bacteriocin-producing LAB are isolated from food origins and are considered to be GRAS (Generally Recognized As Safe) (Woraprayote et al. 2016). These include nisin, a lantibiotic peptide produced by *Lactococcus lactis* and pediocin PA-1 produced by *Pediococcus acidilactici* (Des Field et al. 2018).

Nisin is a protein consisting of 34 amino acids which is stable when autoclaving and efficiently inhibits the development of some important food pathogens such as *L. monocytogenes*, *Staphylococcus aureus* (Popa et al. 2010). Nisin inhibits the bacterial cells by its unique capability of creating pores on the bacterial cell wall (Juturu and Wu 2018). Nisin is a bacteriocin naturally derived from *Lactococcus lactis* subsp. lactis. It has been used in the food industry as a food preservative since 1950 (Khan and Oh 2016). Nisin is harmless, heat-stable, and commercially available for use in various food products. This bacteriocin is usually protected by food ingredients and does not lose its antimicrobial activity after pasteurization, sterilization, and other processing methods. Also, this naturally occurring polypeptide has been widely combined with a range of polymer matrices to produce antimicrobial materials for food preservation purposes (Santos et al. 2018a). Many studies have proved nisin being an efficient antimicrobial agent for food biopreservation such as cheese, meat or vegetables.

Pediocin PA-1 is an antimicrobial peptide produced by *Pediococcus* spp. that has been sufficiently well characterized to be used in food industry as a biopreservative (Diez et al. 2012). Pediocin belongs to class IIa bacteriocin and it's composed of 44 amino acids, containing both cationic and hydrophobic regions and is non-toxic and heat-stable (Verma et al. 2017). *Pediococcus acidilactici* strains have been largely used as fermentation culture starters for a wide range of food products, which include vegetable (e.g., sauerkraut), meat (e.g., sausages), and dairy products (e.g., cheese) and, in fact, pediocin PA-1 producing *P. acidilactici* have been used for many years as starter culture in a variety of food fermentations. Pediocin PA-1, either alone or in combination with other preservation technologies, extends the hurdles to the growth of spoilage bacteria in a number of fermented foods, and offers the advantage of being active at acid pH and acting synergistically with other compounds, such as lactate or organic acids (Diez et al. 2012). Reported effects of nisin or pediocin PA-1 utilization are presented in Table 7.7.

7.4.4 Bacteriophages

Bacteriophages (or phages) are the most abundant biological entities on earth (Choinska-Pulit et al. 2015; Harada et al. 2018), being relatively easy to isolate and propagate, and are highly evolved to inactivate specific bacterial strains, species, or sometimes even genera. Their antimicrobial and therapeutic potential was recognized immediately after their discovery, until the introduction of antibiotics displaced the application of phages shortly after World War II (Kilcher and Loessner 2018). Phages have spent billions of years evolving and developing a powerful protein armamentarium to recognize, infect, and kill bacteria in a very efficient way (Santos et al. 2018b).

Bacteriophages (or phages) are viruses that infect and replicate within bacterial cells while not invading other cells (Janczuk-Richter et al. 2019). According to the International Committee on Taxonomy of Viruses (ICTV), bacteriophages are classified within 13 families, regarding to their morphology and nucleic acid composition. Over 95% of phages taxa represent *Caudovirales* order and includes a vast majority of phages related to foodborne pathogens. Among them, 60% phages are *Siphoviridae* family with noncontractile, long and flexible tails, 25% are *Myoviridae* phages with contractile tails, and 15% are *Podoviridae* with noncontractile, short tails (Choinska-Pulit et al. 2015, Harada et al. 2018).

Phage persistence is an important parameter that can be used as preservative or as a food additive to control growth of foodborne pathogens in food and beverage industries. Bacteriophages are already used in agricultural, food safety and diagnostic applications. The application of bacteriophages represents a promising, safe, chemical-free and environmentally friendly alternative to the use of chemicals or sanitizers on different products (Oliveira et al. 2014). Some characteristics of a given food product could affect the ability of the bacteriophages to remain viable. For example, the structure and

Table 7.7 Nisin or pediocin PA-1 application in food biopreservation

Form of application	Reported effect	Food product	References
Chitosan-coated nisin-silica liposomes	Antibacterial activity against *L. monocytogenes* without altering the sensory properties of samples	Cheddar cheese	Cui et al. (2016)
Starch/halloysite/nisin nanofilms	Significant decrease in the population of *L. monocytogenes* after 1 day comparing to cheeses packed in control films	Minas Frescal Cheese	Meira et al. (2016)
Polyethylene oxide nanofibers containing nisin-loaded poly-g-glutamic acid/chitosan (NGC) nanoparticles	NGC nanoparticles at 5 mg/mL of nisin presented antibacterial activity against *L. monocytogenes*, without altering sensory quality	Cheese	Cui et al. (2017)
Films containing nisin	Significantly decreased the number of *L. monocytogenes* in cheese samples compared to the control group	UF cheese	Divsalar et al. (2018)
Mild heating (63 °C/5 min) in conjunction with nisin (1000 and 1500 IU ml^{-1}) treatment	Total elimination of *L. innocua*	Brined white cheese	Al-Holy et al. (2012)
Encapsulated nisin	Inhibit *L. monocytogenes* growth for 10 days when applied at 400 IU/g and for 24 days when applied at 800 IU/g, while free nisin controlled microbial growth for 4 and 17 days, respectively	Vacuum-sealed, refrigerated beef	Zimet et al. (2018)
Sodium alginate active coating solutions incorporated with nisin	Enhanced shelf life of chicken meat as compared to the control	Chicken meat fillets	Raeisi et al. (2016)
Osmotic solution containing nisin	Osmotic pre-treatment significantly improved quality stability during refrigerated storage, in terms of microbial spoilage, chemical modifications and sensory quality decay	Tuna fish	Sofra et al. (2018)
MAP in combination with Nisin–EDTA treatments	Extension of shelf life by approximately 1 (nisin—500 IU/g; no EDTA added) to 14 days (500 IU/g nisin—50 mM EDTA)	Fresh chicken meat	Economou et al. (2009)
Chitosan-nisin (CS-nisin) microcapsules	Higher efficiency in inhibiting microorganism growth, lipid oxidation and protein degradation; extending the shelf life during storage by 6–9 days, compared to CS alone, nisin alone or the control.	Small yellow croaker (*Pseudosciaena polyactis*).	Wu et al. (2017)
Nisin-based antimicrobial and cold plasma combination treatments	Cold plasma treatment at 40s, followed immediately with nisin-based antimicrobial treatments at 180s and 3600 s led to 2.5 and 4.6 log CFU/g inactivation of *L. monocytogenes*	Granny Smith apples	Ukuku et al. (2019)
Nisin	– Nisin remained stable in fruit juices for at least 30 days and favored the preservation of vitamin C content – The most sensitive microorganism to nisin was *A. acidoterrestris* and the least sensitive was *L. monocytogenes* with a reduction of	Fruit juices	Oliveira Junior et al. (2015)

(continued)

Table 7.7 (continued)

Form of application	Reported effect	Food product	References
	up to 90% of viable cells in peach and mango juices		
Nisin	Reduced and delayed growth of *L. monocytogenes*, without impact on the sensory appearance for 2–5 days	Fresh-cut Iceberg lettuce	McManamon et al. (2019)
Combination of nisin and ε-polylysine with chitosan coating	– Significant inhibition of respiration rate, decline of ascorbic acid and growth of microorganism (yeast and mold, total viable counts, total coliforms count, *Staphylococcus aureus* and *Pseudomonas* spp.) – Increased total phenol content compared with the control after 9-day storage – Strongly effective in inhibiting the white blush of fresh-cut carrots	Fresh-cut carrots	Song et al. (2017)
Electro-activated solutions (EAS) of potassium acetate and potassium Citrate, the EAS were combined with a bacteriocin nisin	– Inactivation efficacy of 5.9–6.1 log CFU/mL against *C. sporogenes* PA 3679. – When testing whole green beans, spore inactivation level was significantly higher reaching 6.5 log CFU/mL.	Green beans puree and whole green beans	El Jaam et al. (2017)
hydrostatic pressure and thermal pasteurization combined with nisin	– Total inactivation of yeast and molds – Nisin with HHP or thermal pasteurization presented a synergistic effect on the inactivation of total aerobic bacteria	Cucumber juice drinks	Zhao et al. (2013)
PA-1 containing fermented cheese whey	– Significant decrease in total viable count, *Staphylococcus aureus* and lactic acid bacterial count, but not in coliform count and yeast-mold count reduction	Raw buffalo milk	Verma et al. (2017)
Poly(lactic acid)/sawdust particle biocomposite film incorporated with pediocin	Inhibition of *Listeria monocytogenes* (99% of total listerial population) during the chilled storage	Raw sliced pork suggests	Woraprayote et al. (2013)

composition of a food product can affect bacteriophages activity by allowing the resistance of the food to the penetration of the bacteriophages into the food matrix, the bacteriophages remaining on the surface of the food, where they may become desiccated (Thung et al. 2017).

Bacteriophages have been used to effectively reduce the viable number of bacterial cells within meat products. The key to the procedure is to ensure the viability of the bacteriophages when applied post-slaughter and post-processing. The bacteriophages have to remain viable during storage in order to reduce the bacterial load. However, once this fact has been achieved, the delivery of viable bacteriophages to the human consumer is not really required (Cooper 2016). A *Salmonella Enteritidis* lytic bacteriophage (SE07) was isolated from retail meat samples in a study performed by Thung et al. (2017). The effectiveness of this phage was determined in different food matrices (beef and chicken meat, fruit juice, fresh eggs) experimentally contaminated with *S. Enteritidis*. The results showed a significant reduction of this bacteria population (about 2 log cycles) in fruit juice and fresh eggs, and also a reduction of bacterial population by 2.1 and 2.0 log cycles on the bacteriophage treated beef and chicken meat samples after incubation at refrigeration temperature for 48 h (Thung et al. 2017). A novel bacteriophage (LPST10) with a high lytic ability and broad host range for *Salmonella* strains was tested on different food products by Huang et al. (2018), demonstrating an inhibitory effect by reducing *Salmonella* counts (0.92–5.12 log CFU/sample) in milk, sausage and lettuce. In 2013, Spricigo et al. (2013) reported that a lytic bacteriophage cocktail composed of UAB_Phi 20, UAB_Phi78 and UAB_Phi87 was able to significantly reduce bacterial load when applied on fresh eggs, pig skin, chicken breasts, and packaged lettuce experimentally contaminated with *Salmonella enterica* serovar *Typhimurium* and *S. enterica* serovar *Enteritidis*. When applied to pig skin and incubated at 33 °C for 6 h, a significant bacterial reduction of >4 and 2 log/cm^2 was observed. Significant decreases in bacterial count were also observed in chicken breasts refrigerated at 4 °C for 7 days, of 2.2 and 0.9 log cfu/g, as well as in lettuce incubated for 60 min at room temperature, respectively 3.9 and 2.2 log cfu/g. However, regarding fresh eggs, only a minor reduction of the bacterial concentration (0.9 log10 cfu/cm^2) was observed (Spricigo et al. 2013). Zinno et al. (2014) demonstrated that a 10^4 UFC/g phage P22 inoculum was effective in host inactivation when applied on different food products, all experimentally contaminated with its host (liquid eggs, whole and skimmed milk, energy drinks, apple juice, chicken mince and chicken breast). Furthermore, wild food strains belonging to the serotypes *Typhimurium*, *Enteritidis*, *Derby Give*, *Newport*, *Muenchen* and *Muenster* were assayed toward phage P22, and only isolates of serotypes *Typhimurium*, *Enteritidis* and *Derby* were inhibited by the presence of this phage. Challenge tests using experimentally contaminated liquid-eggs, chicken breast and chicken mince with mixes of wild *Salmonella Typhimurium* (about 10^4 UFC/g) strains showed a log reduction of 2–3 log cycles after 48 h at 4 °C depending on both mix of strains and the specific food (Zinno et al. 2014). Jun et al. (2018) tested a phage (fHe-Yen9-01) treatment after bacterial inoculation (with *Y. enterocolitica* O:9 strain *Ruokola/* 71) of food samples, including raw pork (4 °C, 72 h), ready-to-eat pork (26 °C, 12 h), and milk (4 °C, 72 h). The results showed decreasing counts by 1–3 logs from the original levels of 2–4 × 10^3 CFU/g or ml. Furthermore, the authors experimentally contaminated kitchen tools and treated them with phages for 2 h. The results showed that the bacterial growth was succesfully inhibited, with bacterial load decreasing by 1–2 logs from the initial levels of about 10^4 CFU/cm^2 or ml (Jun et al. 2018). The effectiveness of phage Listex P100 was investigated by Oliveira et al. (2014) in order to control *L. monocytogenes* growth on different fruits and fruit products (melon, pear and juices and slices of apple) stored at 10 °C. Phage treatment was more effective on melon followed by pear, but no effect on apple products was observed. It was found a reduction of approximate 1.50 and 1.00 log CFU plug^{-1} for melon and pear slices. In apple juice, *L. monocytogenes* was unaltered by phage treatment, the phage decreasing to nearly undetectable counts (Oliveira et al. 2014). Another approach of using bacteriophages as biocontrol agents in food products is represented by the incorporation of viable bacteriophages into absorbent pads which can be placed in direct contact with the food products, permitting the diffusion of phages in the food matrix, keeping it fresh for a longer time (Cooper 2016). Gouvea et al. (2016) studied the effect of an absorbent pad used in refrigerated meat trays, containing a blend of six bacteriophages (BFSE16, BFSE18, PaDTA1, PaDTA9, PaDTA10 and PaDTA1), showing that this blend was able to present antimicrobial activity and could be further used as a biocontrol alternative food packaging (Gouvea et al. 2016). Cellulose acetate films incorporated with a mix of bacteriophages showed antimicrobial activity against *Salmonella Typhimurium* ATCC 14028 according to a study performed by Gouvea et al. (2015). An increase in the lag phase and slower development of microorganisms in the bacteriophages containing medium with the films was observed, compared to control. Bacteriophages were viable for 14 days of evaluation, afterwards they were no longer detected in the film (Gouvea et al. 2015).

7.4.5 Natural Preservatives from Plants

Spices, which are mostly used to amplify the flavor and taste of foods, have been used for a long time due to their preservative and medicinal properties. Several studies about the antimicrobial, antifungal, antioxidant or insecticidal properties of their essential oils have been performed in order to determine their applications as natural food preservatives. Reducing or eliminating food-related microorganisms without affecting food quality can result in an extension of food shelf life and become an appealing option to fight against food borne diseases (Santamarina et al. 2016). Essential oils (EOs) are used in a wide variety of applications in food, pharmaceutical and cosmetics industries due to their flavoring, antioxidant and antimicrobial properties. In particular, the antimicrobial action of essential oils has been attributed to their phenolic compounds and their interaction with microbial cell membranes. They are known to penetrate through the microbial membrane and cause the leakage of ions and cytoplasmatic content therefore leading to cellular breakdown (Salvia-Trujillo et al. 2015).

Antimicrobial extracts from plants have been known as promising alternatives to replace synthetic preservatives in food. Plants are rich in a variety of antimicrobial compounds such as tannins, alkaloids, glycoalkaloids, saponins, alkenyl phenols, flavonoids, sesquiterpenes, terpenoids and lactones (Gyawali et al. 2015a; Pisoschi et al. 2018). Herbs, spices, and plant extracts are considered as being GRAS (Gyawali et al. 2015a; Asensio et al. 2015) products. EOs are produced by more than 17,000 aromatic plant species mostly belonging to families like *Lamiaceae*, *Rutaceae*, *Zingiberaceae*, *Myrtaceae* and *Asteraceae*. They are synthesized and stored in complex secretary structures, glandular trichomes, secretory cavities and resin ducts (Prakash et al. 2015). Clove, oregano, cinnamon and rosemary are some of the most used spices and herbs with great antimicrobial activity. Essential oils obtained from these spices contain several active compounds such as carvacrol, cinnamaldehyde, eugenol, and camphor, compounds that gives them antimicrobial properties (Gyawali et al. 2015b). These compounds are obtained from plants by steam distillation or by extraction with supercritical carbon dioxide (Pisoschi et al. 2018).

Plant polyphenols could be an alternative to EOs. Polyphenols represent secondary metabolites produced by plants, which have potential health properties on human organism, mostly as antioxidants, antiallergic, anti-inflammatory, anticancer, antihypertensive, and antimicrobial activities (Nowak et al. 2016; Harich et al. 2018). Specifically, some phenolic compounds, such as resveratrol, hydroxytyrosol, quercetin, and some phenolic acids, have been reported to have an inhibiting effect on different pathogenic microorganisms. Also, the flavonoids are known to be synthesized by plants in response to microbial infection. Research on polyphenols has shown that they could have a positive impact on lipid oxidation, color stability, and antioxidant activity in meat products (Nowak et al. 2016).

Application in Food

Although the antibacterial properties of EOs have been widely studied, the interest to use them as natural preservatives is a growing trend, in accordance with consumers' increasing health awareness (Espina et al. 2013). Food systems with the minimum number of challenges for antimicrobial application are represented by carbohydrate-based beverages, fresh products, dairy products, and meat, poultry, and seafood products.

A carbohydrate-based beverage is homogeneous and acidic and is composed mostly from water and sugar, having very low content or no proteins or fat. Therefore, antimicrobial agents can be distributed evenly through the product without a substantial amount of antimicrobial loss from the interaction with proteins or fats. Also, the low pH of these products often increases antimicrobial activity. However, the antimicrobial activity could be reduced due to the presence of simple and/or complex carbohydrates in beverages (Davidson et al. 2015). Essential oils of sage, juniper, lemon, and marjoram were shown to have anti-yeast activities against *Geotrichum candidum*, *Pichia anomala*, *S. cerevisiae*, and *Schizosaccharomyces pombe* in apple juice by extending the lag phases of microorganisms (Tserennadmid et al. 2011).

Fruits and vegetables are similar due to their low level of fats and proteins and the presence of contamination at the surface. However, fruits are typically acidic, which is an advantage when applying natural antimicrobials, whereas vegetables have higher pH values. Therefore, the effectiveness of natural antimicrobials in vegetables is often reduced. Natural antimicrobials would most likely be applied to fresh fruits and vegetables in the form of a rinse or spray. Many antimicrobial-incorporated rinse/spray/wash techniques, vapor treatments, and edible coatings have been studied and developed for treatments of fruits and vegetables after harvest in order to improve food safety and spoilage control (Davidson et al. 2015). Thyme, oregano, and lemongrass EOs were found effective against total mesophilic microorganisms. They were tested in order to prolong the shelf life and freshness of cabbage, combined with MAP, and the results demonstrated to be promising for obtaining these properties (Hyun et al. 2015). Oregano and rosemary EOs were combined in order to control *Escherichia coli*, *Listeria monocytogenes* and *Salmonella Enteritidis* in leafy vegetables by de Medeiros et al. (2016). The essential oils combined at sub-inhibitory concentrations were effective in decreasing the counts of these pathogenic bacteria in vegetable broth and in fresh leafy vegetables. Therefore, they represent a good alternative to assure the safety and extend the shelf life of fresh leafy vegetables (de Medeiros et al. 2016). The incorporation of carvacrol nanoemulsions into modified chitosan enabled the development of a bioactive coating to be deposited on green beans, which resulted active against the two tested pathogens, *E. coli* and *S. Typhimurium*, during storage (Severino et al. 2015). The addition of clove essential oil can provide an alternative to chemical preservatives for controlling the fungi in stored rice grains, thus extending their shelf life (Santamarina et al. 2016).

Cheese is a food that is consumed all over the world, being obtained from curded milk by the removal of whey and by curd ripening in the presence of specific microflora. Processed cheese may normally be considered as a stable food product with a satisfactory shelf life (Asensio et al. 2015). There are many challenges for antimicrobial use in dairy products. Their high pH and water activity provide a favorable environment for microorganism growth. The addition of oregano and rosemary essential oils improved the oxidative and fermentative stability of flavored cheese prepared with cream cheese base, prevented lipid oxidation and the development of rancid and fermented flavors, in this way, these essential oils prolonged the shelf life of this product (Olmedo et al. 2013). Organic cottage cheese flavored with oregano EO showed a lower degree of chemical deterioration during storage. Furthermore, this food product with the addition of Cordobes and Criollo EOs preserves longer polyunsaturated fatty acids against oxidation reactions, being an alternative to reduce the production of organic acids during storage, which could be associated with a reduction in microbial activity (Asensio et al. 2015).

Plants are an endless supply of valuable bioactive substances and thus different plant products are being evaluated as natural antioxidants to preserve and improve the overall quality of meat and meat products (Shah et al. 2014). Meat, poultry, and seafood products are likely the most challenging food systems for antimicrobial application due to their properties, such as nonhomogeneous structure, high pH, and high protein and fat content. These factors are unfavorable for most antimicrobials, especially natural compounds, but favorable to the microorganisms (Davidson et al. 2015). Cai et al. (2015) demonstrated that plant essential oil treatment not only maintained fish fillets sensory quality during storage, but also presented the capability to decrease microbial counts and biogenic amines content. Essential oils, such as spearmint oil, have positive effects on antimicrobial properties, delaying protein degradation and nucleotide breakdown, and maintaining hardness, proving to be a potential preservative of the quality and safety of fish and fish products, while extending its shelf life (Cai et al. 2015). It has been found that the addition of nutmeg essential oil influenced the slower lipid oxidation and growth of total aerobic mesophilic bacteria, as well as the improvement of aroma of cooked sausages during long storage period (60 days), prolonging in this way, their shelf life (Sojic et al. 2015). The use of EOs in the production of a dry cured sausage also proved to be an interesting strategy to assure safety against *Salmonella* spp., *L. monocytogenes* and *S. aureus*, but with sensory limitations, that does not allow its use in high concentrations (García-Díez et al. 2016). A possible use of rosemary, thyme and oregano essential oils as food preservatives at a concentration of 0.5% (v/w) or lower could be preferred, since at this concentration they do not modify sensorial characteristics of food, having a bacteriostatic and bactericidal effect against high and low pathogen concentration. These EOs could be used by food industries during the preparation of beef meatballs and other meat ready-to-cook products, having a shelf life of at least 14 days during storage at 4 °C (Pesavento et al. 2015). Salgado et al. (2013) applied clove essential oil to formulations based on sunflower protein concentrates in order to develop edible and biodegradable films, and then they applied the resulted films on refrigerated sardine patties. The sunflower protein films with clove EO allowed to delay their lipidic auto-oxidation and to slightly retard the growth of total mesophilic count (Salgado et al. 2013). The incorporation of cinnamon essential oil into polymeric active package films has been used for extending the shelf life of fish and meat, since they allow the release of antioxidant and antimicrobial compounds (Hu et al. 2015). The addition of cinnamon EO into chitosan nanoparticles and their application on pork meat have been reported to exhibit excellent antioxidant and antimicrobial properties on pork during 15 days of refrigerated storage (Hu et al. 2015).

7.5 Bio-Valorization of Food Waste

Food loss and food waste are often used in scientific literature to identify materials intended for human consumption that are subsequently discharged, lost, degraded or contaminated. Discharge of food material occurs along the entire food supply chain and it involves all sectors of waste management from collection to disposal (Girotto et al. 2015). The development of sustainable solutions for food waste management represents one of the main challenges for our society. These solutions should be capable of exploiting the precious resources represented by food waste to achieve social, economic and environmental benefits. The recovery of food processing wastes as renewable energy sources represents a sustainable option for the substitution of fossil energy, contributing to the transition of food sector toward a low carbon economy (Zhang et al. 2016). Agro-industrial residues and household food waste no longer suitable for human consumption can be used as feedstocks for the production of bioplastics and biofuels together with the extraction of high-value components (Girotto et al. 2015). Biomass wastes, which include solid waste from agricultural residues (rice straw, wet birch pulp), agro-industrial wastes (mushroom waste, cotton cellulose) and liquid waste of food and related industrial wastewater are abundant feedstock for renewable biohydrogen, biomethane and biochemicals productions etc. This technology of waste to energy and biochemicals includes the pretreatment of biomass, subsequently converted to sugars (hydrolysate). Sugars are thereafter transformed into biofuels such as hydrogen, methane, ethanol, and the biomaterial building blocks such as volatile fatty acids: Lactic acid, Acetic acid, Propionic acid, and Butyric acid etc. (Liu and Wu 2016).

7.5.1 Biofuel and Bioenergy Production

Food waste is characterized by a variable chemical composition depending on its origin of production. Food waste may be composed of a mixture of carbohydrates, lipids and proteins, or, if generated from specific agro-industrial sectors, it may be rich in one of these constituents. Different biofuels are therefore produced from food waste using bioprocesses or thermo-chemical processes, depending on their chemical composition (Girotto et al. 2015).

In current applications of anaerobic digestion (AD) systems, organic matter can be converted into a mixture of gaseous compounds, mainly methane (CH_4) and carbon dioxide (CO_2), via acid fermentation and volatile fatty acids degradation, and through the activity acid-forming and methane-forming bacterial biomass. De Gioannis et al. (2017) compared one- and two- stage anaerobic digestion of food waste used for the recovery of methane and hydrogen and methane. The results suggest that a two-stage process where the first reactor is properly operated in order to achieve a significant net hydrogen production, may display a 20% comparatively higher energy recovery yield as a result, mainly, of enhanced methane production as well as of the associated hydrogen production (De Gioannis et al. 2017). In a study performed by Pérez-Camacho et al. (2018), life cycle assessment was used to evaluate life cycle environmental impacts of substituting traditional anaerobic digestion feedstocks with food wastes. The results showed that this substitution conducted to a reduction of greenhouse gas emissions and also can manage the leakage of nutrients to water resources and eliminate eutrophication impacts which occur, typically as the result of field application (Pérez-Camacho et al. 2018). The utilization of organic wastes for biofuels production is considered to be a plausible approach for achieving better energy security, pollution control, process economics, sustainable production, and societal improvements by Stephen and Periyasamy (2018). In their study, after reviewing the scientific literature, they showed the superiority of biodiesel over other liquid biofuels through a comparative assessment of relevant factors. Some of the main constraints for commercial deployment of biodiesel production using organic wastes include higher production costs and higher energy consumption. Furthermore, their study recommends a novel concept for enhancement of biodiesel production from waste cooking oil using coalescer reactor along with a preliminary scrutiny for justifying further research potential of the approach (Stephen and Periyasamy 2018). Yeo et al. (2019) studied the conversion of food waste into an energy resource using naturally occurring fermentative microorganisms embedded in wooden biochips, using a "Smart Food Waste Recycling Bin" (S-FRB) system. The obtained results indicated that the organic content of food waste traded in the system increased from 53% to 72% in the final end-product and achieved a mass reduction rate of about 80%. The heating value of the end-product (3300 kcal/kg waste) confirmed its high potential as a biofuel (Yeo et al. 2019). Rago et al. (2018) studied the feasibility of generating high quality biochar from food waste torrefaction. The fuel properties of torrefied food waste approached those of coal while their high energy yields confirmed their use as potential coal substitutes in thermal conversion systems. Gutierrez et al. (2018) investigated the economic feasibility of a biomethane (produced from food waste) plant using two scenarios: co-digestion of food waste and sewage sludge (1); and co-digestion of food waste and pig slurry (2), both using anaerobic high density polyurethane digesters. The results showed that the economic performance based on net present value (NPV) gave a positive outcome for scenario 1 with 33% of the revenue coming from gate fees, and a negative NPV for scenario 2 (Gutierrez et al. 2018). The clean properties and combustion behavior of hydrochar from food waste were investigated by Wang et al. (2018a). The results showed that the pollutants emissions (HCl, SO_2 and NO) were reduced for hydrochar combustion, but hydrochar from temperature above 220 °C resulted in increased amount of NO, most probably due to the nitro ratio and chemical forms. In another study, Wang et al. (2018b) conducted a co-hydrothermal carbonization of food waste-wood biomass in order to enhance the palletization and hydrochar-fuel properties, resulting in a promising technology for palletization toward solid biofuel production. Citrus waste (CW) contains various polymers of soluble and insoluble carbohydrates that are the ideal raw material for conversion into biological biofuels such as ethanol and biogas. Taghizadeh-Alisaraei et al. (2017) demonstrated that citrus waste represents a good alternative for biofuel production in Iran, the estimated production of ethanol and biogas being of 26.98 million liters and 37.08 million m^3, respectively.

Another approach of food waste bio-valorization in represented by biofertilizers. Du et al. (2018) presented in their study the fact that the conversion of food waste, especially agriculture residues into biofertilizer would reduce its environmental impact, improve nutrition levels of the soil, decrease requirements for synthetic chemical fertilizers and have a direct benefit on food production, confirming that the technology for the conversion of food waste into biofertilizers is viable, but the production efficiency could be improved with better process control strategies, strict quality controls, development of a smart product distribution system and adoption of advanced technologies (Du et al. 2018).

7.5.2 Biomaterials Production

The challenge of finite fossil resources has been addressed by academic and industrial researchers with the development of valuable compounds and polymers based on renewable resources (Girotto et al. 2015). The use of waste materials and by-products in building materials by the incorporation of photocatalytic materials was studied by Saeli et al. (2018). Their results showed the potential of this naturally-derived photocatalyst to be applied in the construction industry, leading to lower atmospheric pollution and environmentally sustainable materials for the preservation of cultural heritage. Blueberry waste from juice processing was valorized to develop starch films through compression molding by Luchese et al. (2018). The developed films showed a homogenous surface, although some pores appeared in the cross-section for the films with the highest blueberry waste content. Results highlighted the use of thermo-mechanical processes such as compression to manufacture sustainable films with enhanced properties through waste valorization by the techniques that are actually employed at industrial scale. Nistico et al. (2017) developed films composed of poly (vinyl alcohol-co-ethylene) and post-harvest tomato plant powder by single-screw extrusion. The films presented good properties and could be competitive for cost, performance and sustainability, depending on the intended application.

7.5.3 Food Ingredients Recovery

Fruit and vegetable processing industries account for the largest segment of food waste which is produced worldwide, thus generating a huge amount of waste. Banerjee et al. (2018) performed a review study about the valorization schemes to extract valuable biomolecules from pineapple on-farm and processing waste for food and therapeutics applications. They concluded that pineapple waste contains a high amount of carbohydrate (55%), being a great substrate for the production of valuable chemicals such as xylitol, lactic acid or succinic acid, which have potential application in the food industry. Also, bromelain enzyme has potential both in therapeutic and food industry (Banerjee et al. 2018). Vong and Liu (2016) determined the composition of okara, having as scope its potential for bio-valorization to obtain bioactive substances for food products. The health benefits and nutritional quality of okara are often enhanced by fermentation, and the fermented okara is also an inexpensive substrate for the extraction of bioactive substances, that could be used in food industry. Another industry that produces a high amount of waste is the winemaking industry Due to the high content of hidroxycinamic acid derivatives, flavonols, tannins, catechins and anthocyanins, winery by-products are postulated as a good sources of natural preservatives whose antibacterial and antioxidant properties can be customize to satisfy the requirements of the diverse food industries (Poveda et al. 2018).

Another approach for food waste use is as animal feed, mostly vegetable wastes. It was found that vegetable by-products are nutritionally and sanitarily appropriate to be used in animal feed. San Martin et al. (2016) tested various drying technologies such as pulse combustion drying, oven and microwave, suitable for vegetable food waste. The obtained meal prototypes were found to comply with all the requirements of the animal feed market.

References

Al-Holy MA, Al-Nabulsi A, Osaili TM et al (2012) Inactivation of *Listeria innocua* in brined white cheese by a combination of nisin and heat. Food Control 23:48–53

Allied Market Research (2016) Prebiotics market

Amara AA, Shibl A (2015) Role of probiotics in health improvement, infection control and disease treatment and management. Saudi Pharm J 23:107–114

Annunziata A, Vecchio R (2011) Functional foods development in the European market: a consumer perspective. J Funct Foods 3:223–228

Asensio CM, Grosso NR, Juliani HR (2015) Quality preservation of organic cottage cheese using oregano essential oils. LWT Food Sci Technol 60:664–671

Banerjee S, Ranganathan V, Patti A et al (2018) Valorisation of pineapple wastes for food and therapeutic applications. Trends Food Sci Technol 82:60–70

Bigliardi B, Galati F (2013) Innovation trends in the food industry: the case of functional foods. Trends Food Sci Technol 31:118–129

Cai L, Cao A, Li Y et al (2015) The effects of essential oil treatment on the biogenic amines inhibition and quality preservation of red drum (*Sciaenops ocellatus*) fillets. Food Control 56:1–8

Caselli M, Cassol F, Calo G et al (2013) Actual concept of "probiotics": Is it more functional to science or business? World J Gastroenterol 19(10):1527–1540

Cencic A, Chingwaru W (2010) The role of functional foods, nutraceuticals, and food supplements in intestinal health. Nutrients 2:611–625. https://doi.org/10.3390/nu2060611

Chen L, Madl RL, Vadlani PV et al (2013) Value – added products from soybean: removal of anti-nutritional factors via bioprocessing, Chapter 8. In: El-Shemy H (ed) Soybean – bio-active compounds. Intech, Croatia, pp 161–180

Choinska-Pulit A, Mituła P, Sliwka P et al (2015) Bacteriophage encapsulation: trends and potential applications. Trends Food Sci Technol 45:212–221

Conik H (2016) Kombucha market. Beverage daily. 18th Feb

Cooper GM (2000) The cell: a molecular approach, 2nd edn. Sinauer Associates, The Central Role of Enzymes as Biological Catalysts, Sunderland, MA

Cooper IR (2016) A review of current methods using bacteriophages in live animals, food and animal products intended for human consumption. J Microbiol Methods 130:38–47

Credence Research, Inc. (2016) Global starter culture market- growth, future prospects and competitive analysis, 2016–2024

Cremon C, Barbaro MR, Ventura M et al (2018) Pre-and probiotic overview. Curr Opin Pharmacol 43:87–92

Cui HY, Wu J, Li CZ et al (2016) Anti-listeria effects of chitosan-coated nisin-silica liposome on Cheddar cheese. J Dairy Sci 99:8598–8606

Cui H, Wu J, Li C et al (2017) Improving anti-listeria activity of cheese packaging via nanofiber containing nisin-loaded nanoparticles. LWT Food Sci Technol 81:233–242

Daniells S (2016) US prebiotics market 'growing strongly' as consumer awareness increases. Nutraingredients-USA. 16th Nov

Davidson PM, Cekmer HB, Monu EA et al (2015) 1 – The use of natural antimicrobials in food: an overview. Handbook of natural antimicrobials for food safety and quality, pp 1–27

De Gioannis G, Muntoni A, Polettini A et al (2017) Energy recovery from one- and two-stage anaerobic digestion of food waste. Waste Manag 68:595–602

de Medeiros Barbosa I, da Costa Medeiros JA, Rima de Oliveira KA et al (2016) Efficacy of the combined application of oregano and rosemary essential oils for the control of *Escherichia coli*, *Listeria monocytogenes* and *Salmonella Enteritidis* in leafy vegetables. Food Control 59:468–477

De Simone C (2019) The unregulated probiotic market. Clin Gastroenterol Hepatol 17(5):809–817. https://doi.org/10.1016/j.cgh.2018.01.018

Delween S (1996) Archaeology of ancient Egyptian beer. J Am Soc Brew Chem 54(1):3–12

Des Field R, Ross P, Hill C (2018) Developing bacteriocins of lactic acid bacteria into next generation biopreservatives. Curr Opin Food Sci 20:1–6

Di Cagno R, Coda R, De Angelis M et al (2013) Exploitation of vegetables and fruits through lactic acid fermentation. Food Microbiol 33(1):1–10

Diez L, Rojo-Bezares B, Zarazaga M et al (2012) Antimicrobial activity of pediocin PA-1 against *Oenococcus oeni* and other wine bacteria. Food Microbiol 31:167–172

Dirar HA (1994) The indigenous fermented foods of the Sudan. Wallingford, Oxon, CAB International, 552 p

Divsalar E, Tajik H, Moradi M et al (2018) Characterization of cellulosic paper coated with chitosan-zinc oxide nanocomposite containing nisin and its application in packaging of UF cheese. Int J Biol Macromol 109:1311–1318

Du C, Abdullah JJ, Greetham D et al (2018) Valorization of food waste into biofertiliser and its field application. J Clean Prod 187:273–284

Economou T, Pournis N, Ntzimani A et al (2009) Nisin–EDTA treatments and modified atmosphere packaging to increase fresh chicken meat shelf-life. Food Chem 114:1470–1476

El Jaam O, Fliss I, Aïder M (2017) Effect of electro-activated aqueous solutions, nisin and moderate heat treatment on the inactivation of *Clostridium sporogenes* PA 3679 spores in green beans puree and whole green beans. Anaerobe 47:173–182

Espina L, García-Gonzalo D, Laglaoui A et al (2013) Synergistic combinations of high hydrostatic pressure and essential oils or their constituents and their use in preservation of fruit juices. Int J Food Microbiol 161:23–30

FAO/WHO (2006) Probiotics in food. FAO Food and Nutrition paper 85

Fazilah NF, Ariff AB, Khayat ME et al (2018) Influence of probiotics, prebiotics, synbiotics and bioactive phytochemicals on the formulation of functional yogurt. J Funct Foods 48:387–399

Felder D, Burns D, Chang D (2012) Defining microbial terroir: the use of native fungi for the study of traditional fermentative processes. Int J Gastron Food Sci 1 (1):64–69. https://doi.org/10.1016/J.IJGFS.2011.11.003

Functional Food Ingredients Market by Type (Probiotics, Prebiotics, Proteins & Amino Acids, Phytochemical & Plant Extracts, Omega-3 Fatty Acids, Carotenoids, Fibers & Specialty Carbohydrates), Application, Source, and Region – Global Forecast to 2023

Functional Foods Market Analysis By Product (Carotenoids, Dietary Fibers, Fatty Acids, Minerals, Prebiotics & Probiotics, Vitamins), By Application, By End-Use (Sports Nutrition, Weight Management, Immunity, Digestive Health) and Segment Forecasts, 2018 to 2024

Gálvez A, Abriouel H, López RL et al (2007) Bacteriocin-based strategies for food biopreservation. Int J Food Microbiol 120:51–70

García-Díez J, Alheiro J, Pinto AL et al (2016) Behaviour of food-borne pathogens on dry cured sausage manufactured with herbs and spices essential oils and their sensorial acceptability. Food Control 59:262–270

Girard AL, Awika JM (2018) Sorghum polyphenols and other bioactive components as functional and health promoting food ingredients. J Cereal Sci 84:112–124

Girotto F, Alibardi L, Cossu R (2015) Food waste generation and industrial uses: a review. Waste Manag 45:32–41

Global Market Insight (2018) Probiotics market 2018–2024

Global Nutraceuticals Industry Report: 2017–2021 – Analysis of the Multi-billion Functional Food and Beverage Industries – Research and Markets, Dublin, Mar. 13, 2017/PRNewswire/. http://www.researchandmarkets.com

Gostin AI, Waisundara VY (2019) Edible flowers as functional food: a review on artichoke (*Cynara cardunculus* L.). Trends Food Sci Technol. https://doi.org/10.1016/j.tifs.2019.02.015

Gouvea DM, Mendonça RCS, Soto ML et al (2015) Acetate cellulose film with bacteriophages for potential antimicrobial use in food packaging. LWT Food Sci Technol 63:85–91

Gouvea DM, Mendonça RCS, Lopez MES et al (2016) Absorbent food pads containing bacteriophages for potential antimicrobial use in refrigerated food products. LWT Food Sci Technol 67:159–166

Gupta S, Abu-Ghannam N (2012) Probiotic fermentation of plant based products: possibilities and opportunities. Crit Rev Food Sci Nutr 52(2):183–199

Gutiérrez TJ (2018) Plantain flours as potential raw materials for the development of gluten-free functional foods. Carbohydr Polym 202:265–279

Gutierrez EC, Wall DM, O'Shea R et al (2018) An economic and carbon analysis of biomethane production from food waste to be used as a transport fuel in Mexico. J Clean Prod 196:852–862

Guzel-Seydim ZB, Kok-Tas T, Greene AK et al (2011) Review: functional properties of kefir. Crit Rev Food Sci Nutr 51(3):261–268

GVR (2014) Starter culture market analysis, market size, application analysis, regional outlook, competitive strategies and forecasts, 2014 to 2020

Gyawali R, Hayek SA, Ibrahim SA (2015a) 2 – Plant extracts as antimicrobials in food products: types. In: Handbook of natural antimicrobials for food safety and quality, pp 31–47

Gyawali R, Hayek SA, Ibrahim SA (2015b) Plant extracts as antimicrobials in food products: mechanisms of action, extraction methods, and applications. In: Handbook of natural antimicrobials for food safety and quality, pp 49–68

Hansen EB (2002) Commercial bacterial starter cultures for fermented foods of the future. Int J Food Microbiol 78:119–131

Harada LK, Silva EC, Campos WF et al (2018) Biotechnological applications of bacteriophages: state of the art. Microbiol Res 212–213:38–58

Harich M, Maherani B, Salmieri S et al (2018) Evaluation of antibacterial activity of two natural bio-preservatives formulations on freshness and sensory quality of ready to eat (RTE) foods. Food Control 85:29–41

Hermann T (2003) Industrial production of amino acids by coryneform bacteria. J Biotechnol 104:155–172

Hoffmann FA (2008) Business considerations in the development of probiotics. Clin Infect Dis 46(2):S 141–S 143

Homayouni A, Alizadeh M, Alikhah H et al (2012) Functional dairy probiotic food development: trends, concepts, and products. In: Rigobelo EC (ed) Probiotics. IntechOpen. https://doi.org/10.5772/48797

Hoover DG, Chen H (2005) Bacteriocins with potential for use in foods, Chapter 14. In: Antimicrobials in food. CRC Press Taylor & Francis Group, pp 429–453

Hu J, Wang X, Xiao Z et al (2015) Effect of chitosan nanoparticles loaded with cinnamon essential oil on the quality of chilled pork. LWT Food Sci Technol 63:519–526

Huang C, Shi S, Ma W et al (2018) Isolation, characterization, and application of a novel specific Salmonella bacteriophage in different food matrices. Food Res Int 111:631–641

Hyun JE, Bae YM, Yoon JH et al (2015) Preservative effectiveness of essential oils in vapor phase combined with modified atmosphere packaging against spoilage bacteria on fresh cabbage. Food Control 51:307–313

Janczuk-Richter M, Marinović I, Niedziółka-Jönsson J et al (2019) Recent applications of bacteriophage-based electrodes: a mini-review. Electrochem Commun 99:11–15

Jun JW, Park SC, Wicklund A et al (2018) Bacteriophages reduce *Yersinia enterocolitica* contamination of food and kitchenware. Int J Food Microbiol 271:33–47

Jung JY, Yun HS, Lee JW et al (2011) Production of 1,2-propanediol from glycerol in *Saccharomyces cerevisiae*. J Microbiol Biotechnol 21(8):846–853

Juturu V, Wu JC (2018) Microbial production of bacteriocins: latest research development and applications. Biotechnol Adv 36:2187–2200

Kearney SM, Gibbons SM (2017) Designing synbiotics for improved human health. In: Microbial Biotechnology. Wiley and Society for Applied Microbiology. https://doi.org/10.1111/1751-7915.12885

Khan I, Oh DH (2016) Integration of nisin into nanoparticles for application in foods. Innov Food Sci Emerg Technol 34:376–384

Khan RS, Grigor J, Winger R et al (2013) Functional food product development–opportunities and challenges for food manufacturers. Trends Food Sci Technol 30 (1):27–37

Khurana HK, Kanawjia SK (2007) Recent trends in development of fermented milks. Curr Nutr Food Sci 3:91–108

Kilcher S, Loessner MJ (2018) Engineering bacteriophages as versatile biologics. Trends Microbiol. https://doi.org/10.1016/j.tim.2018.09.006

Kim EK, An SY, Lee MS (2011) Fermented Kimchi reduces body weight and improves metabolic parameters in overweight and obese patients. Nutr Res 31(6):436–443

Kourkoutas Y, Sipsas V, Papavasiliou G et al (2007) An economic evaluation of freeze-dried kefir starter culture production using whey. J Dairy Sci 90:2175–2180

Kumariya R, Garsa AK, Rajput YS et al (2019) Bacteriocins: classification, synthesis, mechanism of action and resistance development in food spoilage causing bacteria. Microb Pathog 128:171–177

Laulund S, Wind A, Derkx PMF et al (2017) Regulatory and safety requirements for food cultures. MDPI Microorg 5(28):1–14

Leroy F, Verluyten J, De Vuyst L (2006) Functional meat starter cultures for improved sausage fermentation. Int J Food Microbiol 106(3):270–285

Liu C, Wu S (2016) From biomass waste to biofuels and biomaterial building blocks. Renew Energy 96:1056–1062

Luchese CL, Uranga J, Spada JC et al (2018) Valorisation of blueberry waste and use of compression to manufacture sustainable starch films with enhanced properties. Int J Biol Macromol 115:955–960

Marco ML, Heeney D, Binda S et al (2017) Health benefits of fermented foods: microbiota and beyond. Curr Opin Biotechnol 44:94–102

Markowiak P, Slizewska K (2017) Effects of probiotics, prebiotics, and synbiotics on human health. Nutrients 9 (9):1021. https://doi.org/10.3390/nu9091021

McManamon O, Kaupper T, Scollard J et al (2019) Nisin application delays growth of *Listeria monocytogenes* on fresh-cut iceberg lettuce in modified atmosphere packaging, while the bacterial community structure changes within one week of storage. Postharvest Biol Technol 147:185–195

Meira SMM, Zehetmeyer G, Scheibel JM et al (2016) Starch-halloysite nanocomposites containing nisin: characterization and inhibition of *Listeria monocytogenes* in soft cheese. LWT Food Sci Technol 68:226–234

Nistico R, Evon P, Labonne L et al (2017) Post-harvest tomato plants and urban food wastes for manufacturing plastic films. J Clean Prod 167:68–74

Nowak A, Czyzowska A, Efenberger M et al (2016) Polyphenolic extracts of cherry (*Prunus cerasus* L.) and blackcurrant (*Ribes nigrum* L.) leaves as natural preservatives in meat products. Food Microbiol 59:142–149

Oliveira Junior AA, de Araújo Couto HG, Teixeira Barbosa AA et al (2015) Stability, antimicrobial activity, and effect of nisin on the physico-chemical properties of fruit juices. Int J Food Microbiol 211:38–43

Oliveira M, Viñas I, Colàs P et al (2014) Effectiveness of a bacteriophage in reducing *Listeria monocytogenes* on fresh-cut fruits and fruit juices. Food Microbiol 38:137–142

Olmedo RH, Nepote V, Grosso NR (2013) Preservation of sensory and chemical properties in flavoured cheese prepared with cream cheese base using oregano and rosemary essential oils. LWT Food SciTechnol 53:409–417

Paulova L, Patakova P, Branyik T (2013) Advanced fermentation processes. In: Teixeira J, Vincente AA (eds) Engineering aspects of food biotechnology. Taylor & Francis Group, pp 89–110

Pérez-Camacho MN, Curry R, Cromie T (2018) Life cycle environmental impacts of substituting food wastes for traditional anaerobic digestion feedstocks. Waste Manag 73:140–155

Pesavento G, Calonico C, Bilia AR et al (2015) Antibacterial activity of Oregano, Rosmarinus and Thymus essential oils against *Staphylococcus aureus* and *Listeria monocytogenes* in beef meatballs. Food Control 54:188–199

Pisoschi AM, Pop A, Georgescu C et al (2018) An overview of natural antimicrobials role in food. Eur J Med Chem 143:922–935

Polia A, Barbagallo CM, Cicero AFG et al (2018) Nutraceuticals and functional foods for the control of plasma cholesterol levels. An intersociety position paper. Pharmacol Res 134:51–60

Popa M, Niculita P, Mitelut A (2010) Elemente de baza privind biotehnologiile si siguranta alimentelor, 430 p

Poveda JM, Loarce L, Alarcón M et al (2018) Revalorization of winery by-products as source of natural preservatives obtained by means of green extraction techniques. Ind Crop Prod 112:617–625

Prakash B, Kedia A, Mishra PK et al (2015) Plant essential oils as food preservatives to control moulds, mycotoxin contamination and oxidative deterioration of agri-food commodities – potentials and challenges. Food Control 47:381–391

Raeisi M, Tabaraei A, Hashemi M et al (2016) Effect of sodium alginate coating incorporated with nisin, *Cinnamomum zeylanicum*, and rosemary essential oils on microbial quality of chicken meat and fate of *Listeria monocytogenes* during refrigeration. Int J Food Microbiol 238:139–145

Rago YP, Surroop D, Mohee R (2018) Assessing the potential of biofuel (biochar) production from food wastes through thermal treatment. Bioresour Technol 248:258–264

Ranadheera RDCS, Baines SK, Adams MC (2010) Importance of food in probiotic efficacy. Food Res Int 43:1–7

Reis FS, Martins A, Vasconcelos MH et al (2017) Functional foods based on extracts or compounds derived from mushrooms. Trends Food Sci Technol 66:48–62

Ribeiro AR, Altintzoglou T, Mendes J et al (2019) Farmed fish as a functional food: perception of fish fortification and the influence of origin – insights from Portugal. Aquaculture 501:22–31

Rincon-Leon F (2003) Encyclopedia of food sciences and nutrition, 2nd edn. Functional foods. doi:https://doi.org/10.1016/B0-12-227055-X/01328-6

Robinson PK (2015) Enzymes: principles and biotechnological applications. Essays Biochem 59:1–41

Saeli M, Piccirillo C, Tobaldi DM et al (2018) A sustainable replacement for TiO_2 in photocatalyst construction materials: hydroxyapatite-based photocatalytic additives, made from the valorisation of food wastes of marine origin. J Clean Prod 193:115–127

Salgado PR, López-Caballero ME, Gómez-Guillén MC et al (2013) Sunflower protein films incorporated with clove essential oil have potential application for the preservation of fish patties. Food Hydrocoll 33:74–84

Salvia-Trujillo L, Rojas-Graü A, Soliva-Fortuny R et al (2015) Physicochemical characterization and antimicrobial activity of food-grade emulsions and nanoemulsions incorporating essential oils. Food Hydrocoll 4:547–556

San Martin D, Ramos S, Zufía J (2016) Valorisation of food waste to produce new raw materials for animal feed. Food Chem 198:68–74

Santamarina MP, Rosello J, Gimenez S et al (2016) Commercial *Laurus nobilis* L. and *Syzygium aromaticum* L. Merr. & Perry essential oils against post-harvest phytopathogenic fungi on rice. LWT Food Sci Technol 65:325–332

Santos JCP, Sousa RCS, Otoni CG et al (2018a) Nisin and other antimicrobial peptides: production, mechanisms of action, and application in active food packaging. Innov Food Sci Emerg Technol 48:179–194

Santos SB, Costa AR, Carvalho C et al (2018b) Exploiting bacteriophage proteomes: the hidden biotechnological potential. Trends Biotechnol 36(9):966–984

Saxelin M (2008) Probiotic formulations and applications, the current probiotics market, and changes in the marketplace: a European perspective. Clin Infect Dis 46: S76–S79

Settanni L, Corsetti A (2008) Application of bacteriocins in vegetable food biopreservation. Int J Food Microbiol 121:123–138

Severino R, Ferrari G, Dang Vu K et al (2015) Antimicrobial effects of modified chitosan based coating containing nanoemulsion of essential oils, modified atmosphere packaging and gamma irradiation against *Escherichia coli* O157:H7 and *Salmonella Typhimurium* on green beans. Food Control 50:215–222

Shah MA, Don Bosco SJ, Mir SA (2014) Plant extracts as natural antioxidants in meat and meat products. Meat Sci 98:21–33

Silva MA, Albuquerque TG, Alves RC et al (2018) Melon (*Cucumis melo* L.) by-products: potential food ingredients for novel functional foods? Trends Food Sci Technol. https://doi.org/10.1016/j.tifs.2018.07.005

Siro I, Kapolna E, Kapolna B et al (2008) Product development, marketing and consumer acceptance – a review. Appetite 51:456–467

SKLM Commission Secretariat (2010) Microbial food cultures. Deutsche Forschungsgemeinschaft, Bonn

Sofra C, Tsironi T, Taoukis PS (2018) Modeling the effect of pre-treatment with nisin enriched osmotic solution on the shelf life of chilled vacuum packed tuna. J Food Eng 216:125–131

Sojic B, Tomovic V, Kocic-Tanackov S et al (2015) Effect of nutmeg (*Myristica fragrans*) essential oil on the oxidative and microbial stability of cooked sausage during refrigerated storage. Food Control 54:282–286

Song Z, Li F, Guan H et al (2017) Combination of nisin and ε-polylysine with chitosan coating inhibits the white blush of fresh-cut carrots. Food Control 74:34–44

Soni S, Dey G (2014) Perspectives on global fermented foods. Br Food J 116(11):1767–1787

Spahn C, Minteer SD (2008) Enzyme immobilization in biotechnology. Recent Pat Eng 2:195–200

Spricigo D, Bardina C, Cortés P et al (2013) Use of a bacteriophage cocktail to control *Salmonella* in food and the food industry. Int J Food Microbiol 165:169–174

Stephen JL, Periyasamy B (2018) Innovative developments in biofuels production from organic waste materials: a review. Fuel 214:623–633

Taghizadeh-Alisaraei A, Hosseini SH, Ghobadian B et al (2017) Biofuel production from citrus wastes: a feasibility study in Iran. Renew Sust Energ Rev 69:1100–1112

Terefe NS (2016) Emerging trends and opportunities in food fermentation. In: Module in food science. doi: https://doi.org/10.1016/B978-0-08-100596-5.21087-1

Thung TY, Premarathne JM, Chang WS et al (2017) Use of a lytic bacteriophage to control *Salmonella Enteritidis* in retail food. LWT Food Sci Technol 78:222–225

Tserennadmid R, Tako M, Galgoczy L et al (2011) Anti yeast activities of some essential oils in growth medium, fruit juices and milk. Int J Food Microbiol 144:480–486

Ukuku DO, Niemira BA, Ukanalis Z (2019) Nisin-based antimircobial combination with cold plasma treatment inactivate *Listeria monocytogenes* on Granny Smith apples. LWT Food Sci Technol 104:120–127

Utoiu E, Matei F, Toma A et al (2018) Bee collected pollen with enhanced health benefits, produced by fermentation with a Kombucha consortium. Nutrients 10(10):1–24

Verma SK, Sood SK, Saini RK et al (2017) Pediocin PA-1 containing fermented cheese whey reduces total viable count of raw buffalo (*Bubalis bubalus*) milk. LWT Food Sci Technol 83:193–200

Vincentini A, Liberatore L, Mastrocola D (2016) Functional foods: trends and development of the global market. Ital J Food Sci 28(2):338–351

Vong WC, Liu SQ (2016) Biovalorisation of okara (soybean residue) for food and nutrition. Trends Food Sci Technol 52:139–147

Wang T, Zhai Y, Zhu Y et al (2018a) Evaluation of the clean characteristics and combustion behavior of hydrochar derived from food waste towards solid biofuel production. Bioresour Technol 266:275–283

Wang T, Zhai Y, Li H et al (2018b) Co-hydrothermal carbonization of food waste-woody biomass blend towards biofuel pellets production. Bioresour Technol 267:371–377

Wolfe BE, Dutton RJ. Cell(2015) Fermented foods as experimentally tractable microbial ecosystems. Cell 161(1):49–55

Woraprayote W, Kingcha Y, Amonphanpokin P et al (2013) Anti-listeria activity of poly(lactic acid)/sawdust particle biocomposite film impregnated with pediocin PA-1/AcH and its use in raw sliced pork. Int J Food Microbiol 167:229–235

Woraprayote W, Malila Y, Sorapukdee S et al (2016) Bacteriocins from lactic acid bacteria and their applications in meat and meat products. Meat Sci 120:118–132

Wu T, Wu C, Fang Z et al (2017) Effect of chitosan microcapsules loaded with nisin on the preservation of small yellow croaker. Food Control 79:317–324

Yeo J, Oh J, Cheung HHL et al (2019) Smart Food Waste Recycling Bin (S-FRB) to turn food waste into green energy resources. J Environ Manag 234:290–296

Zhang Z, O'Hara IM, Mundree S et al (2016) Biofuels from food processing wastes. Curr Opin Biotechnol 38:97–105

Zhao L, Wang S, Liu F et al (2013) Comparing the effects of high hydrostatic pressure and thermal pasteurization combined with nisin on the quality of cucumber juice drinks. Innov Food Sci Emerg Technol 17:27–36

Zimet P, Mombrú AW, Faccio R et al (2018) Optimization and characterization of nisin-loaded alginate-chitosan nanoparticles with antimicrobial activity in lean beef. LWT Food Sci Technol 91:107–116

Zinno P, Devirgiliis C, Ercolini D et al (2014) Bacteriophage P22 to challenge *Salmonella* in foods. Int J Food Microbiol 191:69–74

Creating Products and Services in Bioinformatics

8

Radu Cristian Toma, Laura Dorina Dinu, and Camelia Filofteia Diguta

Abstract

In this chapter, bioinformatics is defined, emphasizing the interdisciplinary aspects. There is detailed information regarding the bioinformatics fields such as database design and data mining, sequencing, gene and protein expression, structural bioinformatics, phylogenetic tree construction, biological networks, and their practical application. In addition, tools provided by bioinformatics are described (open-sources and web-based services in bioinformatics, educational programs, and training platforms in bioinformatics). The chapter analyze the path of a bioinformatics student toward entrepreneurship in the US context versus the European context.

Keywords

Bioinformatics · Databases · Health sciences · Precision medicine · Translational bioinformatics · Pharmacoinformatics · Microbiomics · Oncology · Biomedical computing

8.1 Bioinformatics: Definition, Evolution, and Interdisciplinary Aspects

Broadly speaking, *computational biology* means applying in a mode interdisciplinary tools and techniques in several branches of science, such as computer science, statistics, and mathematics, to solve biology issues.

Computational biology covers a wide range of branches of biology, for example, genetics, biophysics, cell biology, biochemistry, and molecular biology.

Computational biology deals with molecular data analysis, such as biosensors (DNA sequences, RNA, or proteins), three-dimensional structures of proteins, data on the expression of different genes, or network molecular biology. With these data, a wide variety of solutions can be analyzed and problems solved such as identifying disease-causing genes to develop new cures, reconstitution of the evolutionary path of species, etc.

Elements of computational biology are *bioinformatics* and *statistics*. Bioinformatics addresses the development of software tools, algorithms for collecting, storing, and analyzing biological data and statistics, through using models and statistical methods, analyzing these data biologically.

Bioinformatics can be seen as an interdisciplinary domain in which techniques and computerized technologies are developed and

R. C. Toma (✉) · L. D. Dinu · C. F. Diguta
Faculty of Biotechnologies, University of Agronomic Sciences and Veterinary Medicine of Bucharest, Bucharest, Romania
e-mail: Radu.Toma@biotehnologii.usamv.ro; laura.dinu@biotehnologii.usamv.ro

F. Matei, D. Zirra (eds.), *Introduction to Biotech Entrepreneurship: From Idea to Business*,
https://doi.org/10.1007/978-3-030-22141-6_8

applied to study and answer questions in the fields of molecular biology, genetics, biomedicine, etc.

In a way, bioinformatics is a technology that is used to manage, query, and analyze large amounts of data in life sciences. In another way, bioinformatics is a holistic and systemic approach that finds new models, and discovers new elements.

The word "Bioinformatics" appeared in 1970 in a Dutch paper. The definition of "Bioinformatics" was "*the study of informatics processes in biotic systems*" (Hesper and Hogeweg 1970). In 1978, Pauline Hogeweg described in a scientific paper her activity as "bioinformatics."

The development of bioinformatics and its evolution as an interdisciplinary field of biosciences and informatics is in close relation to the formation and development of computers, molecular biology, and genetics starting in 1950.

After 1953, many theoretical concepts and technologies in genetics were set (double helix structure of DNA in 1953 by Watson and Crick). Also, both the first protein sequences (insulin) and the first protein structures (of myoglobin) were discovered. At the same time, in informatics, various technologies were developed, for example, electronic computers, programming languages (the first programming language was FORTRAN, 1953). The concepts of today's internet and TCP protocol and in 1991 the World Wide Web were then invented.

Some of the basic questions in bioinformatics appeared in the late 1960s/early 1970s (in 1977, F. Sanger published a paper about Sanger DNA sequencing) and evolved during the 1990s when the US Department of Energy, later run by the NIH (US National Institutes of Health), started the Human Genome Project (HGP) aiming at sequencing the human genome.

Today, we can say that the development and growth of bioinformatics is in an important way associated with the HGP.

In 2005, a new stage began: next-generation sequencing (NGS). Nowadays, next-generation sequencers can sequence multiple complete genomes in less than 24 hours.

8.2 Applications of Bioinformatics in Life Sciences

Life sciences researchers refer to collecting and analyzing a vast of scientific data on biological molecules such as nucleic acid and protein sequences, 3D protein structures, microarray gene expression, and biological pathways. In the last few years, bioinformatics has progressed very quickly and plays an important role in the development of life sciences. Many publicly available resources are developed to study, explore, and understand the genomes of organisms (genomics), RNA molecules (transcriptomics), proteins (proteomics), including their properties, functions, and interactions involved in the biological process. Correct use of the bioinformatics tools and databases can help scientists and researchers to manage, analyze, and interpret the enormous amount of data and to reduce the number of costly experiments.

The European Nucleotide Archive (ENA; https://www.ebi.ac.uk/ena), provided from the European Molecular Biology Laboratory's European Bioinformatics Institute (EMBL-EBI) (Hinxton, UK) is a freely available database from life sciences covering the spectrum of molecular biology, including the life science research literature (http://europepmc.org) (Harrison et al. 2019). The European Bioinformatics Institute (EMBL) supports and provides over 40 data resources (https://www.ebi.ac.uk/services/), which cover the entire range of biological sciences (Cook et al. 2019). The centrality of the ENA to the bioinformatics infrastructure has been recognized in the ELIXIR Core Data Resource (https://www.elixir-europe.org/platforms/data/core-data-resources) and ELIXIR Deposition Database (https://www.elixir-europe.org/platforms/data/elixir-deposition-databases) (Harrison et al. 2019). EMBL-EBI is member of the International Nucleotide Sequence Database Collaboration (INSDC; http://www.insdc.org/), together with the GenBank at the National Center for Biotechnology Information (NCBI), National Library of Medicine, National Institutes of Health in

Bethesda, MD, USA (https://www.ncbi.nlm.nih.gov/) (Benson et al. 2018) and the DNA Data Bank of Japan (DDBJ: http://www.ddbj.nig.ac.jp/) in Mishima, Japan (Kodama et al. 2018). These three major public sequence databases are freely available on the Internet.

The Ensembl project (https://www.ensembl.org) represents a fundamental resource for creating, maintaining, and updating reference genome annotation and comparative genomics resources available to the entire scientific community without restrictions (Cunningham et al. 2019).

SILVA (http://www.arb-silva.de) provides a comprehensive web resource for up-to-date, quality-controlled databases of aligned ribosomal RNA (rRNA) small (16S/18S, SSU) and large subunit (23S/28S, LSU) ribosomal RNA (rRNA) sequences for all three domains of life (Bacteria, Archaea and Eukarya) and supplementary online services including specialized tools and databases to perform classification searching sequence analysis and phylogenetic reconstructions (Quast et al. 2013; Glöckner et al. 2017). SILVA is an ELIXIR Core Data Resource (https://bio.tools/silva). Another useful bioinformatics website is ExPASy (https://www.expasy.org/), which provides access to scientific databases and software tools in different areas of life sciences, including proteomics, genomics, phylogeny, evolutionary biology, population genetics, transcriptomics, and more (Artimo et al. 2012).

Multiple sequence alignment is an essential technique in many bioinformatics applications. Clustal Omega (https://www.ebi.ac.uk/Tools/msa/clustalo/) is a multiple sequence alignment program for identifying regions of similarity that can indicate structural, functional, and/or evolutionary relationships between molecular sequences (nucleic acid or protein). Other useful freely available alignment algorithms are MUSCLE (https://www.ebi.ac.uk/Tools/msa/muscle/) and T-Coffee (tree-based consistency objective function for alignment evaluation http://www.tcoffee.org/) to multiply align proteins, RNA, and DNA sequences (Edgar 2004, Taly et al. 2011).

In addition, databases have been developed to support protein sequences, three-dimensional structures, protein functions, and protein–protein interactions. The Universal Protein Resource knowledge base (UniProtKB) is developed and maintained by the UniProt consortium, a collaboration between the SIB Swiss Institute of Bioinformatics, the European Bioinformatics Institute (EMBL-EBI), and the Protein Information Resource (PIR) (Gabella et al. 2018). The UniProtKB is composed of UniProtKB/Swiss-Prot—the reviewed section (manually annotated) containing the data of protein sequence and functional information (mainly protein name or description, taxonomic data, cross-reference, and literature citation) and UniProtKB/TrEMBL—the unreviewed section (automatically annotated) with computational analyses. UniProt can be accessed via the website at http://www.uniprot.org/ (Gabella et al. 2018). Other popular online databases include the PIR (https://proteininformationresource.org/)—protein sequence database; wwPDB (worldwide Protein DataBank) (https://www.wwpdb.org/)—protein databases; InterPro (www.ebi.ac.uk/interpro)—database for protein sequence analysis and classification into families that predicts the presence of functionally important domains and sites; PRIDE (www.ebi.ac.uk/pride)—PRoteomics IDEntification database (Gabella et al. 2018; Mitchell et al. 2018). These databases that integrate different bioinformatics methods offers a platform for data mining, annotation by sequence similarity, visualization of the protein interactions, metabolic pathway mapping, and generation of the information for kinetic simulations. STRING, as part of the ELIXIR Core Data Resources, is available online at https://string-db.org/ and provides a critical assessment of knowing and predicting protein–protein interactions including direct (physical) and indirect (functional) associations (Szklarczyk et al. 2019). The new version of STRING (11.0) covers to 5090 organisms (Szklarczyk et al. 2019). The IntAct Molecular Interaction Database provides a freely available (https://www.ebi.ac.uk/intact/home.xhtml), open source database system and analysis tools for molecular interaction data. The Database of Interacting Proteins (DIP; http://dip.doe-mbi.ucla.edu/) catalogs and searches the protein–protein interaction network of the target organism.

Phylogenetic analyses are important aspects of bioinformatics. The choice of nucleotide or protein sequences depends on the properties of the sequences to obtain a correct phylogenetic tree. The Molecular Evolutionary Genetics Analysis (MEGA) software contains many sophisticated methods and tools for phylogenomics and phylomedicine (Kumar et al. 2018). MEGA provides tools for sequence alignment (nucleic acid and protein), construction, and visualization of phylogenetic trees, estimating sequence similarities and divergence, rates of molecular evolution, determining genetic relatedness in addition to exploring online databases (Kumar et al. 2016). The new version is MEGA X, available in two interfaces (graphical and command line) and can be downloaded from www.megasoftware.net free of charge (Kumar et al. 2018). The popular nucleotide substitution models used are the Jukes–Cantor and Kimura models. The neighbor joining is the simplest algorithm for building a tree from a distance matrix.

With the increasingly common use of technologies, such as DNA and RNA sequencing, NGS technologies, microarray analysis, and high-throughput proteomics and metabolomics, comes the need for novel methods to turn these new types of data into new knowledge. The Sanger method for DNA sequencing was an important step toward developing modern sequencing methods.

Over the past several years, NGS technologies have revolutionized life sciences. These technologies are now widely available, can generate large volumes of sequence data, and reduce the cost of DNA sequencing. The main NGS technologies are 454 sequencing, Illumina sequencing, SOLiD sequencing, available for second-generation sequencing; HeliScope Single Molecule Real Time (SMRT) sequencing technology, nanopore sequencing by Oxford Nanopore, available for the third generation sequencing. Whole-genome sequencing (WGS) is currently a rapid alternative to phenotypic methods. Transcriptome analysis (using RNA sequencing) or gene expression microarray data represents a valuable tool for exploring gene expression, followed by a pathway analysis.

The high-throughput microarray technologies are an efficient method for the simultaneous detection and identification of microorganisms (using DNA, RNA (cDNA) and protein sequences) in one assay, generating large amounts of genomic data. Microarrays have been used largely for gene expression studies, detection, and characterization of microorganisms, and polymorphism profiles. Microarrays produce vast amounts of information requiring approaches and algorithms to interpret the data. Thus, bioinformatics has become an important tool for managing and analyzing microarray data. There are free image-processing software programs available on the Internet. WebArray (http://webarraydb.org/webarray/index.html) is a web platform for the analysis of microarray data. The Bioinformatics Array Research Tool (BART) is available for download from https://bitbucket.org/Luisa_amaral/bart, and enables scientists with no bioinformatics knowledge to perform their own state-of-the-art customized analyses on a variety of microarray experiments using an intuitive interactive interface.

A summary of databases described in Sect. 8.2 is presented in Table 8.1.

8.2.1 Application of Bioinformatics Start-ups in Life Sciences in Europe

Although most of the synthetic biology market is concentrated in North America, there are many interesting European genomics companies. *Genedata* (https://www.genedata.com/) with headquarters in Switzerland is a company that provides software solutions and consulting solutions that support large-scale, experimental processes in life science research, in addition to academia worldwide. *Genomatrix* (https://www.genomatix.de/index.html), with headquarters in Germany, is a company that provides technologies for analyzing and interpreting genomic data. *BaseClear* (https://www.baseclear.com/services/bioinformatics/) offers a large range of bioinformatics solutions, especially in the area of

Table 8.1 The bioinformatics databases used as tools in life sciences

Database Name	Link address	Description
ENA	www.ebi.ac.uk/ena	A resource for nucleotide databases, sequence assembly information, and functional annotation in Europe
DDBJ	www.ddbj.nig.ac.jp	A resource for nucleotide sequence databases in Japan
Ensembl	www.ensembl.org	Genomic data sets
ELIXIR	www.elixir-europe.org	Databases and software tools
SILVA	www.arb-silva.de	Ribosomal RNA databases
ExPASy	www.expasy.org	Scientific databases and software tools
Clustal Omega	www.ebi.ac.uk/Tools/msa/clustalo	Multiple sequence alignment
Muscle	www.ebi.ac.uk/Tools/msa/muscle	Multiple sequence alignment
T-Coffee	www.tcoffee.org	Multiple sequence alignment
UniProtKB	www.uniprot.org	A resource for protein databases and annotation data
MEGA X	www.megasoftware.net	Construction of phylogenetic trees to study evolutionary history
WebArray	http://webarraydb.org/webarray/index.html	Web platform for the analysis of microarray data
BART	https://bitbucket.org/Luisa_amaral/bart	Microarray data and analysis tools

NGS data analysis, comparative genomics or metagenomics and expression analysis; the online viewing and analysis platforms are extremely handy tools. ecSeq (https://www.ecseq.com), with headquarters in Germany, is a company that provides solid expertise in the analysis of high-throughput sequencing data. Other companies have offices in Europe. *Seven Bridges Genomics* (https://www.sevenbridges.com/), with European headquarters in the UK, Serbia, and Turkey, provides end-to-end bioinformatics solutions, including access to datasets, analytic workflows, algorithms, cloud-computing infrastructure, and scientific support. *Illumina* (https://www.illumina.com/), located in the UK and Netherlands, is a company that develops integrated systems for the analysis of genetic variations and biological functions; provides the products and services that serve sequencing, genotyping, gene expression, and proteomics. *GeneBio* Geneva Bioinformatics SA (http://www.genebio.com/) with headquarters in Switzerland commercializes intellectual property and associated services and resources developed at the SIB Swiss Institute of Bioinformatics. *Genevia technologies* (https://geneviatechnologies.com/), located in Finland, offers a larger portfolio of bioinformatics products that help life science researchers (academic institutes and industrial companies). *Bioinformatics Consultants* (https://www.bioinformaticsconsultants.com/), based in Stockholm (Sweden), offer a wide range of tailored bioinformatics and biostatistics services to academia and industry.

Bioinformatics today plays a significant role in life sciences research, providing new ways and approaches for the assessment of valuable data.

Generally, the research workflow consists in data acquisition and storage, predictive modeling, classification, and computational analysis. Standardization and validation are crucial for the efficient use of software tools and data resources. A software platform contains various software tools, enabling data resources to be more accessible to researchers with no knowledge of bioinformatics. Bioinformatics is widely used among researchers to manage and analyze the large data sets obtained and permits databases to deliver user-friendly solutions that are easy to interpret. Thus, it offers data science a promising future for exploring molecular biology and controlling R&D costs, with an impact on research productivity.

8.3 Application of Bioinformatics in Health Sciences

Biomedical research is progressing rapidly aided by new innovative technologies and the vast amount of data generated that needs to be maximally shared, integrated, and exploited using efficient bioinformatics tools. Thus, a relatively new field, translational bioinformatics, has arisen to develop "storage, analytic, and interpretive methods to optimize the transformation of increasingly voluminous biomedical data, and genomic data, into proactive, predictive, preventive, and participatory health," as defined by American Medical Informatics (https://www.amia.org/applications-informatics/translational-bioinformatics).

Bioinformatics is fundamental to precision medicine (PM), a model of predictive healthcare that considers the genetic background of the individual and his/her environment, selecting the optimal medical decision to prevent, diagnose or treat a disease. PM can improve people's health and wellbeing prescribing personalized therapies; therefore, several countries are implementing programs to support research and applications of PM, such as the Precision Medicine Initiative started in 2016 in the USA.

8.3.1 Bioinformatics and "Omics" Technologies for Biomedical Research

Nowadays, there are many companies that are closely integrating genomics technology with diagnostics and services, especially in the field of health care. Moreover, the development of high-throughput technologies in biomedical research, such as microarrays and NGS, offers major opportunities to improve public health. As the cost of these technologies decreases, there is an exponential increase in the amount of data produced; therefore, to effectively use this "big genomic data" strong and efficient bioinformatics systems are required.

A pioneering company that makes secure software solutions to manage genomic data sets and platform integration for customers engaging in genomics-based clinical research and medical applications is *Station X, Inc.* Their flagship product is a cloud-based software GenePool® for analyzing, visualizing, and managing complex genomic data and for sharing data with collaborators for interactive investigations (https://angel.co/station-x). *DNAnexus* is a leading provider of solutions for management, storage, and scaling in genomics, providing a "cloud-based platform as a service" and a global network for sharing the data (https://www.dnanexus.com). They already work with top biopharmaceutical companies, leading genome centers, and diagnostic test manufacturers, and in 2018 announced DNAnexus Apollo, a powerful platform for clinical genomic data science exploration, analysis, and discovery of the new targets and biomarkers of disease progression and therapy response. In Europe, some start-ups have developed tools for interpretation of NGS data and mechanisms for data sharing in genomics for science and biomedical applications. Founded in 2014, *Repositive* has since grown to the biggest portal for accessing public genomic data sources and is one of the contributors to the National Institutes of Health Data Commons in the USA. *Desktop Genetics*, a biotechnology company based in London, enables the CRISPR gene-editing revolution with powerful bioinformatics tools to help discover and treat human diseases. The gene-editing platform DESKGEN AI incorporates modern strategies for improving the precision editing rate and has the largest database of genome editing data in the world, helping pharma, tech, and academic customers working in drug discovery and functional genomics (https://www.deskgen.com). Another example is the SeqHepB system, a unique bioinformatics genomic sequencing software that rapidly and accurately identifies those viral mutations and sub-mutations that lead to drug resistance and multi-drug resistance and helps clinicians to make an informed decision for each individual patient infected with hepatitis B virus (http://www.seqhepb.com).

8.3.2 Analysis of Bioinformatics and Microbiomics

Although scientific research around microbiota is currently booming, many experts are working to exploit its potential as the next source of therapeutic agents and to develop medical applications. In a report published by Zion Market Research in 2017, the global human microbiome market was valued at around USD721.63 million and is expected to grow around USD1.3 billion by the end of 2024, mainly as a result of the increase in research yielding new applications of the human microbiome (https://globenewswire.com). Therefore, a US company, *Second Genome*, developed the first therapeutic database and a Microbiome Discovery Platform with microbiome profiling and analytical informatics technologies to discover the role of microbes in disease and to identify the bioactive molecules that have positive disease-modifying effects (https://www.secondgenome.com). Potential therapeutic candidates identified (bacterial strains, metabolites, peptides and proteins) can then be tested using traditional drug development processes and results are analyzed using bioinformatic tools to improve the drug design process. In Europe, *Eagle Genomics* applies commercial open source principles to genomics and helps to industrialize microbiomics with a powerful software system for microbiome analysis that has already been involved in obtaining innovative products in the microbiomics sector (https://www.eaglegenomics.com).

8.3.3 Bioinformatics and the Discovery of Cancer Drug and Biomarkers

Cancer is a disease of the genome caused by the accumulation of mutations occurring in critical genes, such as oncogenes and tumor-suppressor genes, that drive unregulated cell growth and differentiation. During the last two decades, understanding the effect of mutations in key regions of the genome was enabled by high-throughput technologies and unlocked new possibilities for cancer therapy. Nowadays, tumor genomic profiling may be used as a clinical tool in some types of cancer for diagnosis, prognosis or to guide treatment choices (Servant et al. 2014; Singer et al. 2017).

Epic Sciences, Inc. (USA) was founded in 2008 to develop novel diagnostics to personalize and advance the treatment and management of cancer. At the beginning, the company provided analysis services for companies developing drugs to identify and characterize based on image analysis software circulating tumour cells from blood samples. In 2016, *Station X* announced that *Epic Sciences* will use a customized version of GenePool® pre-loaded with standard clinical reporting data about up-to-date findings and tumour profiling for biopharmaceutical and clinical testing (https://www.epicsciences.com). The National Cancer Institute and the *SevenBridges* bioinformatics company launched in 2014 the Cancer Genomics Cloud (CGC), an open-source cancer informatics pipeline populated with The Cancer Genome Atlas (TCGA) data, one of the most complete genomics databases to which more than 11,000 patients have contributed their own medical data related to 33 different tumor types and subtypes (http://www.cancergenomicscloud.org). The company *Elypta* (Sweden) developed a liquid biopsy platform based on the discovery that a panel of 19 metabolic biomarkers are specifically deregulated across several cancer types, and then using a learning algorithm generating a quantitative score that may be used as a tool to aid prognosis or to monitor treatment. The technology has shown unprecedented accuracy, from 92.7% to 100% when blood and urine measurements were combined to diagnose and monitor the severity of renal cell carcinoma, the most common form of kidney cancer. Further investigations are aimed at developing a test for all types of cancer, based on Elypta's biomarkers approach (https://elypta.com/science).

8.3.4 Pharmacoinformatics

The drug discovery process is a critical issue in the pharmaceutical industry as it is a very costly and time-consuming process to produce new drug potentials and enlarge the scope of diseases incurred. "In silico" drug designing is an expression used for the identification of new drugs and target molecules by employing computer and various bioinformatics tools for structure-based drug discovery (SBDD). This inventive process of finding new medications is based on the knowledge of the molecules that bind to the biological target of interest (generally proteins/enzymes), and on the knowledge of the three-dimensional structure of the biological target obtained through methods such as homology modelling, NMR spectroscopy, and X-ray crystallography. Bioinformatics approaches accelerate the process of drug candidate screening, drug target identification and validation or protein modeling, in addition to facilitating the characterization of side effects and predicting drug resistance. Mainly, bioinformatic tools store and manage available information about drugs and targets and provide strategies and algorithms to predict biologically active candidates and new drug targets. Molecular docking is an automated computer algorithm that attempts to find the best match between two molecules, which is a computational determination of binding affinity between molecules. Docking can be performed using bioinformatics tools that are able to search a database containing molecular structures and retrieve the molecules that can interact with the query structure using virtual high-throughput screening, building up chemical and biological information databases about ligands and targets/proteins to identify and optimize novel drugs. Also, it uses in silico filters to calculate drug likeness or pharmacokinetic properties for the chemical compounds prior to screening to enable early detection of those compounds that are more likely to fail in the clinical stages and further to enhance the detection of promising compounds.

The number of static docking software programs currently available is high, but in the last few years there has been a rapid development of molecular dynamics simulations. Recently, an American company developed Desmond, a software package for performing high-speed molecular dynamics simulations of biological systems on conventional and supercomputers. Desmond is available as commercial software through *Schrödinger, Inc.*, but is free of charge for universities or other non-profit research institutions (https://www.deshawresearch.com). The *GTN* company, founded in 2017 by Noor Shaker, with knowledge in the field of drug discovery, and her co-founder Vid Stojevic, a theoretical physicist, developed a software that screens a huge amount of possible drugs, discovering molecules entirely hidden from view using quantum computing. Nowadays, this European company collaborates with the ten largest global pharmaceutical companies (https://gtn.ai/index.html.). Another company from Spain, designs and develops third-generation drugs for cancer and neurological diseases using an in-house technology platform, IPROTech, which allows the development of innovative peptidomimetic drugs. Several innovative drugs developed by the company cross the blood–brain barrier and the gastrointestinal tract to attack hard-to-reach targets and have proven efficacy in the preclinical phase. Moreover, *Iproteos* participated in a project co-funded by the European Union Horizon 2020 to develop a therapeutic strategy against epilepsy (http://www.iproteos.com).

8.3.5 Biomedical Computing

In the last few years many business opportunities have arisen in the field of clinical and healthcare bioinformation systems such as IT support systems for health care decision-making or computer-controlled medical devices. For instance, the Conversation Platform™ developed by *Conversa Health, Inc.*, is a digital check-up platform that deliver to patients personalized messages based on data from electronic health records, biometric monitoring devices or general health data derived from digital check-up responses. This helps to monitor and manage large patient populations delivering flexible and fully-automated virtual care based on a library of over 500 "clinically intelligent conversations" that covers chronic care, pre- and post-surgical care, or prevention and wellness advice (https://conversahealth.com). Another bioinformatics development, PriorAuthNow software, uses data analytics and interoperability standards to accelerate the process of submitting, monitoring, and completing prior authorization and enables provider groups (large hospitals, diagnostic laboratories, specialty pharmacies) to reduce turnaround times to obtain these agreements from the payer to cover specific services before the service is performed (https://priorauthnow.com). *Xcode Life Sciences* from Asia is a personal genomics company that uses bioinformatics tools to design and analyze genetic testing panels for preventive health care. Their tests include genetic testing for nutrition to understand food intolerances and sensitivities or the ability to metabolize different food groups, and pharmacogenetic testing to customize the therapeutic strategy and fitness genetics to optimize training and achieve top performance (https://xcodelife.co/).

8.4 Application of Bioinformatics to Food and Drug Safety

Within regulatory science for food and drug safety, regulatory bioinformatics is considered a priority area, for both regulatory agencies and industry. Regulatory bioinformatics is aimed at establishing the best practices to understand and interpret data from innovative technologies, such as microarrays and NGS, and to ensure the safe introduction and use of these applications (Healy et al. 2016). This area of bioinformatics has great potential to improve public health and will bring many beneficial applications for the safety of medical products and food.

In the last few years, innovative bioinformatics tools with applications in food safety have been developed for foodborne disease surveillance and outbreak investigations. A major area in this field is the development of robust and user-friendly genomics-based prediction software to perform advanced pathogen characterization in molecular epidemiology to support public health investigations. Predictive serotype systems for foodborne pathogens *Salmonella in Silico* Typing Resource (SISTR) is an open web-accessible platform with a large database of over 4000 publicly available genomes and bioinformatics tools that rapidly perform and predict (~95% accuracy) the *Salmonella* serotype using analysis of the pathogen draft sequence (Yoshida et al. 2016). SISTR or *SerotypeFinder system* for *E. coli* provides faster and cheaper data than conventional procedures, such as pulse field gel electrophoresis or multi-locus sequence typing (MLST), and it is expected to replace traditional serotyping in food regulatory and public health laboratories (Taboada et al. 2017). Similarly, the commercial software *Ridom SeqSphere+* performs whole-genome microbial typing (cgMLST) based on data from NGS or Sanger sequencing and can be used for outbreak analysis or real-time surveillance (https://www.ridom.de/seqsphere/). Other bioinformatics applications focus on the prediction of antimicrobial resistance from whole-genome sequence data, such as *Comprehensive Antimicrobial Resistance Database* (CARD), a platform with a database of resistance genes, their products, and associated phenotypes, and *Resistance Gene Identifier* (RGI) software (https://card.mcmaster.ca/).

8.5 Developing Bioinformatics Tools

8.5.1 Software

At first, sequence analysis was a huge problem in bioinformatics. The more complicated the structure of the protein the more challenging the assemblage. *COMPROTEIN* is the first computer program for sequence alignment. It was developed by Margaret Dayhoff and Robert Ledley in 1962. Initially, the software was run on a IBM7090 machine at Johns Hopkins University.

In 1970, Saul Needleman and Christian Wunsch published research in the "Journal of Molecular Biology" about a *Dynamic Program algorithm* to find the optimal alignment of protein or nucleotide sequences.

Also in 1978, after Margaret Dayhoff analyzed the substitution rate among different pairs of amino acids from homologous proteins, the *PAM substitution matrices* were constructed. They are the best matrices to use for sequence alignment even today. Although in 1992, the *BLOSUM matrices* were published and became the new default, the PAM matrices continued to be used.

After the discovery of introns in 1981, Temple Smith and Michael Waterman published a research work in the "Journal of Molecular Biology" providing an adapted solution to sort out the alignment of DNA sequences in the presence of exons and introns, *the Smith–Waterman algorithm*, which performs local sequence alignment of two strings of nucleic acids or protein sequences.

In the 1980s, more protein and nucleic acids sequences were solved, so a new problem occurred. The Needleman–Wunsch and Smith–Waterman algorithms become too slow to manage the newly created databases of sequences. Researchers needed to find if there were known sequences similar to the one that they were studying. Thus, in 1990, Steve Altschul published a paper describing the *BLAST algorithm* (https://blast.ncbi.nlm.nih.gov/Blast.cgi) in the "Journal of Molecular Biology." By indexing known sequences and sacrificing optimal alignment in some cases, BLAST increased the querying speed of the database. Moreover, in 1997, they published in "Nucleic Acids Research" a paper about new algorithms: *Gapped BLAST algorithm* and *PSI-BLAST algorithm* (https://www.ebi.ac.uk/Tools/sss/psiblast/).

In the late 1990s, when the gene prediction and whole-genome alignment became areas of huge interest for more and more research and researchers, various algorithms were developed for gene expression data analysis. An example is the *GenScan* software (http://genes.mit.edu/GENSCAN.html) based on the Hidden Markov Model developed by Christopher Burge and Samuel Karlin in 1997.

Other tools related to the bioinformatics field are:

The *bioinformatics.org* (http://www.bioinformatics.org) site is an information portal and they also host relevant bioinformatics tools and projects.

EBI Tools Index (http://www.ebi.ac.uk/Tools/) provides a good preview of various tools from EBI.

DAVID (https://david.ncifcrf.gov/) is a user-friendly tool for pathway analysis.

Biocourseware (http://www.biocourseware.com/) is the e-learning development group of the TouchApp Limited providing interactive and engaging learning experiences for users of iPads and Android tablets. A few of their apps are *Genetic Decoder*, a handy genetic code converter for students and researchers in biological sciences. Simply click to change DNA or RNA codons and related amino acid information will be displayed. *DNA Mapping* is a free web tool to locate and map enzyme cutting sites of your DNA sequence. *DNA Sequence Editor* is a free web tool for analyzing and editing your DNA sequence.

8.5.2 Databases

As more protein sequences were obtained, it become a big problem to compare the similarity between two or more sequences. Such evaluation requires aligning the sequences. The number of sequences exceeded the manual alignment capability.

In the 1960s, Margaret Dayhoff created a database of protein sequences. In 1965, she started to publish an annual "Atlas of Protein Sequence and Structure." At first, it included all 65 protein sequences that were known at the time, and the number started to increase every year as more and more sequences were added. That was the first *bioinformatics database*. In 1983, it evolved into the *PIR*, which was incorporated into *UniProt* (https://www.uniprot.org/).

In 1982, a new database was created to store nucleotide sequences: *GenBank* (https://www.ncbi.nlm.nih.gov/genbank/). It grew exponentially, doubling every 2 years.

To provide a systematic assessment of the growing number of prediction methods, John Moult started Critical Assessment of Structure Prediction (CASP13) in 1994, which has been running for over 25 years now (http://predictioncenter.org/casp13/index.cgi).

With the arrival of the NGS technologies (starting in 2005), the *Sequence Reading*

Sequence (SRA) (https://www.ncbi.nlm.nih.gov/sra) was created in 2007 to store the still increasing quantity of read sequences. Now, it has over 1900 billion nucleotides, doubling every 6 months. As it grew too fast, in 2011, the NIH (National Institutes of Health) considered interrupting the SRA owing to budgetary constraints, but this has not happened fortunately.

Other very useful databases are:

ChEMBL (https://www.ebi.ac.uk/chembldb/) is used to introduce small molecule databases.
PubChem (https://pubchem.ncbi.nlm.nih.gov/) is similar to ChEMBL, a way of introducing small molecule databases.
STRING (https://string-db.org/cgi/input.pl)—known and predicted protein–protein interactions.
BioGPS (http://biogps.org/#goto=welcome)—provides a quick overview of genes of interest from multiple databases. It can be used to introduce the concept of plugins and to ask the students to create their own plugins, layouts, etc.
Online Bioinformatics Resources Collection (OBRC) (http://www.hsls.pitt.edu/obrc/) is a good resource, a collection of many databases.

Nowadays, more and more algorithms and databases have been created to analyze data on gene expression by a number of start-ups worldwide.

A summary of databases described in Sect. 8.5 is presented in Table 8.2.

8.6 e-Learning Platforms for Educational Programs and Training in Bioinformatics

Modern technologies also streamline the process of continuous training. Their use eliminates the inconvenience of interrupting the professional activity of the trainees. Travel difficulties are also overcome. Thus, a modern discipline such as bioinformatics is found in many digital educational programs using e-learning platforms for teaching. Online platforms are also used to train specialists, from various related fields, in the field of bioinformatics.

According to the Cambridge dictionary, the term *e-learning* is defined as home-based learning and using computers and Internet-based courses.

To accomplish its function, e-learning is provided through a virtual learning environment (VLE). It is defined as a Web-based platform called the *e-learning platform*, which delivers the course content in digital format and is usually integrated within the educational institutions. Typically, a VLE allows participants to be organized into groups and roles. Also, a *virtual course* is structured into resources, activities, and interactions, providing for the various stages of

Table 8.2 The main databases used as tools in bioinformatics analysis

Database name	Link address	Description
UniProt	https://www.uniprot.org/	Protein database
GenBank	https://www.ncbi.nlm.nih.gov/genbank/	Nucleotide database
CASP13	http://predictioncenter.org/casp13/index.cgi	Protein structure modeling
Sequence Reading Sequence	https://www.ncbi.nlm.nih.gov/sra	Nucleotide database
ChEMBL	https://www.ebi.ac.uk/chembldb/	Small molecule databases
PubChem	https://pubchem.ncbi.nlm.nih.gov	Small molecule databases
STRING	https://string-db.org/cgi/input.pl	Database with both known and predicted protein–protein interactions
BioGPS	http://biogps.org/#goto=welcome	Provides a quick overview of genes of interest from multiple databases
OBRC	http://www.hsls.pitt.edu/obrc/	Collection of more databases

evaluation in addition to reporting on participation.

Without claiming an exhaustive list, some of the most well-known and learned educational and training platforms in bioinformatics are listed below.

Definitely one of the most popular and frequently used online platforms with resources in the field is *Rosalind* (http://rosalind.info/about/), a platform with bioinformatics resources ranging from solving computational biology problems in molecular biology to providing an educational learning platform on bioinformatics topics. The platform is accessed by creating an account. As the authors of this platform describe it, Rosalind's name commemorates Rosalind Franklin, whose X-ray crystallography with Raymond Gosling facilitated the discovery of the DNA double helix by Watson and Crick. The platform offers a wide variety of resources for learning bioinformatics (http://rosalind.info/about/).

Another very useful learning resource in the online environment is *Bio.Info* (https://edu.t-bio.info/), a portal offering many online courses and specializations in a wide range of disciplines such as epigenetics, metagenomics, machine learning, genomics, etc.

Bio.Info "is an educational website maintained by Pine Biotech, a tech transfer company from the University of Haifa that specializes in solutions for Bioinformatics (https://edu.t-bio.info/).

Bioinformatics Barcelona Association (BIB) (http://www.bioinformaticsbarcelona.eu) is a non-profit association for the provision of education and training, the promotion of advanced research, knowledge and technology transfer, the stimulation of competitiveness and innovation within the industrial sector, and the provision of greater visibility as an international node in the field of bioinformatics.

A well-known platform that offers MOOC-type courses for both individual training and training of large and very large company employees is *Coursera.org*. Coursera provides access to education, partnering with universities and organizations offering courses online. When a course is completed, an electronic course certificate is received (https://www.coursera.org/).

Some of the best beginner and intermediate level courses or specializations on this platform are: Bioinformatics Specialization (offered by San Diego University) and Bioinformatics: Introduction and Methods, offered by Peking University (https://www.coursera.org/specializations/bioinformatics).

Bioinformatics Specialization includes courses on Finding Hidden Messages in DNA; Genome Sequencing; Comparing Genes, Proteins, and Genomes; Molecular Evolution; Genomic Data Science and Clustering; Finding Mutations in DNA and Proteins (https://www.coursera.org/).

As mentioned above, this list is by no means an exhaustive one, but it give a picture of the educational and training offer for bioinformatics.

8.7 Bioinformatics Start-ups in the EU: European vs US Context

In terms of start-ups in the bioinformatics field, the European and US ecosystems are very different.

The disparity is notable given that the EU economy is comparable with that in the USA, and also like the USA, Europe creates many top mathematicians, geneticists, biologists, computer scientists, and software developers.

One of the most important reasons why Europe is behind the rest of the world is funding. Raising significantly larger amounts of money in the USA allows young companies to fulfil their needs to keep pace, unlike their competitors in Europe who are having trouble raising funds for start-up incubators.

Another important difference in start-up approaches in Europe compared with the USA is: first income or first growth? American start-ups compete in a huge market and need a high degree of market penetration to gain a competitive edge over potential imitators. Therefore, growth and traction are the main success factors for US new businesses, unlike their European counterparts.

The slower growth rate of European start-ups is also due to the diversity of European markets.

Considering the need to translate products into several languages, and to build partnerships in different countries, economic growth is locked locally. In short: it is impossible for a market, a social network or other platforms to have access to all European countries at the same time. Because they tend to be multibillion-dollar opportunities, again, it becomes clear why there are fewer unicorn companies in Europe.

A few examples of the most notable start-ups in the European field of bioinformatics are:

1. *Repositive.io* (https://repositive.io) founded by Fiona Nielsen, a specialist in computer science in the UK, is based on the concept of a social enterprise (https://repositive.io).
2. *DNADigest* (http://dnadigest.org/index.html), also founded by Fiona Nielsen in the UK in 2013, was aimed at promoting best practices for data sharing in genetics/genomics research through events, online communication, and research publications (http://dnadigest.org/index.html).
3. *DeskGen* (https://www.deskgen.com/), founded by *Victor Dillard, Riley Doyle, and Edward Perello in the UK*. The goal of Desktop Genetics is to discover, understand, and treat the genetic causes of human diseases. At the core of their products and services lies the DESKGEN AI. The DESKGEN AI is a machine-learning platform for designing CRISPR experiments with maximum efficiency, while minimizing unwanted side-effects (also known as "off-target" effects) (https://www.deskgen.com/landing/#/).
4. *Bison SeqTech* (http://www.bison-seqtech.dk), founded by Bent Petersen, Simon Rasmussen, and Thomas Sicheritz-Pontén in Denmark, is a company offering a wide range of customized services such as raw data quality control, data cleaning, bacterial genome assembly, eukaryote genome assembly, human genome assembly, genome annotation, comparative genomics, metagenome assembly and annotation, taxonomical annotation, and functional annotation (http://www.bison-seqtech.dk).
5. *Genestack* (http://www.genestack.com/) was founded in 2012 by Dr. Misha Kapushesky, previously a researcher at the European Bioinformatics Institute in Cambridge, UK. Genestack is focused on multi-omics data management, offering a platform and analysis tools (http://www.genestack.com/).
6. *Eagle Genomics* (https://www.eaglegenomics.com) was founded by Abel Ureta-Vidal in 2008. The company is named after the Eagle Pub in Cambridge, where Francis Crick and James Watson announced their discovery about DNA in 1953 (https://www.eaglegenomics.com).
7. *Genebox* (https://www.genebox.me) is a genomics start-up company incubating at a technology center hosted by the University College Dublin in Ireland and it is aimed at providing genome sequencing, cloud storage data, and a genome data interpretation service (https://www.genebox.me).
8. *Genialis* (https://www.genialis.com/) was founded in 2013 in Slovenia. Its CEO, Nejc Skoberne, makes a succinct description about the history of this startup: "*Back then, two students of computer science were deciding to found a company. We did not completely know what we wanted to do, but then a pharmaceutical company that wished to have a certain software approached the Bioinformatics Laboratory. Then we said to each other "ok, considering we want to have a company, we will just create this software". That is how it started. Before that, I never even worked in bioinformatics, that was something completely new to me, but now we are all in it.*" (https://www.genialis.com/, https://startup.si/en-us/newstext/324/nejc-koberne-genialis-work-with data-represents-a-large-part-of-research-in-biology).

References

Aerobic Capacity. https://xcodelife.co/. Accessed 2 Feb 2019

Amia. https://www.amia.org/applications-informatics/translational-bioinformatics. Accessed 25 Feb 2019

Angel. https://angel.co/station-x. Accessed 27 Feb 2019

Artimo P, Jonnalagedda M, Arnold K, Baratin D, Csardi G, de Castro E, Duvaud S, Flegel V,

Fortier A, Gasteiger E et al (2012) ExPASy: SIB bioinformatics resource portal. Nucleic Acids Res 40: W597–W603
Benson DA, Cavanaugh M, Clark K, Karsch-Mizrachi I, Ostell J, Pruitt KD, Sayers EW (2018) Genbank. Nucleic Acids Res 46:D41–D47
Bison. http://www.bison-seqtech.dk. Accessed 23 Feb 2019
Cancer Genomics Cloud. http://www.cancer-genomicscloud.org. Accessed 24 Feb 2019
CARD. https://card.mcmaster.ca. Accessed 2 Feb 2019
Conversa. https://conversahealth.com. Accessed 2 Feb 2019
Cook CE, Lopez R, Stroe O, Cochrane G, Brooksbank C, Birney E, Apweiler R (2019) The European Bioinformatics Institute in 2018: tools, infrastructure and training. Nucleic Acids Res 47:D15–D22
Coursera. https://www.coursera.org/. Accessed 23 Feb 2019
Coursera. https://www.coursera.org/learn/bioinformatics-pku. Accessed 21 Feb 2019
Coursera. https://www.coursera.org/specializations/bioinformatics. Accessed 2 Feb 2019
Cunningham F, Achuthan P, Akanni W, Allen J, Amode MR, Armean IM, Bennett R, Bhai J, Billis K, Boddu S, Cummins C, Davidson C, Dodiya KJ, Gall A, Girón CG, Gil L, Grego T, Haggerty L, Haskell E, Hourlier T, Izuogu OG, Janacek SH, Juettemann T, Kay M, Laird MR, Lavidas I, Liu Z, Loveland JE, Marugán JC, Maurel T, McMahon AC, Moore B, Morales J, Mudge JM, Nuhn M et al (2019) Ensembl 2019. Nucleic Acids Res 47:D745–D751
DE Shaw Research. https://www.deshawresearch.com. Accessed 24 Feb 2019
DeskGen. https://www.deskgen.com. Accessed 27 Feb 2019
DeskGen. https://www.deskgen.com/landing/#/. Accessed 21 Feb 2019
DNAdigest. http://dnadigest.org/index.html. Accessed Feb 2019
DNAnexus. https://www.dnanexus.com. Accessed 27 Feb 2019
EagleGenomics. https://www.eaglegenomics.com. Accessed 25 Feb 2019
EagleGenomics. https://www.eaglegenomics.com. Accessed 27 Feb 2019
Edgar RC (2004) MUSCLE: a multiple sequence alignment method with reduced time and space complexity. BMC Bioinform 5:113
Edu.t. https://edu.t-bio.info/. Accessed 22 Feb 2019
Elypta. https://elypta.com/science. Accessed 24 Feb 2019
Episciences. https://www.epicsciences.com. Accessed 27 Feb 2019
Gabella C, Durinx C, Appel R (2018) Funding knowledge bases: towards a sustainable funding model for the UniProt use case [version 2; referees: 3 approved]. F1000Research 6(ELIXIR):2051
Genbox. https://www.genebox.me. Accessed 25 Feb 2019
GeneStack. http://www.genestack.com/. Accessed 24 Feb 2019
Genialis. https://startup.si/en-us/newstext/324/nejc-koberne-genialis-work-with-data-represents-a-large-part-of-research-in-biology. Accessed 25 Feb 2019
Genialis. https://www.genialis.com/. Accessed 25 Feb 2019
Globe. https://globenewswire.com. Accessed 27 Feb 2019
Glöckner FO, Yilmaz P, Quast C, Gerken J, Beccati A, Ciuprina A et al (2017) 25 years of serving the community with ribosomal RNA gene reference databases and tools. J Biotechnol 261:169–176
GTN Ltd. https://gtn.ai/index.html. Accessed 24 Feb 2019
Harrison PW, Alako B, Amid C, Cerdeño-Tárraga A, Cleland I, Holt S, Hussein A, Jayathilaka S, Kay S, Keane T et al (2019) The European nucleotide archive in 2018. Nucleic Acids Res 47:D84–D88
Healy MJ et al (2016) Regulatory bioinformatics for food and drug safety. Regul Toxicol Pharmacol 80:342–347
Hesper B, Hogeweg P (1970) Bioinformatica: een werkconcept [Bioinformatics: a working concept]. Kameleon 1(6):28–29
iProteos. http://www.iproteos.com. Accessed 24 Feb 2019
Kodama Y, Mashima J, Kosuge T, Kaminuma E, Ogasawara O, Okubo K, Nakamura Y, Takagi T (2018) DNA data bank of Japan: 30th anniversary. Nucleic Acids Res 46:D30–D35
Kumar S, Stecher G, Tamura K (2016) MEGA7: molecular evolutionary genetics analysis version 7.0 for bigger datasets. Mol Biol Evol 33(7):1870–1874
Kumar S, Stecher G, Li M, Knyaz C, Tamura K (2018) MEGA X: molecular evolutionary genetics analysis across computing platforms. Mol Biol Evol 35 (6):1547–1549
Mitchell AL, Attwood TK, Babbitt PC, Blum M, Bork P, Bridge A, Brown SD, Chang H-Y, El-Gebali S, Fraser MI et al (2018) InterPro in 2019: improving coverage, classification and access to protein sequence annotations. Nucleic Acids Res. https://doi.org/10.1093/nar/gky1100
Prior. Auth now. https://priorauthnow.com. Accessed 2 Feb 2019
Quast C, Pruesse E, Yilmaz P, Gerken J, Schweer T, Yarza P, Peplies J, Glockner FO ((2013)) The SILVA ribosomal RNA gene database project: improved data processing and web-based tools. Nucleic Acids Res 41: D590–D596
Repositive. https://repositive.io. Accessed 21 Feb 2019
Ridom. https://www.ridom.de/seqsphere/. Accessed 2 Feb 2019
Rosalind. http://rosalind.info/about/ Accessed 2 Feb 2019
SecondGenome. https://www.secondgenome.com. Accessed 27 Feb 2019
Seq. http://www.seqhepb.com. Accessed 27 Feb 2019
Servant N et al (2014) Bioinformatics for precision medicine in oncology: principles and application to the SHIVA clinical trial. Front Genet. https://doi.org/10.3389/fgene.2014.00152

Singer J et al (2017) Bioinformatics for precision oncology. Brief Bioinform. doi:https://doi.org/10.1093/bib/bbx143

Szklarczyk D, Gable AL, Lyon D, Junge A, Wyder S, Huerta-Cepas J, Simonovic M, Doncheva NT, Morris JH, Bork P, Jensen LJ, Mering CV (2019) STRING v11: protein–protein association networks with increased coverage, supporting functional discovery in genome-wide experimental datasets. Nucleic Acids Res 47:D607–D613

Taboada EN et al (2017) Food safety in the age of next generation sequencing, bioinformatics, and open data access. Front Microbiol. https://doi.org/10.3389/fmicb.2017.00909

Taly JF, Magis C, Bussotti G, Chang JM, Di Tommaso P, Erb I, Espinosa-Carrasco J, Kemena C, Notredame C (2011) Using the T-Coffee package to build multiple sequence alignments of protein, RNA, DNA sequences and 3D structures. Nat Protoc 6:1669–1682

Yoshida C et al (2016) The *Salmonella In Silico* Typing Resource (SISTR): an open web-accessible tool for rapidly typing and subtyping draft *Salmonella* genome assemblies. PLoS One. https://doi.org/10.1371/journal.pone.0147101

9 Start-Up and Management Features in Biotech Business

Daniela Zirra, George Căruţaşu, Gabriel Eugen Garais, Gabriela-Lucica Mărgărit, and Alexandra Perju-Mitran

Abstract

The start-up of new businesses is strongly influenced by the features of the economic environment, the diversity of business opportunities, the variety and availability of funding sources, and finally the available resources and economic and institutional infrastructure. As in any field, starting a business in biotech involves several steps. From this point of view, we mention that there are no significant differences in this area compared to other fields of activity. We do, however, have to say that the entrepreneurial process in biotech presents several specific elements that every investor must consider. Thus, the innovative feature of any biotechnology approach is crucial to the development of a successful business. At the same time, if the managerial process is the one applied to all types of business, in biotech it must consider the types of strategies that can be applied, the technological level, or the types of partnerships that can be established. Therefore, a further clarification of the elements of the entrepreneurial process in biotech is always useful and necessary.

Keywords

Biotech business · Biotechnology entrepreneurship · Business start-up · Managerial strategies · Innovation process · Business development · High-tech industry

9.1 Introduction

Currently, there are two main industries that are considered new and emerging. These are IT & C (especially the software products sector) and biotechnology industries (Acs et al. 2009: 29). Within them, many small businesses are operating, making the intensity of entrepreneurial activities much higher in these industries than in other areas of the economy. The source of inspiration and knowledge for small businesses, especially start-up ones, is the result of R & D activities carried out in universities, institutes, and performing companies.

At the same time, the factor that generates high economic performance is a quality entrepreneurial activity that is capable of harnessing new

D. Zirra (✉)
Department of Commerce, Economic Integration and Business Administration, Romanian-American University, Bucharest, Romania

G. Căruţaşu · G. E. Garais
Faculty of Computer Science for Business Management, Romanian-American University, Bucharest, Romania

G.-L. Mărgărit
Faculty of Biotechnologies, University of Agronomic Sciences and Veterinary Medicine of Bucharest, Bucharest, Romania

A. Perju-Mitran
Faculty of Management-Marketing, Romanian-American University, Bucharest, Romania

F. Matei, D. Zirra (eds.), *Introduction to Biotech Entrepreneurship: From Idea to Business*,
https://doi.org/10.1007/978-3-030-22141-6_9

knowledge and imparting a high rate of innovation in the actions undertaken. In the global economy, competitive advantage is given by the ability of economies and firms to adapt to rapid business changes, but comparative advantage can only be obtained if firms in a national economy have rapid and rhythmic access to new knowledge. The entrepreneurial process in biotechnology is particularly complex and has a special place in the sustainable development strategies of the most developed countries of the world. Any incentive approach to the biotech industry must be based on a wide range of factors, which can be easily grouped into two main categories.

The first category concerns the microeconomic aspects, namely their clients and their increasingly complex needs; the large number of stakeholders active in this industry and which can sometimes influence the development of the business; employees with outstanding performance in their fields of research, which are the most important "treasure" of the firm; the intensity and quality of the competitive activity in this field; the general public, which can be generally assimilated by the media; new knowledge and technology providers without which no start-up would be possible; and a well-developed cluster network that favors quality partnerships.

Macroeconomic issues are also very important for imprinting the pace of industry development. The main categories of macroeconomic factors that influence biotech industry are demographic developments, the general framework and level of economic development of the country, the natural environment and the available resources, the pace of spreading and assimilation of technological progress, the level of development of society and cultural aspects, and the political and legislative influences and framework.

Entrepreneurship in biotechnology is out of reach for most people because the biotech-entrepreneur can only be a researcher or scientist with notable performance in his field of activity, usually with a high level of international visibility. A new business in biotech industry can only be opened within an innovative cluster, and often the business idea is generated by the existence of a business incubator that has been developed and operates under the wing of a prestigious university. Another particular element for this industry is the existence of very well-developed networks, not only within the cluster but also outside it, with international ramifications. The role of the network is to facilitate communication between scientists themselves on the one hand and between them and support companies (for managerial and commercial activities), potential donors, associations and foundations, innovative knowledge and technology providers, and, last but not least, regulatory bodies in this highly specialized field.

The easier it is to open a new firm, and multi-directional support is more consistent, the more the entrepreneurial phenomenon in biotech industry can grow, and the performance will be more consistent. In this respect, the existence, functionality, and quality of IT & C support tools are vital, such as an E-platform that can be created and developed both to stimulate the biotech start-up and to increase the chances of success of the new business. Taking into account all the issues listed, it is obvious that innovation, new knowledge, and advanced technologies are the key to this industry, which is so special and necessary in the current stage of global economic development. Therefore, innovation management in biotech industry plays an essential role in its growth and development.

9.2 Micro- and Macro-Factors that Stimulate the Opening of a Biotech Business

Micro-factors consist of several actors that are situated close to the company that have the possibility to affect its operation and/or its ability to serve the market. Micro-factors are seen as forces that the company can influence, while macro-factors are beyond the company's sphere of influence.

The most important factors in the company's microenvironment, customers, can be either individual consumers or businesses. Reaching the customer with a value proposition is the purpose of the entire value delivery network, while at the same time, consumer acceptance is the "ultimate obstacle" to the extension of biotechnology in Europe, with consumers having the right to also influence its regulation (Vilella-Vila et al. 2005).

The main reason why European agriculture is still largely free of genetically modified (GM) crops (Aerni et al. 2015) is due to the force of publics with an actual or potential interest in or influence on a company's ability to achieve its objectives. Although the European Commission and many European leaders have been supportive of the further growth of biotechnology, green biotechnology has turned out to be a much more controversial issue in Europe than expected when a coalition of environmental NGOs, consumer groups, and small traditional or organic farmers whose sentiments also found expression in public opinion surveys started an anti-biotech movement (Kurzer and Cooper 2007).

In considering the competitive environment for biotech companies, the key concerns are:

1. The effect of industry structure on competition in the industry. New entrants are generally less likely to enter an industry as dynamic as biotech, where the more established companies have a role in defining industry standards.
2. Differences in the approaches of competing companies, such as price competition, service offers, and aggressive advertising campaigns.
3. Market strategies and strength of market leaders and their closest rivals. This might reveal why some biotech rivals are more successful than others.

A major force shaping competition within an industry is the threat of new entrants. The threat of new entrants is a function of both barriers to entry and the reaction from existing competitors. There are several types of entry barriers:

Economies of scale may require the company to enter the market on a large scale, risking strong reactions from the existing players, or come in on a small scale by accepting a cost disadvantage. Economies of scale refer to the decline in unit costs of a product or service (or an operation, or a function that goes into producing a product or service).

Product differentiation may become an issue by forcing entrants to allocate more money and time than forecasted to overcome existing customer preferences. The capital costs of entering the biotech industry in certain countries can be so large that it discourages market entry.

Independent of scale, existing companies can have cost advantages that are not available to potential entrants regardless of the entrant's size. Such advantages may include access to cheap resources, patent ownership, technological and market know-how, assets obtained earlier at lower costs, more favorable locations for business development, and lower loan costs.

Customers of biotech products can be either companies or individual consumers. When switching from one company to another, the move may incur a switching cost, more precisely a one-time imposition that may not necessarily be financial that the buyer has to deal with if they switch from one company's products. To overcome this situation, new entrants may have to offer first-time buyers a price cut or service, or differentiate themselves by offering higher quality, at the risk of lower profit margins upon market entry.

New entrants may also have to persuade suppliers and distribution channels by providing extra incentives. Suppliers will influence overall value delivery by providing the resources needed by them to produce the company's goods and services. With regard to supplier issues, managers must be vigilant in monitoring supply availability and costs, as shortages or delays, natural disasters, strikes, and other events will negatively influence sales in the short term or force price increases and lower customer satisfaction in the long term.

Partnering with resellers on the European market has become more difficult. European manufacturers don't have many small, independent resellers; they have to partner with large and growing organizations with enough power to impose terms or even restrict access for smaller manufacturers to certain markets.

Intermediaries can be physical distribution companies, marketing services agencies, credit companies, insurance companies, banks, as well as businesses that help the company optimize its own performance.

Biotech companies must consider their own size and position within the industry compared to their competition. Large companies can take strategic measures that smaller firms cannot afford, while small companies can develop strategies that give them better rates of return than large companies.

Micro-factors also include stakeholders and publics with an actual or potential interest in or influence on a company's ability to achieve its objectives.

Before addressing the macro-factors affecting the entrepreneurial process, we will make some assessments of recent developments in the economy as a whole. As we know, globalization is a phenomenon that has a strong impact on all segments of economic and social life, because among other effects, it generates rapid and significant changes both in the way activities are carried out and in the decision-making process. In the current context, influenced by the speed of technological progress, entrepreneurship has, at the same time, become a simple but also more complicated process: simple, because there are virtually no major global barriers to opening and growing and developing new businesses almost anywhere in the world, and complicated, because businesses have to cope with a stronger and more aggressive competitive environment.

It is interesting to look at what is happening in the context of globalization in the high-tech areas of the economy, often called knowledge-based or intense-scientific activities, as we notice an increase in entrepreneurial activity, associated with revenue growth in the field and with a strong clustering process, referring here to biotechnology and communications and information technology industries (Acs et al. 2009: 14).

There is no generally accepted definition for entrepreneurship. Moreover, there is a wide variety of definitions developed in various stages of the development of the entrepreneurship science. If we want to synthesize the different approaches to defining entrepreneurship (Kruger 2004), we can say that it is a specific way to run a business because it considers the following five key aspects: creation—starting a new business or setting up a new company; general management—establishing the main management directions of the new business, together with the allocation of the necessary resources for the business; innovation-based—launching new products, services, processes, working methodologies, technologies, etc. on the market (including the process of creating them); high level of risk associated—the willingness to accept a high level of risk (the higher the degree of novelty of the products or services, or even the markets that enter it, the more pronounced is the risk of failure); and performance-based—achieving high levels of growth and/or profit by creating and operating planned activities.

Measuring developments and performance in the entrepreneurial sphere is also difficult to achieve (Verheul et al. 2002a) since there is no coherent set of indicators, generally accepted and used worldwide, to allow for the development of relevant comparative analyses on the one hand and the rhythm and quality of statistical reporting in the field is deficient on the other hand. Static and dynamic methods of assessing entrepreneurial dynamics are considered, as well as indicators for the business owners and self-employed. One of the most complex approaches of the factors that influence or determine the entrepreneurial activities to evolve in the favorable direction of their development or, on the contrary, to their negative stagnation or very slow evolution is based on six main sections. The theory promoted by authors takes into account both the factors that influence the supply and demand of entrepreneurship and those related to the entrepreneurs themselves, the balance of the "entrepreneurial market," the influences from the governmental sphere, and the factors specific to the analyzed countries.

In terms of entrepreneurship demand, it is based on the existence of a significant number of entrepreneurial opportunities. As a rule, the structure of a national economy along with the growth and diversification of demand for goods and services has a strong impact on entrepreneurship.

The entrepreneurial offer is influenced by many aspects (push and pull aspects), which are often fundamentally different from one country to another, such as demographic aspects (population growth rate, its density in different regions, age structure), the degree of involvement of women/men in entrepreneurial activities, degree of urbanization of the country, development gaps between urban and rural areas, unemployment rate, income level, and the phenomenon of immigration.

We all know that the decision to become an entrepreneur is not easy. A person who decides to enter into entrepreneurial activity relies on the cumulative existence of three groups of factors: the existence of opportunities (factor objective of nature) and the ability to perceive them and fructify them (subjective nature); the appropriate individual profile or traits; and the desire or disposition to accept the risk and the ability to manage it.

The balance of the entrepreneurial market is very easy to understand if we analyze it through the rate of entrances and exits from the entrepreneurial sphere (business ownership rate). If the market is in the equilibrium area, the birth rate (entry rate) will be situated slightly above the death rate (exit rate) of new companies; it means that the entrepreneurial phenomenon is on the right track, and if the trend persists for long periods, it means that the entrepreneurial activity is developing and progressing.

Regarding influences from the government sphere, things are a little more complicated because the range of interventions targets a very diverse set of issues. Among these, we can enumerate: stimulating the growth of opportunities, especially in high-tech areas, by supporting the development of the leading industries, encouraging entrepreneurship through a wide range of incentives, promoting a value system that paves the way for performing entrepreneurial activities, etc. In a simplified manner, government interventions take place in the following segments of the entrepreneurial process:

- Policies elaborated at macroeconomic level, with reference to the tax system and regulations, the labor market (employment contracts, labor force mobility, taxation and other obligations related to salary payments, income level, etc.)
- Policies that regulate the opening and closing of firms (including transition through difficult times and bankruptcy)
- Policies applied in the competition field
- Compliance policies with legislative changes; policies that regulate the financial market and increase the accessibility of the necessary funds
- Policies that provide direct support to entrepreneurial activities
- Sectoral, regional, and local development policies and entrepreneurship stimulating policies in disadvantaged groups on the labor market
- Policies in the field of education to stimulate the increase of the general training level, the training of highly qualified personnel, and the acquisition of entrepreneurial skills.

In terms of cultural aspects, they refer to the cultural values promoted at national level and the status of the entrepreneur in society. At the US economy level, entrepreneurial activities are seen with respect and great openness, and entrepreneurs are supported on multiple levels. From a cultural point of view, business failure on the North American continent is not blamed, but it is seen as an opportunity to gain experience, develop entrepreneurial skills, and then have the determination and capacity to start a new business. At European level there still is serious reluctance to begin entrepreneurial activities: on the one hand, because in some regions or countries, and especially in academic domain it's not usual to have your own business, and, on the other hand, because there is no clear, coherent, and unitary structure that encourages and supports entrepreneurial activities and potential entrepreneurs, especially in high-tech areas, such as the biotechnology industry.

According to specialized studies (Aldrich and Martinez 2010: 418), the web community only needed 3 years to attract the attention of public sector decision makers, while the biotechnology field only did the same after 20 years. But once the biotechnology industry has gained a central stance in all politicians' concerns, the volume of investments has reached billions of dollars. In addition, this star industry status has further strengthened the legitimacy of companies, which has led both to the growth of entrepreneurial activities and to the increasing involvement of world-class universities such as Harvard, Stanford, and the University of California-San Francisco.

9.3 Characteristic Features of the Entrepreneurial Process in Biotechnology

In general, entrepreneurship has attracted the attention of many researchers and economists, and entrepreneurship in innovative industries is more exciting. In this respect, the main sphere of concern addresses some important issues, for example (Acs 2010: 165): the role entrepreneurs have in creating innovative, high-tech new firms; the profile of entrepreneurs in this field and to what extent they are different from entrepreneurs in other areas; what are the public policies to stimulate entrepreneurial activity; and how to stimulate entrepreneurs to achieve performance. In the biotech industry, as in any knowledge-based or science-based industry, entrepreneurs can be considered the most important factors contributing to the transition of innovation as a result of research activities, to the market, and to the end users of new and innovative products. They are responsible and have the ability to identify new business opportunities and to feel the needs of the market.

The current business environment is characterized by rapid changes, driven largely by the speed of technological progress, an increasingly accentuated complexity, the amplification and diversification of risks, and an increasingly aggressive competitive environment. To all these features, we have to add the extraordinary impact of globalization on the activity of the firms and the decision-making processes that take place within them.

Under these circumstances, firms need to adapt quickly if they want to survive and then grow in a sustainable way. This process of adaptation can take place if a few basic conditions, namely innovation, flexibility, a prompt response to change, and a proactive behavioral outlook on business developments in both the national and international contexts, are fulfilled cumulatively. In such a context, the companies that best meet the new requirements are those that operate in high-tech industries such as biotechnologies.

It is obvious that only economic areas that can be characterized by knowledge-based activities or science-based activities can always be up-to-date in relation to change. In addition, as a global industry, biotechnology, with its three segments (red—health, green—agricultural, and white—chemical and environmental), can provide the right response to a series of major problems that the contemporary world is facing.

Another important aspect is the size of the companies. We appreciate that only small entrepreneurial firms, grouped around a new knowledge collection basin, can meet the above-mentioned requirements (innovative, flexible, rapid, proactive). They can use and apply all the ingredients needed for fast, efficient, harmonious, and internationalized growth.

High-tech entrepreneurs (new venture entrepreneurs) can impress their companies with an appropriate rhythm of development through innovation, originality, creativity, and excellence, thus generating a driving phenomenon around them. In turn, the training effect contributes to the enhancement and diversification of entrepreneurial activities in the field in which they operate.

In a global economy, internationalization is one of the keys to success and the vector of achieving high performance. The specific features of the biotechnology industry provide the necessary conditions for a highly internationalized activity: clustering around key knowledge providers, in the most favorable regions, within well-established international collaboration networks.

In order to develop and grow efficiently, the entrepreneurship-based economy needs a favorable framework, focusing on two main directions. On the one hand, we are referring to the whole process through which the conditions that stimulate the elaboration and capitalization of new knowledge can be created. On the other hand, we refer to the development and implementation of policies, instruments, and measures designed to encourage, accelerate, deploy, and develop a quality, highly creative, and profitable entrepreneurial activity. Decision makers in the US economy have understood this very well, so the main concerns are focused on developing and implementing the best policies in this area. Thus, some relevant examples (Acs and Szerb 2006: 112–117) in this regard are as follows: account is taken of the directions in which the global economy evolves, given the developments on the commercial plane, the characteristics of the global migration phenomenon, the creation and dissemination of new knowledge, and elements of technological progress; the foundation of growth and economic development policies is the entrepreneurial phenomenon and its requirements, which is supported by a broad and stimulating range of provisions on taxation, education, research and development activities, the legislation governing the establishment and development of new firms, the right of intellectual property, etc.; the implementation of specific policies aimed at regional development through the formation and development of small innovative companies; and focus on entrepreneurs, to simplify their efforts to set up a new firm, access the necessary funding, protect their rights, etc.

The development of high-tech, highly innovative industries, such as IT&C and biotechnology, has grown rapidly, particularly thanks to the small business (Wuebker et al. 2010: 473). For this reason, one of the most important measures that decision makers can apply is to reduce the costs of setting them up. Substantial efforts are needed to develop clusters dedicated to biotechnology, around the best-performing new knowledge providers, with world-renowned well-known researchers.

The excellent performance of the US biotechnology industry is based on a multilevel approach to entrepreneurship from the public sector (federal governments, at national level, and in the area of concern of local authorities), at which it was decided to place the industry in a central position in the sphere of economic and development policies and to support it through several mechanisms (Ahn and Meeks 2008: 26): research grants, subsidies, capabilities (offices, research laboratories in scientific parks), development of higher education in the field, tax exemptions, development funds, etc. Pennsylvania, for example, decided that "Biotech Industry is one of the key drivers" of economic development, which shows once more the importance it has.

For an accelerated development of highly innovative entrepreneurship, the USA has fully complied with the OECD's recommendations to create and develop a pipeline-type network between universities and other research entities on the one hand and to set up new firms in areas called hot industries, such as biotechnology, nanotechnology, and advanced manufacturing. The main effect of the implemented measures is the growth and development of high-tech entrepreneurship, with a strong impact on the process of transforming new knowledge into marketable innovative products to meet consumer needs.

Biotechnology is primarily an entrepreneurial industry characterized by a high level of uncertainty and risk and by a strong dynamic of the number of entrants and exits of companies in this field of activity (firm's birth rate and death rate).

In the biotechnology industry, the most important values of the firm (being considered as "assets") are researchers and scientists attracted to work without being considered employees with a contract of employment (in the classical sense) in different ways: usually they are employees of universities (or laboratories or research institutes) that collaborate with the company; founder members of the newly created firm; members in scientific councils/scientific advisory board; and Presidents of the Scientific Council (Verheul et al. 2002b: 226).

9.4 Specific Elements of Starting Up a Business in Biotech

Due to the significant impact of high-tech entrepreneurship on economic development at local, regional, and national level, there are some traits of this category of activities that we must mention (Acs 2010: 166): innovation is promoted at all levels; human capital is highly qualitative, and human resources are well-prepared, with a valuable research background; a large share of the capital allocated is venture capital; business growth and development is endogenous as it is a very good vehicle for spreading new knowledge; they are centered mainly on the creation of new and non-service products, which address a large number of consumers; a relatively small initial capital is needed and exposure to the market is higher than in other areas due to the fact that business originators come from the research environment and the dissemination of the results of activities is at an expanded or even global level in many cases.

The decision to open a company in the biotech industry (Kolchinsky 2004: 3) is taken after careful consideration of the following: the size of the costs associated with the development, and then the marketing of a new product; a market need—consumers—the opportunity; an expanded or large but fast-growing market; a stimulating level of competition; the existence of the necessary human resources (researchers and scientists); and the necessary funds and the degree of accessibility of the necessary funds. In addition, the mechanisms underlying the technological transfer and the level of access to the new knowledge providers are also taken into consideration (Nurmemmedov 2004: 9). The success of such an entrepreneurial approach is strongly influenced by the time needed for a newly developed product to become marketable, ownership rights, and an appropriate business model.

Entrepreneurship with high or increased impact, as defined by Acs (2010), is based on three main elements:

- The entrepreneur
- Technological changes
- Venture capital

According to the theories launched by Schumpeter, Knight, and Hayek in different periods of time, the existence of an opportunity is an objective element, and noticing it is influenced by subjective aspects. Entrepreneurs are the ones who have the ability to observe and then to capitalize on an opportunity, and entrepreneurship is the process through which this is done.

In the biotech industry, business is carried out in clustering organizations without which performance cannot be achieved, because of both the complexity of the field of activity and the difficulties that might be encountered in identifying and obtaining the necessary resources.

Generally, for any start-up intention, there is a need first and foremost for an opportunity in the market that the potential entrepreneur decides to exploit.

The decision to open a new business in the biotech industry is never simple, but there are a few elements that need to be emphasized. First of all, it is about some general issues, related opportunity, and the decision to enter into the entrepreneurial process. Secondly, we will refer to support tools that can be used in biotech entrepreneurship to support and stimulate entrepreneurial start-ups.

9.4.1 General Aspects

In general, business opportunities can be analyzed from at least three points of view, namely how the business idea can be put into practice, how the new business will work, and the category in which the opportunity falls, in relation to various classification criteria (Santos et al. 2010: 31). In a simplistic way, by operationalization we can understand the introduction of new goods and/or services, different raw materials, or methods of planning and organizing the activity that generates additional profit, or which can be traded at a price higher than their cost of production.

But equally important is identifying a situation as an opportunity. American researchers in the entrepreneurial domain claim that opportunities exist in the environment, and whoever is interested should only discover and take advantage of them. An opposite direction is the one promoted by European researchers who argue that entrepreneurial opportunities are generated by how individuals can perceive, analyze, and understand what is happening in the environment.

However, we can observe that regardless of how the problem is approached, most opinions tend to think that opportunities are not created but exist, and the difference lies in how individuals get to notice them (to discover them or to understand what is going on around them and, in this way, to identify them).

After an in-depth analysis of the opportunities area, Baron and Ensley (2006) identified five features that make an opportunity be seen as a business opportunity as follows: by deploying it, you can find the answer to a problem that a customer has (there is a potential market for the new product or service); following the implementation of the business idea, a positive cash flow can be obtained, i.e., profit; the associated risk level can be accepted and managed; the new product or service is better than the old one, which was designed to meet the same category needs; and by opening the new business, positive changes can be made in the field.

Regarding the conditions that must be met for an individual to make the decision to open a new business, the same authors (Baron and Ensley 2006) specified that they are the following: a favorable financial forecast was produced, resulting from well-founded calculations; existence of favorable analyses and assessments by experts in the field; the idea has a high degree of novelty; the market exists and is ready to accept the new product or service; and the individual "feels the market" and understands that his idea will be successful.

We draw attention to the fact that entrepreneurial activities in the biotech industry do not just mean the opening of a new business. Entrepreneurial initiatives must also take place in the situation of companies already existent on the market, regardless of their "age." Strengthening, developing, and eventually internationalizing a business can only be achieved in optimal conditions by creating and continuously implementing new ideas and creating an environment conducive to the emergence of entrepreneurial initiatives.

In a continuously changing environment, such as the one today, a biotech company can only gain strong competitiveness through innovation and good quality entrepreneurship.

Being successful in a field of activity, no matter which one, is not easy for anyone. Success in the biotech industry is gained through perseverance, assiduous work, assimilation of new knowledge, transit of difficult times, breaking down of barriers, periods of recovery after failures, and the desire to win. You will say that if you do all this, it's simple, but things are a little different.

To have entrepreneurial success in this industry requires having some features that differentiate the regular entrepreneurs from the ones who really are successful. In decision-making process in entrepreneurship in biotechnologies, regardless of its nature, motivational factors play a very important role. Entrepreneurial activity is provocative and difficult at the same time, so the decision to enter this field is influenced by both contextual factors and personal factors and by the motivation anyone can have to become an entrepreneur.

A particularly interesting model which seems best suited to companies active in the biotech industry, especially start-ups, is focused on three coordinates, namely focus-locus-modus. The main idea that emerges from the proposed model (Onetti et al. 2010) is that the business owner should focus on a quality management of the business and the resources attracted to its enterprise, choosing the most appropriate location for quick access to new knowledge, attracting high-quality employees, and the necessary capabilities. We add the focus to internationalization and activation in a well-developed network.

9.4.2 The Need and Structure of an E-Platform to Simulate a Biotech Start-Up

Computerized information technology is needed nowadays to accompany every social and economic activity. The use of IT offers tremendous support in conducting processes of data analysis and data management. Well-organized data and optimized automated data processing is a basic need and demand for any activity that comes in relation to other activities. By defining the relationship that is established between different levels and categories of data generates lists and reports that offer support to any decision making for the decisional apparatus of a social-economic entity.

E-platforms are now not only an important tool but also indispensable. The main purpose of an E-platform is to store and manage all the information that a company or an individual need to handle at its activities. E-platforms can, from a technical point of view, be as application or as web-based application. Desktop application has now not so many uses anymore because it functions off-line and the need of information and access to collaborative features must be done in real time with online connectivity to the Internet. So, most of the E-platforms are now using web-based technology.

The use of Internet as main transport utility for needed data structures has brought and maximized the efficiency in information management and access. Web-based application offers a real-time access to updated versions of information through advanced data management and data mining tools.

For students and entrepreneurs who want to learn and test economic interactions activity, a protected and simulated economic environment can be used by using special designed E-Platforms that can function through the concept of gamification. In such a way an online platform for even new start-ups can be used to better understand economic interaction between entities and its key activities.

Such a tool gives the advantage of engaging in virtual data optimizations and a large spectrum of analysis over tasks and actions that must be taken in parallel in a real social and economic entity through a simulated E-platform environment that gathers in a game or play type interactions of many simulates competitors economic.

As regards the structure and functionality of the sections they contain, online platforms should meet the following requirements:

- Database structure design must suit the data flexibility for generating complex reports
- Platform general technical sections:
  - Responsive web interface design adaptive for different displays
  - User management with different complex user roles
  - Forms with flexible and versatile design
  - Lists design with multiple filter possibilities
  - Designing internal forms for storing specific simulated entrepreneurial documents to upload: setting up the start-up; recruiting and hiring employees; business plan; quality assurance and marketing; commercial activities; research and development; periodic evaluation of employees; financial reporting and accounting
  - Security modules for preventing IT attacks
  - Import/export data for backup of data
  - Multilingual capabilities of the platform

The needed elements of innovation expected impact and transferability potential that must be included in the E-platform are related to the active and joint interoperability among the following elements:

- Connecting start-up accounts to communicate and evaluate each other's uploaded documents
- Filtering information based on accounts roles from within
  - One start-up
  - Comparing documents uploaded by multiple start-ups
- Joining users from different countries to work as a team to operate a start-up

The platform must provide an interactive framework for all users involved in the simulated project on all legal and linguistic peculiarities to all procedures implemented. In this regard adopt a spiral development cycle specific to IT projects with the following steps.

After sizing the number of potential users and identifying the hardware infrastructure available, the platform administrator references architecture for the client/server application type having following levels: database level, prior studies showing the use of SQL, followed by the interrogating to be achieved through web programming language interface by assuring low cost of developing the application using open source tools. In the design phase the formats will be described in detail and their relationship will influence the forming of the relational database system and tools necessary for load processing.

Implementation of the system will consist in developing effective software product in accordance with the requirements of the detailed design resulting in functional modules of the system. In terms of functionalities, the platform includes the following modules:

- General registration required to access the private content of the application being used as identification in all activities of users on the platform managed by the system administrator.
- Module for managing of roles through which registered users can assign different permissions in the platform such as tutors, entrepreneurs in simulated enterprises, entrepreneurs, and start-ups simulated enterprise employees. The module for simulated enterprises and start-ups will support the general description of the business areas and the associative component, simulated enterprises, or start-up managed system.
- The administration of specific activities undertaken by companies aiming simulated start-up and providing access to reference documents in the field, patterns of achievement of the business plan, and various other documents depending on the specific legislation of each partner. Management module for business recruitment and hiring consultants for simulated enterprises that will support specific activity by providing access to reference documents and templates localized for each partner regarding the evaluation of professional work experience and employment.
- Module for employees and individual projects being used for highlighting activities of each employee of a company or simulated start-up by conducting technical-economic documentation in the following areas: production and services, acquisitions, sales, human resources, financial reporting, and accounting and research/development.
- A financial reporting module that provides support and access to key reference documents on financial reporting methods and localized examples in line with national partners. Individual assessment module is needed which will centralize the evaluation forms that can be used for each activity, specific reference documents, and managing results obtained by the participants in the platform.

It emphasizes that the platform offers a centralized knowledge management in innovation and entrepreneurship containing reference documents that can be used in specific activities related to enterprise simulated and start-ups giving also access to located resources according to the national character and specific legislation for each of the users.

The innovative character of the E-Platform is emphasized through central management of system resources knowledge management providing a full rating system for individual activity and task of simulated enterprises and start-up with the possibility of interaction between entrepreneurs and tutors and providing selective access to documents by activity and country in which it operates.

9.5 Business Managerial Strategies in the Biotech Area

An extremely interesting and useful theory, promoted by Audretsch et al. in 2002 (The eclectic theory of entrepreneurship), was designed to identify the factors that contribute to the stimulation and development of entrepreneurship and the coordinates of the general framework of entrepreneurial activity on the one hand (from the economic goods or product market perspective) and the entrepreneur offer (from labor market perspective). The authors made a wide-ranging analysis of entrepreneurship on three levels: micro-level (entrepreneurial phenomenon as a way of increasing self-employment), meso-level (the entrepreneurial activities and their dynamics), and macro-level (factors influencing the entrepreneurial process through policies, technologies and their transfer, the state of the economy, and the cultural context).

The US economy is more favorable to the entrepreneurial phenomenon than to Europe taken as a whole. On the European continent there are a multitude of countries, that are very different in terms of the level of development and entrepreneurship. The pronounced heterogeneity of European countries makes the analysis of the entrepreneurial phenomenon as a whole difficult to achieve, and the creation of a unitary framework and the determinants of entrepreneurship is almost impossible. At best, general guidelines at European level on entrepreneurship development, together with a general framework and development policy at the level of the European Union, can be mentioned.

In addition, it can be said that the transition to the US entrepreneurial economy has been facilitated by some key strategic issues (Acs and Szerb 2006: 111): simplifying and reducing barriers to entry in highly technological areas, an affordable price level in communications and transport, a friendly taxing system, advantageous financing conditions, and legislation that boosts marketing related to the new and innovative products that are produced by firms and that have been developed as a result of easy access to new knowledge provided by research entities (universities, research institutes, and laboratories).

Another major strategic aspect is the location for the entrepreneurial activities of biotech firms. A study (Ahn and Meeks 2008: 27), carried out in 18 countries and which highlighted 600 companies from the biotech industry, revealed some significant aspects of the elements underlying the decision to choose the best location for its opening and development. The most relevant aspects refer to the proximity of a high-class research entity (university, institute, laboratory), easy access to highly qualified staff, access to funds (venture capital, grants, etc.), a high level of quality of life (schools, services, housing, transport, cultural and sports facilities, etc.), accessible and advantageous areas for activity, a dynamic and performing entrepreneurial environment, access to quality services that supports and fosters the creation of new products (innovative), financial and fiscal support, the availability of patents facilitating technology transfer, support services for commercialization and market access, etc.

Biotech area is a very vast domain containing many resources, ideas, and techniques. To be an entrepreneur in biotechnology requires competence, skills, passion for work, scientific knowledge, and managerial spirit.

In the beginning of entrepreneurial activity, the most important is the planning activity. To reach the target, first of all, is what goal a manager sets for himself, what alternatives he has to get there, and which one is the perfect solution. Peter Drucker said: "Plans are only good intentions unless they immediately degenerate into hard work."

Business managerial strategies are, as a package, an important tool in a biotech entrepreneur activity. Regarding the stage of the company, strategies are different, but the common root is to obtain the specific wanted goal.

If a company is in the beginning, main strategies are the creation of a prototype, or implementation of a concept, a patent, etc. with the identification of supply sources, distribution channels, and achieving one or more partnership. The main goals could be projecting, planning, and organizing activities inside the biotech company, aside from maintaining costs at a low level.

Different approaches are necessary for the growing stage when managers are controlling production flow, the balance between costs and financial returns, managing new products and testing them on the market, and receiving feedback from customers and partners. So, we can talk about:

- Creating R&D activity that focuses on generating cash flow

- Executing a good financing strategy to fund R&D projects
- Researching and developing compelling science products
- Securing intellectual property, technologies, and business ideas

In specific cases, business managerial strategies are focused on identifying project or projects with the highest strategic value through multiple actions like:

- Different profile of product or technology
- The unique mechanism of action
- Top efficiency
- The potential for establishing a new standard of care in the pharmaceutical domain or new standard of production conforming with HACCP in human alimentation
- Sales benefits depending on market size or customers' needs

In many cases, life cycle models are representative of entrepreneurial work. For instance, Kazanjian and Drazin (1990) described the growth model in four phases:

Phase 1—Concept and development, where the main target is focused on the invention and development of product or service. As managerial strategies we need to follow the idea: creating a product, developing it in the prototype, testing the prototype, and searching the financial resources to bring the prototype into large-scale production.

Phase 2—Developing the prototype at the level of large-scale production and preparing the introduction on the market. In this phase, the manager must take into consideration planning and organizing the activities inside his company and all the production stages. For organizing the production flow, it is necessary to solve all technical problems which can appear in the production zone. After that, one big problem remains: how to make the entry of the product on the market.

Phase 3—Growing. How slow or fast is growing the new product on the market is strongly influenced on how well the attention on the market movements is focused. Some mistakes appeared because the company does not make the correlation between clients' necessities and the quantity of the products launched on the market. At the same time, clients need to be assured about the quality of the product, so guaranteeing the quality is very important. The third main problem regarding market is how to properly advertise for attracting the attention of the client for the product. And not the least important issue is how to be solve the personnel problems inside the company.

Phase 4—Stability stage. Strategies are focused on consolidating the market position with the developed product, but also preparing future products.

Making a balance between costs and sales, R&D activities, major opportunities in market penetration, development of a biotech company depend on top manager's strategic vision.

In the last stage, the decline of a company, the managerial team takes actions about the following aspects:

- Finding errors in production
- Finding gaps in technologies cycles, low-qualified executive staff instead of highly qualified
- Finding funds for supplementary capital

In this stage, the company must be evaluated frequently because the level of sales is decreasing, and the next R&D products can no longer be produced.

9.6 The Role of Innovation Management in Biotech Business Development

Innovation is present in all biotech industry segments, and innovation management in this industry plays a vital role in sustainable growth and development. In this chapter, our goal is not to develop the concept of innovation management. The goal we are pursuing is to highlight a few key aspects related to the elements that, from the point of view of innovation management, are favorable factors for this industry. Generally speaking, innovation management means promoting innovative aspects in a company so that it can generate positive effects on its performance.

In short, innovation and its management include generating new products that can reach a wide range of consumers; upgrading and developing existing products in response to novelty items introduced by other competitors; continuous improvement of business processes and activities, both in order to increase performance and to reduce costs and save available resources; identifying and developing that business model that best suits current and future developments in the external economic and social environment.

From our point of view, a leading innovation management in the biotech industry addresses several aspects: creating an industry-wide framework to stimulate and facilitate the transfer of new knowledge and innovative technologies from their generator (universities, research institutes, and laboratories) to companies that are established or active in the field; creating efficient structures in which to create an environment conducive to an intensive process of generating innovative ideas (incubators and dedicated clusters); and creating tools to favor the successful creation and development of knowledge-based companies, scientific-based companies, and innovative-based activities.

Knowledge is an important input in any field of activity, but in high-tech areas such as the biotechnology industry, they are the most important production factor, alongside highly skilled human resources. It is already a well-known fact that for the generation of new knowledge used in economic activities, some basic ingredients (Acs et al. 2009: 28) that should exist in any economy are needed. The first is to have an intense research and development activity (R&D), both in traditional institutions for this category of activities (universities, research institutes, and laboratories) and at the level of companies. The second concerns the availability and possibility to attract to the R&D activities a significant number of highly qualified people with high potential and with a solid experience in the field. In other words, a very high-quality human capital is needed. A third important ingredient is a high number of researchers and scientists who can set a vigorous rhythm of generating new knowledge that is so necessary in today's globalized and highly technological economy.

Biotechnology industry is understood as entrepreneurial activity of scientists. The intensity and importance of this activity in the US economy is revealed by the official statistics by the fact that, in 1980, about half of Massachusetts Institute of Technology's biology staff was involved in starting or participating in one way or another in entrepreneurial activities, and informally the whole staff (with the exception of a single person (Etzkowitz 1998: 823). The phenomenon is special, because of the number of new firms, having as a starter for opening the activity within the university.

Many studies have centered on the value chain elements in the process of producing knowledge in bioscience. An especially important element in this field is related to the way in which new knowledge can be exploited through marketing, especially the entrepreneurial activity devoted to biotechnology. Mainly, there are two main developments that stimulate the performance of the entrepreneurial process in the biotech industry, namely, incubating new businesses (stimulating the opening of new companies in innovative clusters) and co-incubating new businesses, opened as a result of the expansion of networks outside the borders. Then commercialization of new knowledge (their market capitalization) is essential for sustainable development of the biotech industry. In this regard, Cooke et al. (2006: 117) emphasize that an important factor in stimulating the marketing of knowledge (classified as anticipatory, participatory, and precipitating) is the way in which "new knowledge" is disseminated.

As the access to new knowledge is earlier in the production process, the faster the spread rate and the chances of shortening the period from the moment of knowledge emergence until their marketing is reduced. Thus, knowledge can be accessed in three main ways: sequentially, as research advances, not as a final result; in real time, through access to research infrastructure and highly qualified human resources (researchers, scientists); and early access to inventions, innovations, and local breakthroughs.

But to get to the market, knowledge passes through various stages: exploration knowledge (from the core research area), exam knowledge (the first-stage knowledge that has been used for tests or other processes so that they are able to provide feedback on the quality level and their value to future consumers), and exploitation knowledge (knowledge of the second stage, brought into marketable form, able to meet the needs of a wide range of consumers).

Another notable point is that, for example, in high-tech American entrepreneurship, rapid and important steps have been taken to consolidate the process of incorporating the activities of generating new knowledge (Etzkowitz 1998: 824), with those intended to be marketed, in a unitary and well-regulated mechanism. This development has had a favorable impact on biotech entrepreneurial science evolution (that can be assimilated like the result of the amplification of its cross-disciplinarity) as well as the evolution of the biotechnology industry (by intensifying the use and application of research results in industrial processes).

What is important is that as biotech industry has developed and achieved notable performance in some US states, the number of innovative and high-tech clusters has increased (Acs et al. 2009), and competitive research has become increasingly important. Also, in order to further stimulate industry, more emphasis is placed on precompetitive research activity, in order to provide an extra chance for the process of producing and capitalizing on new knowledge and innovative products.

At European level, Germany is one of the countries that pays particular attention to stimulating entrepreneurial activities generated by universities and public research institutes or laboratories. The main objectives of this attention are to stimulate the development of entrepreneurial culture in high-tech areas; to encourage the economic (market) recovery of the results obtained in the scientific research activities; to boost entrepreneurship by opening new innovative start-ups, beneficial to both biotech industry and the national economy. An important effect of these efforts is the intensification of high-tech clustering in biotechnology in regions like Rhein-Ruhr, Dresden, Thueringen, Karlsruhe, and Stuttgart.

Clusters focused on high technology and innovative entrepreneurial firms have been the engine behind the US economy's comparative advantage in knowledge-based era, through the successful production and marketing of new knowledge (Verheul et al. 2002b: 220). Porter (2001: 11) shows that a decisive aspect for an innovative and performing process is the quality of the regional business environment, respectively: available and accessible inputs (workforce, physical infrastructure, research and environment infrastructure, human resources, administrative elements, new knowledge that can be attracted and actually used, information, and capital); the local context that stimulates investment and the pace of knowledge renewal and the intensity of competition; relationship with companies operating within or near the cluster; and structure, characteristics, and dynamics of new product demand.

We want to mention three highly innovative American clusters with outstanding performance: California's Silicon Valley, Massachusetts' or Boston's Route 128 (Boston Innovation District), and North Carolina's Research Triangle. The first two were heavily funded by federal research funds and the third (focused on biotechnology) was formed mainly as a result of North Carolina's efforts, benefiting to a small extent from federal funding. In 1981, North Carolina Biotechnology Centre was established (especially for pharmaceutical activity), through which various grants and services were made available to firms to support the development of biotech companies. In the same area, Microelectronics Centre of North Carolina was established, developed as a "technopolis" to perform research development activities, supported by a powerful business incubator.

The Silicon Valley and Route 128 clusters began to develop around strong university centers, respectively Stanford University and Harvard University together with Massachusetts Institute of Technology. The Research Triangle, with strong state support, had the chance to grow around three powerful universities, also with highly performing research and development activities, like the ones mentioned before, respectively: North Carolina State University, in Raleigh; University of North Carolina, in Chapel Hill; and Duke University, in Durham.

An extremely important step in the development of the biotechnology industry (even before the formation and development of the highly innovative clusters so often found in the US economy) is the business incubators. And in this area, the US economy has overtaken Europe, bearing in mind that the creation and development of incubators has been started long before (Aernoudt 2004: 127). According to the definition given by the American National Business Incubation Association (NBIA), incubators play a significant role in stimulating new business start-ups, whose work is centered on the creation, development, and marketing of innovative products (knowledge-based and innovative-based economic activities). Another very important role of incubators is to impress an increased dynamic of the entrepreneurial process. In other news, the services offered to potential entrepreneurs are very varied: providing space for start-ups; legal advice both in the field of business in general and also specific to the field in which the new business opens; counseling and managerial expertise, which in many cases can make a difference between company survival and development, on the one hand, and their death or failure, on the other hand; easier access to a wide range of sources of funding; providing expertise in the operational process; and facilitating access to markets and consumers.

The more a business incubator is performing, the more impact it will have, generating positive effects on increasing the number of newly established firms, increasing firms' survival chances while they are on their own after leaving the incubator, increasing the chances of sustainable business development, etc.

Another very important aspect for effective innovation management is the protection of intellectual property rights that is vital to biotechnology industry and other high-tech industries. US decisional makers quickly understood this, so they currently have one of the most powerful mechanisms in the world (Acs and Szerb 2006: 111). The more intellectual property rights are protected, the more venture firms are advancing, and the number of new high-tech firms increases. In other news, the USA has developed a two-pronged protection system: one at the national level and one at the level of the states, in parallel with the implementation of a taxation system that encourages intense-innovative entrepreneurship. The last key element is ensuring fast access to the funds needed to open businesses and develop their business within a very high-performing financial system.

Biotechnology industry, for which the fundamental source of value is innovation, is by its nature entrepreneurial and has some defining features: cooperative, competitive, and strong association relationships (proof is the network of innovative clusters, in which small firms in this industry are set up, grow, and develop); strong orientation toward creating and capitalizing on new knowledge; innovation has a systemic character (Cooke 2001: 960).

9.7 Conclusions

Biotechnology industry is an area that is predominantly developed on the basis of the results of the scientific research activity (Acs et al. 2009: 25). The novelty elements and the high pace of innovation in this industry are the key to a high-quality entrepreneurial process and intense entrepreneurial activity. We note that, due to the high degree of sophistication of biotechnology, the clustering phenomenon of companies is particularly significant. Thus, we can observe that the companies active in the biotech industry group their activity in regions that ensure their harmonious development and fast access to new information and knowledge and other very high-quality resources (such as infrastructure and human resources). Clustering poles are, as a rule, universities and/or private companies that invest heavily in research and development (R&D). In fact, they represent the locomotive of the entrepreneurial economy in the regions in which they operate.

In a simplified manner, the workflow that has favored achieving a high level of performance in US biotech is based on the integration of academic activities (universities, laboratories, etc.) those carried out in companies (Aldrich and Martinez 2010). Scientists, academics, and students (doctoral and postdoctoral) have been encouraged to conduct research internships in companies, so strong connections have developed between new knowledge creators and those with the skills needed to market the results. Thus, two main effects have occurred. On the one hand, we can see that, over time, a parallel labor market for scientists has emerged, outside the demand and supply of work in academia and research. On the other hand, highly qualified academic and research staff have acquired and developed entrepreneurial skills and many of them have opened their own firm. To this process we add strong support from public decision makers, investors, and NGOs.

Biotechnology industry is not only important in the US or European economies. Since 2008, countries like India, Singapore, or China have decided to make the creation and development of clusters in biotechnology a priority (Wuebker et al. 2010: 469). For this reason, investments made with venture capital have as their primary objective the activities of software industry, biotechnology industry, and the third place was occupied by the industry focusing on the production of renewable energy.

Due to its complexity, the entrepreneurial process is characterized by multidisciplinarity (Acs and Szerb 2006: 119), which is why its correct and efficient approach is not available to anyone. This is all the more relevant as in the countries with a high level of development the relationship between entrepreneurship and economic development is directly proportional and in developing or underdeveloped countries the relationship is inversely proportional. Finally, we want to bring the American biotech industry model back to the readers' attention, because it is currently the world's most performing and therefore worthy to follow.

There are some features of US biotech entrepreneurship that we want to mention here (Verheul et al. 2002b: 230), of which the most important are:

- Entrepreneurial culture is very well developed—entrepreneurship is encouraged very early through proper education (for the economy in general, for business, entrepreneurial knowledge, entrepreneurial attitude); high-performing American universities attract excellent students from all over the world; successes in the field are recognized and awarded; potential failure is not a hindrance but a chance for a new start; and efforts to open a new business are not a barrier.
- The labor market is very flexible and the contracts are less stringent (formal), which ensures a high mobility of highly qualified staff and collaborations in various forms with firms and other research entities.
- Access to venture capital is ensured through appropriate policies and mechanisms, and the capital market for innovative entrepreneurship is simpler and more transparent (e.g., through Angel Capital Electronic Network funds ranging from \$ 250,000 to \$ 5 million can be accessed).
- Intellectual property rights are carefully protected, being considered extremely important for a performing activity, because in this way the creative-innovative process is stimulated (ideas, knowledge, and new products)—the costs for small US firms are reduced in the biotech industry (as in other high-tech industries) compared to other business areas.
- Reduced costs and simplified bureaucracy for SMEs, with regard to the tax regime, efforts made by the need to adapt to changes, and legislative requirements in the field.

Finally, based on the above-mentioned elements, we can conclude that the most important key factors for stimulating entrepreneurial activity (Shimasaki 2014) for the sustainable growth and development of biotech industry anywhere in the world are appropriate education (and we refer here both in terms of adopting entrepreneurial activities as a profession and in terms of increasing the share of highly qualified staff in scientific research in biotechnology); flexibility of contractual labor market relationships to support researcher mobility; creating a capital market structure that favors investment and simplifies access to finance for businesses in this industry (newly created and existing ones); de-bureaucratization and cost reduction associated with the establishment, operation, and development of firms; and a more flexible and well-structured system in the field of intellectual property protection.

References

Acs ZJ (2010) High impact entrepreneurship. In: Acs ZJ, Audretsch DB (eds) Handbook of entrepreneurship research. An interdisciplinary survey and introduction, 2nd edn. Springer, pp 165–182. https://www.researchgate.net/profile/Sharon_Gifford/publication/226441377_Risk_and_Uncertainty/links/53e28e910cf275a5fdd9ea2e/Risk-and-Uncertainty.pdf#page=412

Acs ZJ, Szerb L (2006) Entrepreneurship, economic growth and public policy. In Small business economics (2007), Number 28, pp 109–122. Springer, doi:https://doi.org/10.1007/s11187-006-9012-3., https://search.proquest.com/docview/220909353?pq-origsite=gscholar

Acs ZJ, Aldrich HE, Audretsch DB, Baumol WJ, Boko S, Gatewood EJ, Johannisson B, Reynolds PD, Sabel C, Thurik RA, Lundström A (eds) (2009) The role of SMEs and entrepreneurship in a globalised economy. Expert report no. 34 to Sweden's Globalisation Council. https://www.government.se/49b731/contentassets/8efd3c3a4c844f88883513fa451760bd/the-role-of-smes-and-entrepreneurship-in-a-globalised-economy

Aerni P, Gagalac F, Scholderer J (2015) The role of biotechnology in combating climate change: a question of politics? Sci Public Policy 43(1):13–28

Aernoudt R (September, 2004) Incubators: tool for entrepreneurship? Small Bus 23(2):127–135. https://search.proquest.com/docview/220943444?pq-origsite=gscholar

Ahn MJ, Meeks M (2008) Building a conducive environment for life science-based entrepreneurship and industry clusters. J Commer Biotechnol 14:20–30. https://doi.org/10.1057/palgrave.jcb.3050076. https://link.springer.com/content/pdf/10.1057%2Fpalgrave.jcb.3050076.pdf

Aldrich HE, Martinez MA (2010) Entrepreneurship as social construction: a multilevel evolutionary approach. In: Acs ZJ, Audretsch DB (eds) Handbook of entrepreneurship research. An interdisciplinary survey and introduction, 2nd edn. Springer, pp 387–430. https://www.researchgate.net/profile/Sharon_Gifford/publication/226441377_Risk_and_Uncertainty/links/53e28e910cf275a5fdd9ea2e/Risk-and-Uncertainty.pdf#page=412

Audretsch DB, Thurik R, Verheul I, Wennekers S (eds) (2002) Understanding entrepreneurship across countries and over time. In: Entrepreneurship: determinants and policy in a European-US comparison. Economics of science, technology and innovation, vol 27, pp 1–10. Kluwer Academic, Dordrecht. https://books.google.ro/books?hl=en&lr=&id=zjQ1MGthzesC&oi=fnd&pg=PA2&dq=Business,+Biotechnology,+Entrepreneurship,+European,+paradigm,+American,+success,+Perspective,+Key+elements,+Gap&ots=n5VO9YQ1cw&sig=RL3NYj9CWly7wDZJMN97ntX2qV8&redir_esc=y#v=onepage&q&f=false

Baron R, Ensley M (2006) Opportunity recognition as the detection of meaningful patterns: evidence from comparisons of novice and experienced entrepreneurs. Manag Sci 52:1331–1344. https://doi.org/10.1287/mnsc.1060.0538

Cooke P (2001) Regional innovation system, clusters and knowledge economy. Ind Corp Chang 10(4):945–974. http://www.ie.ufrj.br/intranet/ie/userintranet/hpp/arquivos/101120164301_Cooke2001RegionalInnovationSystemsclustersandknowledgeeconomy.pdf

Cooke P, Kaufmann D, Levin C, Wilson R (2006) The biosciences knowledge value chain and comparative incubation models. J Technol Transfer 31:115–129. http://citeseerx.ist.psu.edu/viewdoc/download?doi=10.1.1.472.1798&rep=rep1&type=pdf

Etzkowitz H (1998) The norms of entrepreneurial science: cognitive effects of the new university-industry linkages. Res Policy 27:823–833. http://www.oni.uerj.br/media/downloads/1-s2.0-S0048733398000936-main.pdf

Kazanjian RK, Drazin R (1990) A stage-contingent model of design and growth for technology based new ventures. J Bus Ventur 5:137–150

Kolchinsky P (2004) The entrepreneur's guide to a biotech startup, 4th edn. https://www.ctsi.ucla.edu/researcher-resources/files/view/docs/EGBS4_Kolchinsky.pdf

Kruger ME (April 2004) Creativity in the entrepreneurship domain, University of Pretoria. https://repository.up.ac.za/bitstream/handle/2263/27491/Complete.pdf?sequence=11

Kurzer P, Cooper A (2007) Consumer activism, EU institutions and global markets: The struggle over biotech foods. J Publ Policy 27(2):103–128

Nurmemmedov E (2004) Bio-entrepreneurial partnership – a tool for biotechnology transfer. A systems of innovation approach. Master Thesis, Lund University International Master's Program in Environmental Science. https://www.lumes.lu.se/sites/lumes.lu.se/files/nurmemmedov_elmar.pdf

Onetti A, Zucchella A, Jones MV, McDougall-Covin PP (10 August, 2010) Internationalization, innovation and entrepreneurship: business models for new technology-based firms. Springer Science+Business

Media. doi https://doi.org/10.1007/s10997-010-9154-1. https://s3.amazonaws.com/academia.edu.documents/43010284/Internationalization_innovation_and_entr20160224-19946-1x4c2ch.pdf?AWSAccessKeyId=AKIAIWOWYYGZ2Y53UL3A&Expires=1549643076&Signature=XO7Y3Z25gdiANF%2FKMOVPDKWirX4%3D&response-content-disposition=inline%3B%20filename%3DInternationalization_innovation_and_entr.pdf

Porter ME (2001) Research triangle. Harvard Business School, Council of Competitiveness, Monitor Group, ontheFRONTIER. https://www.hbs.edu/faculty/Publication%20Files/COI_ResearchTriangle_cfb27f08-1329-4637-b6b0-d185d9f418ee.pdf

Santos SC, Curral L, Caetano A (2010) Cognitive maps in early entrepreneurship stages. From motivation to implementation. Entrep Innov 11(1):29–44. https://repositorio.iscte-iul.pt/bitstream/10071/7860/1/post_print_Santos,%20Curral%20%26%20Caetano,%202010%20IJEI.pdf

Shimasaki CD (ed) (2014) Biotechnology entrepreneurship: starting, managing, and leading biotech companies. Elsevier Inc. https://www.ott.nih.gov/sites/default/files/documents/pdfs/Ferguson-Chapter14-BiotechEntrepreneur-2014.pdf

Verheul I, Wennekers S, Audretsch DB, Thurik R (2002a) An eclectic theory of entrepreneurship: policies, institutions and culture. In: Audretsch DB, Thurik R, Verheul I, Wennekers S (eds) Entrepreneurship: determinants and policy in a European-US comparison. Economics of science, technology and innovation, vol 27. Kluwer Academic, Dordrecht, pp 13–81. https://books.google.ro/books?hl=en&lr=&id=zjQ1MGthzesC&oi=fnd&pg=PA2&dq=Business,+Biotechnology,+Entrepreneurship,+European,+paradigm,+American,+success,+Perspective,+Key+elements,+Gap&ots=n5VO9YQ1cw&sig=RL3NYj9CWly7wDZJMN97ntX2qV8&redir_esc=y#v=onepage&q&f=false

Verheul I, Bosma N, von der Nol F, Wong T (2002b) Determinants of entrepreneurship in the United States of America. In: Audretsch DB, Thurik R, Verheul I, Wennekers S (eds) Entrepreneurship: determinants and policy in a European-US comparison. Economics of science, technology and innovation, vol 27. Kluwer Academic, Dordrecht, pp 209–245. https://books.google.ro/books?hl=en&lr=&id=zjQ1MGthzesC&oi=fnd&pg=PA2&dq=Business,+Biotechnology,+Entrepreneurship,+European,+paradigm,+American,+success,+Perspective,+Key+elements,+Gap&ots=n5VO9YQ1cw&sig=RL3NYj9CWly7wDZJMN97ntX2qV8&redir_esc=y#v=onepage&q&f=false

Vilella-Vila M, Costa-Font J, Mossialos E (2005) Consumer involvement and acceptance of biotechnology in the European Union: a specific focus on Spain and the UK. Int J Consum Stud 29(2):108–118

Wuebker R, Acs ZJ, Florida R (2010) The globalization of innovation and entrepreneurial talent. In: Acs ZJ, Audretsch DB (eds) Handbook of entrepreneurship research. An interdisciplinary survey and introduction, 2nd edn. Springer, pp 457–484. https://www.researchgate.net/profile/Sharon_Gifford/publication/226441377_Risk_and_Uncertainty/links/53e28e910cf275a5fdd9ea2e/Risk-and-Uncertainty.pdf#page=412

Financing and Investment in Biotechnology

10

Cristian Paun

Abstract

Capital is the most important production factor. In the absence of the capital, the other production factors remain unused. Capital accumulation is the key element for any business initiation and development. The limited access to finance is the common problem of any start-up or SME. However, innovative sectors (such as biotech) have more financing options in the early stages (such as venture capital, business angels, investment funds, mezzanine financings). This chapter is discussing the main challenges for initiating a business in the biotech sector from this perspective of attracting more capital that can be provided by classical 3Fs (known as "family, friends and other fools"). The chapter is presenting also the alternatives for later stages when the business became mature and is able to attract more long-term capital from capital markets (such as initial public offering—IPO, leasing, supplier credit facility, or buyer credit facility). The financing of innovative sectors is also considered as a priority for many governments of developed and emerging countries. This chapter is introducing some of the specific public financing schemes and state-aid mechanisms that are available now for innovative sectors such as the biotech sector.

Keywords

Capital · Financing · Investments · Financial markets · Financial institutions · Financial instruments · Cost of capital · Financial risks

Capital is the most important production factor among the others (natural resources, labor, or entrepreneurship. Without capital, any entrepreneurial endeavor cannot start or is limited to early stages. Genuine capital is generated by capital accumulation through saving it from consumption. Capital is the result of those individuals that postpone their consumption for later for various reasons (e.g., future is uncertain and saving reserves for that time is a wise decision). Entrepreneurship supposes always to deal with market disequilibrium: to identify an unexploited consumption needs (that are, in fact, unlimited) and to allocate limited resources to produce goods or services that are addressed to them (Kirzner 1978). Entrepreneurship cannot be detached from its important feature: the complete assuming of uncertainty derived from the permanent action in the future (von Mises 2010). There is a clear distinction between *the entrepreneurs* (investors assuming the whole business uncertainty) and *the capitalists* (those who are saving and lending their capital to the entrepreneurs without

C. Paun (✉)
The Bucharest University of Economic Studies, București, Romania
e-mail: cristian.paun@rei.ase.ro

F. Matei, D. Zirra (eds.), *Introduction to Biotech Entrepreneurship: From Idea to Business*,
https://doi.org/10.1007/978-3-030-22141-6_10

assuming the ownership, decisions, and business uncertainty) (Rothbard 2007).

Before any business starts, the entrepreneurs are trying to find enough capital for their production process from various sources. So, before being able to produce and to sell something to the market, you should be able to sell your business ideas to third parties (investors and capitalists) and to try to get their capital by issuing financial instruments (Corelli 2018). The access to capital is depending on the market experience (for start-ups is more difficult than for mature businesses), the type of the business (how innovative, how technology intensive), the existence and the quality of real assets, the credibility of initial investors, the existence of additional security arrangements before asking for capital (like contracts signed in advance), etc. (Mishkin and Eakins 2017). Moreover, the success or failure of a business financing strategies is also highly influenced by the development and sophistication of financial system in the host country (Cifci et al. 2017; Deltuvaitė and Sinevičienė 2014). The lack of financing alternatives is increasing the cost of capital and the financing risks.

10.1 Financing and Investment in the Business Initiation and Development

When entrepreneurs wants to initiate a business their access to capital will be very limited due to the following reasons (at least): (1) the business experience cannot be proven or is limited to the initiator of the business (as initial equity investors providing the seeding capital to the business); (2) the small size of a start-up (the insignificant value of real fixed assets—machineries, vehicles, buildings—is limiting the creditworthiness of the business; the lack of collaterals); (3) the market for the business is not clear and sustainable (the lack of clients; the potential of the business is limited to the personal beliefs of initiators); (4) the business is not yet a bankable entity (despite the fact the initiator could be); (5) the asymmetry of information; and (6) the access of start-ups to more sophisticated financing operations is very limited (the access to capital markets, for instance) (Munro 2013).

The seed capital for a business is merely provided by the individual investors that are starting that business and the capitalists that are directly closed to them—informal investors generally recognized as "family, friends, and other fools" and that are accepting a high risk of any early stage of any business (Freeman 2012). The financing sources for any business are *internal* (generated by the business itself) and *external* (provided by third parties). Among internally generated financing sources of capital we can identify reinvested profits, depreciation and amortization of fixed assets, and any further increase of equity capital subscribed by former shareholders. Internal financing resources are very limited (as volume), are depending on the size of the assets (of the business), and are more expensive than external ones. When you are looking internally for financing your business, you should not be transparent with third parties outside of it (providing financial statements, for instance) or you will not alter your ownership and decision (by including external stakeholders). External financing resources suppose the involvement of external third parties providing capital to the business by lending (credit and bonds) or by investing capital (equities). Due to their mentioned features, *start-ups and early-stage businesses are more focused on internal financing resources* than external ones (Rossi 2018). A business can access the financial resources *directly* and *indirectly* (by involving financial markets and institutions). The direct access of capital supposes a required knowledge (to know where and how this capital can be accessed), a better understanding of risks, a limited volume of capital, and a very limited type of financing schemes and instruments (Saunders and Cornett 2018). The indirect access of capital means the involvement of financial intermediaries in two ways: the financial institution is limited to sell the financial instrument issued by a company sharing the risks of this mechanism (the financial institution does not alter the financial instrument) and the financial institution is previously selling its own financial instruments (bonds, equities, deposit certificates) and is including in his own portfolio the financial assets issued by companies (including start-ups, early-stage businesses and SMEs). Unfortunately, financial intermediation is very important but is not so accessible by early-stage businesses (with minor exceptions) (Chesini et al. 2018). The success of any start-ups is highly depending on its capacity to be connected to the financial markets and institutions intermediating local and international capital. When we are discussing about business financing, the distinction between short-term and long-term capital is important too. Long-term capital is involved in the development of any business, and it is redirected toward the acquisition of fixed assets directly involved in the production process (Atche 2017). Short-term capital is needed for business's operations (commercial credit, financing of inventories of raw materials or finished production, etc.). For a business, financing long-term financing needs (buying machineries) by involving short-term capital (and the opposite is the same) is very inefficient. Long-term capital is provided by capital markets (bonds and equities) and less by commercial banks (loans). A start-up or an early-stage company will be limited to its own capital or reinvested capital that is assigned to the long-term capital. In this case, short-term financing needs (working capital) will be covered by long-term capital. Due to that, any business initiation will face with additional financing costs compared with mature businesses, until the company will become a bankable and credible one for capitalist and investors (OECD 2015).

In conclusion, financing is one of the most sensitive for any business initiation and growth. New started businesses will encounter high difficulties to attract important capital from the beginning. Therefore, the start-ups will be limited to families, small investors closed to the initiators of the business. This problem is more sensitive when initiation refers to capital-intensive production processes (such as is the case of biotechnology). Additional specific capital providers (business angels, venture capital funds, private investors) are needed. The public support (grants provided by the government) is also vital, in this case.

10.2 The Specific Case of Financing and Investing in the BioTech Businesses

Financing the seed capital for starting a business implies special features: the lack of proven previous experience, the lack of required security arrangements (contracts signed in advance with suppliers or purchasers), the higher uncertainty of this sector (derived from new technologies proposed for production process, from new products proposed to the market, etc.), or the higher innovative status than regular businesses.

In theory, there is a direct link between stages of a business and the financing strategies available. Innovative sectors (including biotech) have more opportunities for early stages and expansion phase of the business's lifetime than non-innovative ones (Cardullo 1999).

In the case of *non-innovative* sectors, the value added is much reduced, the competition is very high, and the finished products are not so differentiated between the competitors. In such economic sectors, the returns are very low. The competitors' market strategies are focused on costs and prices not on the products' differentiations or markets' diversification. This very reduced profitability rate forces such businesses to be concentrated on a very narrow number of financing strategies after the capital is seeded into the start-up by family, friends, and other "fools" (closed people to the initiators of the business). After the business becomes a bankable one, the loans are merely used to finance the short- and the long-term financing needs. Later, if the business will become big enough, the capital markets can be a reliable alternative to provide needed long-term capital (through corporate bonds). At the maturity of this business, equity financing is finally accessed by non-innovative businesses.

In the case of *innovative economic sectors*, such as biotech, the financing strategies are completely different, especially for the early and expansion stages (see Fig. 10.1). For the high-tech and very innovative businesses the early-

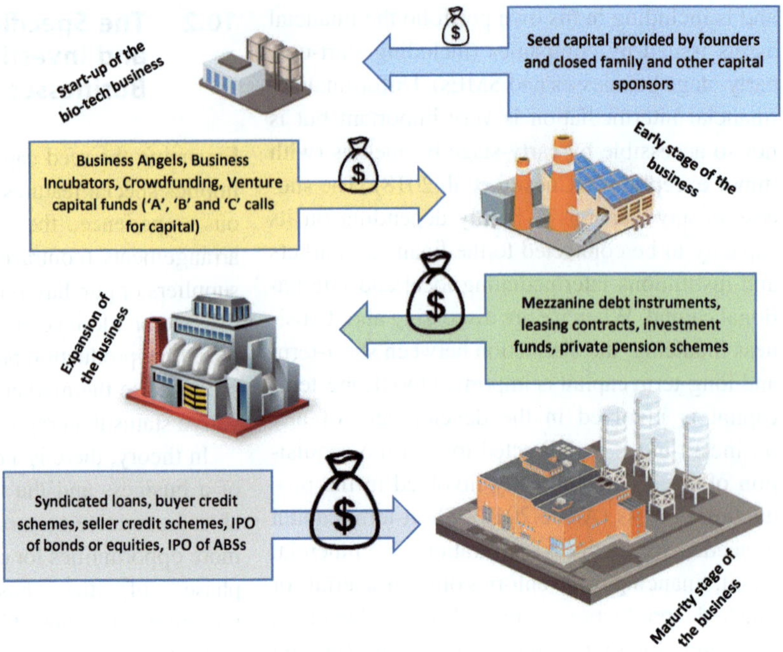

Fig. 10.1 The comparative analysis of relationship between available financing channels and business's stages—the biotech innovative sector case

stage funding is classified into "A," "B," and "C" stages. In the "A" early-stage phase of the biotech business, the most important capital providers are merely looking to the product or service provided to the market, to the already existing base of the clients and the potential of this incipient market, and to the potential of any other markets for the proposed products. The capital raised in this stage is over 20 million $ and often takes 8–12 years to exit from the investment; for very innovative biotech businesses, this amount can be slightly higher. The presence of innovation is boosting a lot the value of the biotech business in the early stages and during the expansion phase. This potential significant increase of the value of the business is explaining the additional financing alternative and the interest of more capital providers to participate and to assume corresponding higher risks that are associated with such activity sectors. The value of capital provided to the business is increasing with each stage of the business (OECD 2004).

The funding of "A," "B," and "C" early stages of a biotech business is based on the capital provided by the following type of investors: biotech *business angels*[1] (private individual investors or grouped investors acting as a network and assuming a very high business risk and requiring a corresponding high expected return and capital gain), biotech *business incubators*[2] (a structure focused on improving the access to loans, guarantees, public financing schemes; on helping with financial management and accounting and helping with the access to other potential early-stage capital providers), biotech *crowdfunding schemes*[3] (a targeted financing scheme based on a social network and providing specific benefits for participating capital investors), and biotech *venture capital funds*[4]

[1] Some of the relevant business angels for biotech sector are Life Science Angels, WINGS—The Medical Technology Angels, Cambridge Business Angels.

[2] Such as the Babraham Bioincubator Concept (UK), Eurasanté Bio-Incubator (France), or Umeå Biotech Incubator (Sweden).

[3] Examples of biotech Crowfunding Schemes: Capital Cell (UK), DavinciCrowd (France).

[4] As the following: BioDiscovery 5 (France), Complexa (Germany), Erytech (France), LSP Health Economics Fund 2 (the Netherlands), and Medicxi Growth 1 (UK).

(private equity individual and/or institutional investors providing capital for early-stage investments and assuming more risks and responsibilities for a very limited time). Almost all investors involved in these stages will exit at the end of the expansion phase by selling their equity participation to the regular individual/institutional investors (Bryant 2014; Cumming and Hornuf 2018).

In the later stage of an innovative business's expansion, *investment funds, pension funds, the leasing mechanisms*, or the *mezzanine debt* (such as convertible bonds into equity financing at the maturity) could be more involved to provide additional capital to this biotech business.

When the business reaches the maturity stage more complex financings can be involved such as *syndicated loans, buyer loan facility, seller loan facility, IPO of bonds and equities*, and *IPO of depositary receipts.*

10.3 Costs and the Risks of Biotech Businesses' Financings

Financing of biotech business supposes costs and risks. The cost of capital is essential for estimating the opportunity of any investment: the estimated future net cash flows of investments are discounted with this cost of capital and the calculated *net present value* should be positive and higher than all other investment opportunities; the *internal rate of return* of any biotech investment should be higher than this estimated cost of capital. In the calculation of the cost of capital, we are using the time value of money and discounting principles. Internal rate of cost is that discount rate determined by equalizing to 0 of the net present value associated with any financing scheme. The investors will always search *to minimize* this cost of capital and *to optimize* the capital structure of their project (the proportion of debt and equity in the capital budget). The financing plan is presenting all selected financing alternatives based on their cost and risk. The cost is estimated for each of them and included in the weighted average cost of capital (WACC) (Baker and Martin 2011; Bierman 2002) (Fig. 10.2).

The investment projects in the biotech sector are often characterized by higher value of the investment project than regular businesses, the higher required recovered time, and, therefore, higher uncertainty regarding the economic value added by these investments. The seeding capital and early-stage financings are mainly focused on equity financings. Debt financing is accessed later, at the end of the expansion stage and during the business's maturity. The cost of equity is higher than the cost of debt due to the tax shield: the dividends are paid after taxation and the interest rates (coupon rates) are paid before taxation. The cost is only one of the pillars in the financing decision. The other one is the risk. The financing of a business involves specific risks not only for creditors but also for debtors. The risks commonly associated with financing schemes are (the debtors' perspective): *interest rate risk* (for

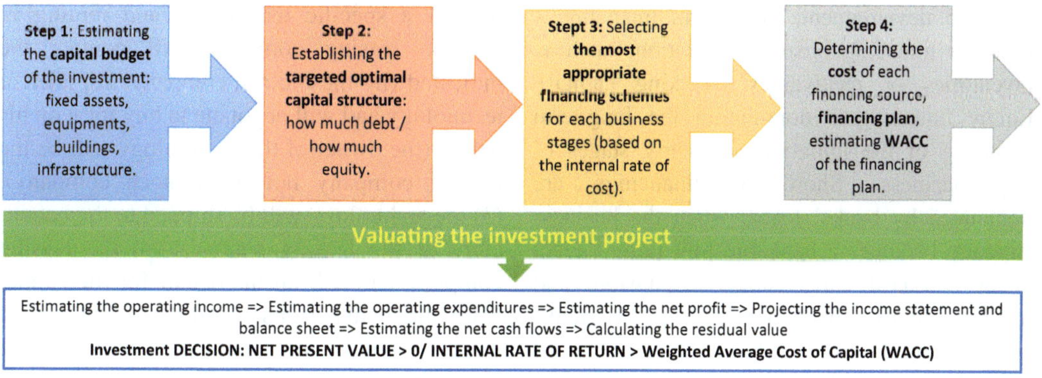

Fig. 10.2 The main steps to estimate the cost of capital for a biotech investment project. Source: Own graphical representation based on Baker and Martin (2011) and Bierman (2002)

a debtor that borrowed capital with variable interest rate, any increase in the interest rate will produce a loss and a higher cost) and *foreign exchange rate risk* (if the capital is borrowed in a foreign currency, the depreciation of local currency will increase the cost of capital producing an extra loss for any company borrowing money from international financial markets). The creditors are additionally exposed to default risk (the risk that any credit would not be reimbursed on time by the debtor), systemic risk (the default of the debtor is produced by the government or by the governance not by its internal problems), and the liquidity risk (due to a mismatching of maturities in their financing operations). All these risks associated with financial flows are influencing the level of interest rate (the cost of capital) and the financing conditions. Finally, we should mention that the quality and the development of the financial system and financial institutions are important for improving the cost and the risks of financing schemes. The competition among venture capitalists, the financial market diversification, and the sophistication of financial products and services improve the risk and the cost of financing mechanisms.

10.4 Financing Alternatives Available for Expansion and Development of the Biotech Business

Long-term financing is used for business expansion and development. Short-term financing is used for financing business operations such as inventories or receivables (working capital). Equity financing is dominating the long-term financing of the initiation and early stages of any business. Short-term financings are dominated by banks' loans, when the business structure becomes a bankable unit (after "death valley"). Latest early stages, expansion, and maturity involve much debt and credit. Banks are reluctant to involve their capital obtained from shorter term banking deposits (up to 2–3 years) into long-term financings due to existing liquidity risk.

In this section, we focus on the main private financing mechanisms available for the expansion and maturity stages of a biotech business:

- *Expansion phase*: leasing financing, syndicated loans, buyer credit, seller credit.
- *Maturity phase*: initial public offering (IPO) of bonds and equities, IPO of depositary receipts (asset-backed securities—ABS).

10.4.1 Leasing Financing of a Biotech Project

Leasing mechanism is a non-banking financing mechanism appropriate for technology-intensive economic sectors (including biotech), for companies that are in the early stages (without so clear market experience, still not well-defined product, etc.). Leasing financing supposes that an independent financial institution (that is not a bank) is involved for technology transfer. This leasing company will buy the indicated technology and will rent it to the biotech company. This decision is taken by the leasing company after an economic and financial analysis of the business plan and of the biotech company (including market perspective, business idea, quality of equity investors involved until that stage, existing security arrangements, etc.).

The mechanism is very simple (Fig. 10.3): (1) The biotech company is signing a contract with a leasing company regarding the acquisition of a specific technology and renting it; (2) the leasing company will contact the indicated (by the biotech company) supplier of technology and will negotiate to buy for it on his behalf (the owner of this technology will be the leasing company not the biotech company); (3) the technology will be shipped to the beneficiary of the leasing contract; (4) the leasing company will ask on his behalf for the capital from the banks (loans) and from the capital markets (bonds and equities); (5) the leasing company will pay on demand the supplier of the technology; (6) the biotech company will install the technology and will start to produce

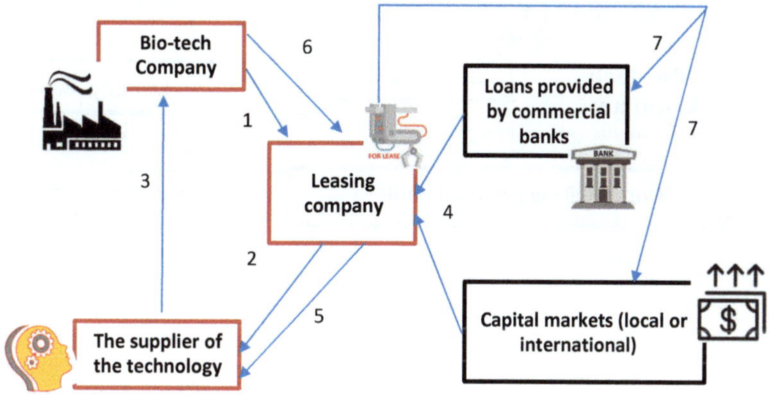

Fig. 10.3 Leasing mechanism for biotech business

and to sell to the markets; from these revenues, the leasing company will pay for the rentals to the leasing company; and, finally, (7) the leasing company will pay back for the debts and loans or will pay for dividends (the case of equity financings) (Blackstaff 2001).

The benefits of this long-term financing scheme are the following: the leasing company will share the business risks together with biotech companies (the moral depreciation of the technology, market risk, etc.). Moreover, this financing alternative is suitable for early-stage businesses when fixed assets (used as real guarantees) are not so consistent. This financing scheme is also very convenient due to the available options at the end of the contract: to buy the equipment to its residual value, to continue to rent the equipment, or to exit from this contract without costs and penalties. The contract has also the option to include the depreciation and amortization costs on the beneficiary's operating expenditures (in this case the contract is covering all the lifetime of the equipment and the three options at the end are missing).

10.4.2 Syndicated Loans and Biotech Sector

Long-term business loans are very complicated for commercial banks due to the source of their capital (term deposits that have a shorter maturity than long-term business credit). Due to this potential liquidity risk, the commercial banks involved in the long-term crediting operations for business sector development are syndicating the loans by involving more banks in them. These banks can be located in different countries and are ranked based on their experience in participating/coordinating such complex financing schemes (the BANKSCOPE system, for example, can be used to identify the most experienced bank to syndicate such loans in a specific region, sector, or country).

The mechanism of syndicated loan is the following (Fig. 10.4): (1) the biotech company (as beneficiary of this long-term loan) will contact a leadership bank (arranging bank) that will coordinate the whole financing scheme; (2) if the amount of capital or the maturity is too demanding for this leading bank, an additional coordinating group of banks will be created by leadership (based on its own connections); (3) the management group of banks will analyze the financial situation of the biotech company and will establish the loan conditions (interest rate, maturity, reimbursement schedule, grace period, sharing of risks and obligations, etc.); (4) an additional syndicate of banks is created in order to increase the financing capacity and to improve the risk sharing; (5) if the amount of capital cannot be completely covered by the banks from the first two syndicates, a credit memorandum will be addressed to local or international financial markets; based on it, more banks can join to this complex crediting operation (6).

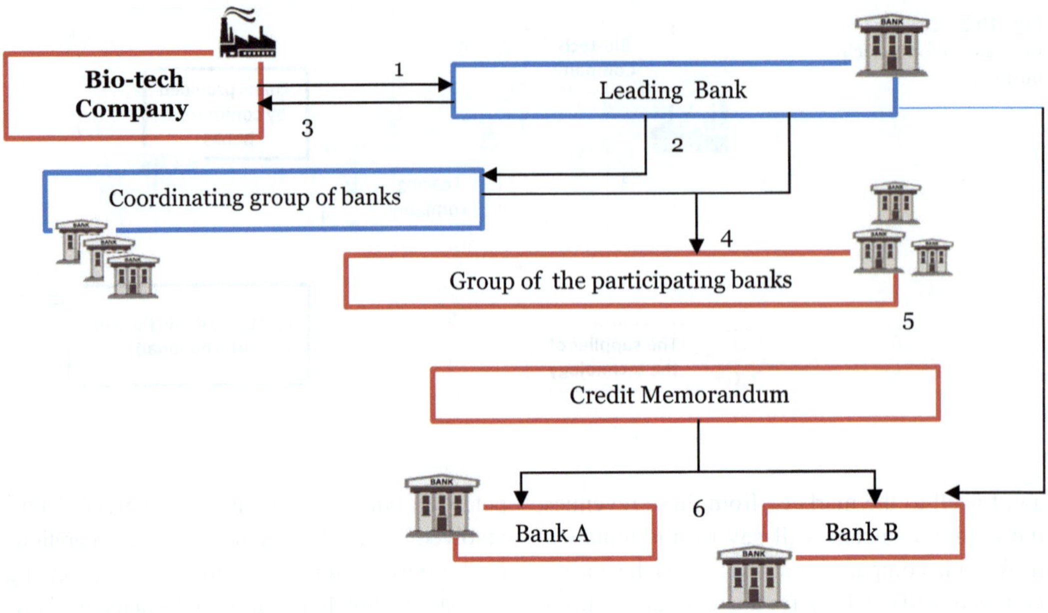

Fig. 10.4 The mechanism of the syndicated loan

The majority of capital is provided by leading bank (and coordinating group)—2/3 of the capital in this case. The remaining capital is provided by the group of the participating banks and, if this group will fail to provide that remaining 1/3 share, the banks that subscribed to the syndicated loan memorandum. This long-term loan facility is also suitable for mature biotech businesses and for project financings (public and private partnerships).

10.4.3 Supplier and Buyer Loan Facility

These two long-term loan facilities are suitable for biotech businesses when technology is imported from abroad. Supplier and buyer credit are linked to high-tech exports of complex machineries and equipment. The higher risk for banks involved in a long-term crediting scheme is solved this time not by the syndication of banks but by the involvement of governmental agencies in the export crediting operations (these agencies are partially supporting the risk and the cost of financing by providing special refinancing conditions to the banks involved in such operations).

The mechanism of the supplier loan facility is the following (see Fig. 10.5): (1) the company that is exporting the technology is contacted by a biotech company for delivering the technology and a guarantor bank is asked to provide a guarantee letter that will cover the default risk of this loan; (2) the biotech will sign a contract for transferring the technology with the international supplier and the technology is shipped according to the negotiated terms; (3) the exporter's bank will provide a long-term loan to the exporter (this loan can be subsidized by governmental agencies); (4) the exporter will pay the debt service; and, finally, (5) the biotech will pay for the import of the technology on the due date (after 3–5 years).

The chief advantage of this mechanism consists in the fact that the biotech will obtain the desired technology without involving its own capital from the beginning. The technology will be paid by the sales of the company to the market. The cost of capital will be included by the exporter of the technology in its price invoiced to the biotech company. Moreover, the exporter will provide its support for finding the necessary

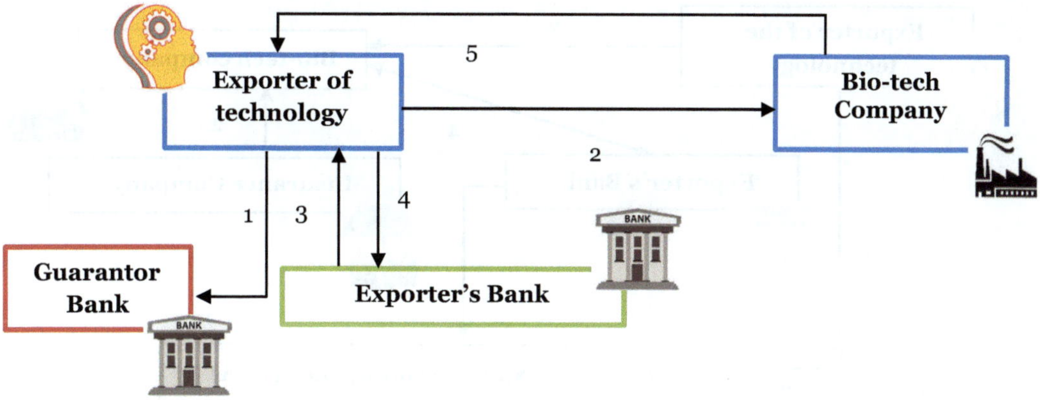

Fig. 10.5 The mechanism of the supplier loan facility

financing in his country, due to the interest to promote the exports abroad, especially high-tech exports.

The other financing facility available, when the technology is imported by biotech company, is the buyer loan facility (see Fig. 10.6). In this case, the exporter of the technology will apply for a guarantee letter that will cover the credit default risk of the involved banks (1); an insurance company from the biotech company's country will provide an insurance policy for political risk (covering the case when the loan will not be repaid due to political problems in the debtor's country) (2); the technology will be contracted and transferred to the biotech company (3); the biotech company will receive a buyer loan facility from the exporting country (interested to stimulate high-tech exports) (4); and will use it to pay on demand for the import of the technology. Later, the loan will be paid back by the biotech company from its sales on the market and according to its business plan. Additionally, an export credit agency (ECA) can refinance the exporter's bank (Willsher 1995).

Again, the biotech company is importing the technology with the financial support of the exporter's bank. Normally, high-tech exports have a financing scheme available (due to the features of this type of contracts—high value, high risk) backed by governmental agencies (such as ECAs; some of them being involved also in the guaranteeing and insuring the financing mechanism). The difference between this mechanism and seller loan facility is the higher risk (the political risk/systemic risk) and, therefore, the higher cost.

10.4.4 Initial Public Offering (IPO) of Bonds/Equities on the Capital Markets (Local or International)

The initial public offering (IPO) of bonds (convertible bonds, bonds with warrants, simple corporate bonds) is used at the end of the early stages to consolidate the existing capital and to prepare the business for its growth and maturity stage. The issuance of equities (combined with bonds and/or with the subscription of bonds against equities) is mainly used after the early stages. The IPO of equities does not provide significant capital for supporting the growth of the business but is a reliable exit mechanism for the previous investors in the early stages of the business (business angels, venture capital investors, investment funds, etc.).

The financing mechanism supposes few distinctive steps: the biotech company interested to finance its business's growth and development will contact a leading bank (an experimented investment bank) that will coordinate the whole IPO; if the amount of capital or the maturity is very demanding for a single coordinating bank, the leadership bank can ask few other investment banks to join the coordination group (this group

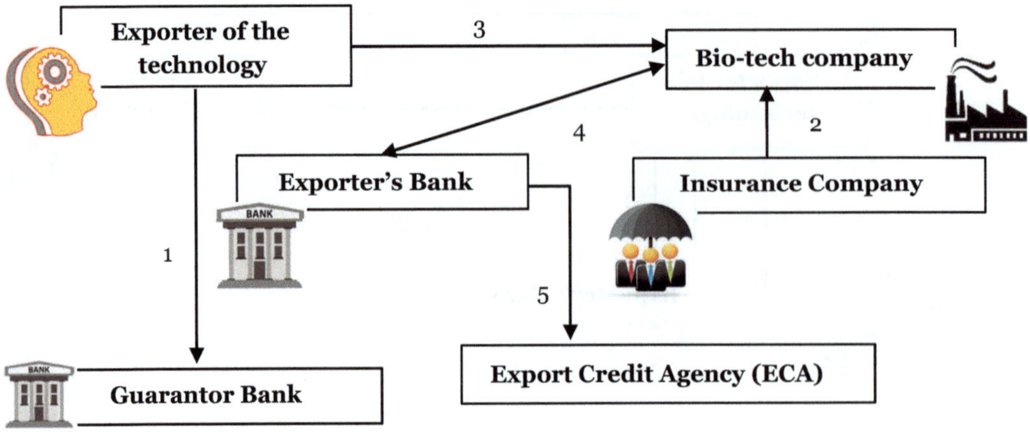

Fig. 10.6 The mechanism of the buyer loan facility

can ensure the underwriting for unsubscribed securities) (Phase 1); in the next phase, the coordinating group of banks (including leadership bank) will analyze the financial situation of issuer of bonds/equities and will establish the IPO's conditions that will be published in an IPO's prospectus—the technical document describing in detail the financing mechanism (Phase 2); the coordinating group will contact a number of relevant investment banks included in the underwriting group (this banks will sell 2/3 of the securities with an undertaking regarding unsubscribed bonds or equities to be included in their own portfolios) (Phase 3); the remaining 1/3 share of securities is sold through selling group of banks without any commitment in terms of including these shares in their own portfolios (the unsubscribed shares will be returned to the leadership bank) (see Fig. 10.7).

The major problem for biotech company is the situation when the IPO is partially unsubscribed by investors and required capital cannot be attracted from capital markets. In this case, there are few strategies that can improve the situation (mainly due to the reduced interest of investors in this project): to promote better the IPO at the level of the potential investors by sending the best sales consultants to them (the roadshow strategy); the adjustment of the issuance conditions (a discount when the securities are issued, a premium when the securities are called back by issuer—the case of bonds only, etc.); extra payment to convince the coordination group to accept an underwriting commitment from the beginning of IPO (in this case, the coordinating group will have also an underwriting function); and, finally, to convince the issuer to accept the IPO with a lower subscribed value and to help the beneficiary to find additional long-term financing scheme (Minnesota Department of Employment and Economic Development 2005; Daxhammer and Resch 2017).

Compared with previous long-term financing schemes (syndicated loan, leasing financing, buyer loan facility, and seller loan facility), in the case of IPO, the role of financial intermediaries is limited to the selling of the financial securities to the investors from the capital markets and (eventually) to share the risks of this issuance (the securities to be unsubscribed by investors). Therefore, the cost is different (lower in this case compared with leasing or loans) covering only the sale operations of the securities. The cost of capital is directly negotiated with the buyers of these securities (bonds/equities) and can be significantly improved by the financial institutions involved.

10.4.5 The Issuance of Depositary Receipts (DR)

Another recommended long-term financing scheme for the growth stage of a biotech company

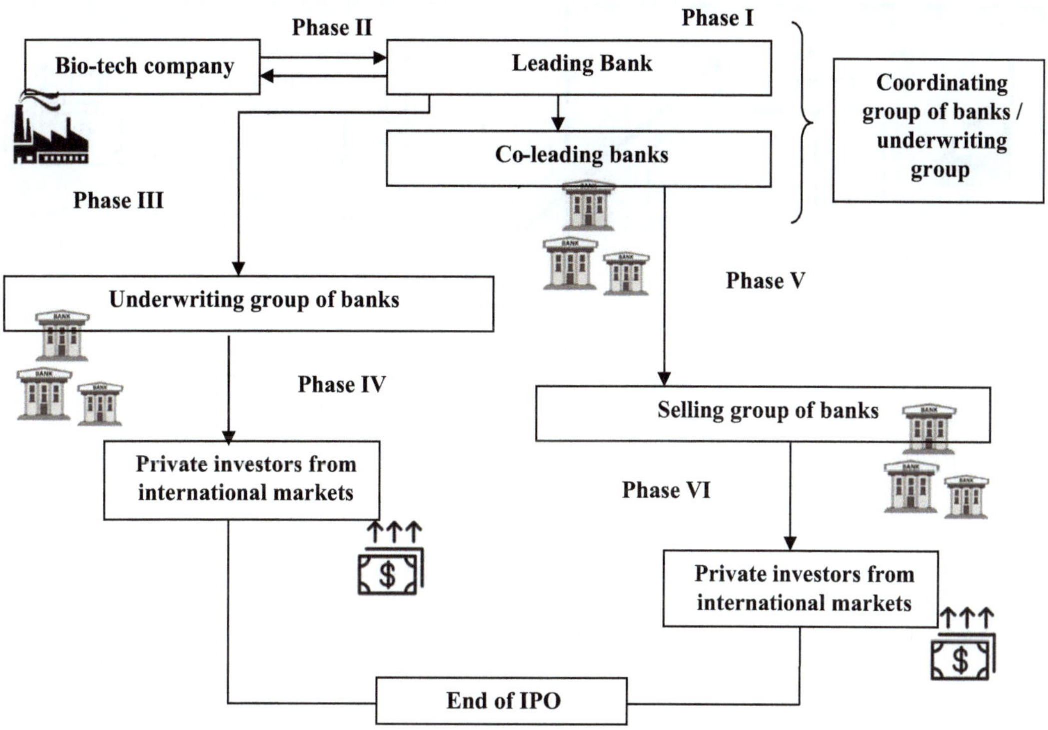

Fig. 10.7 The mechanism initial public offering (IPO) of bonds/equities on local/international capital markets

is the initial public offering (IPO) of depositary receipts. This scheme is suitable for the companies that are located in very risky countries, the companies that have difficulties to meet the stock exchange's listing conditions (e.g., minimum capital requirements) or the companies that are not too known by potential international investors. Normally, this financing scheme has an international dimension (can be globally addressed—the case of GDR or regionally addressed—the case of ADR or EDR).

To access capital markets for mature biotech companies but not big enough to be accepted as listed companies there, the issuance of DR can be the alternative (Fig. 10.8): the biotech company will contact an "AAA" investment bank with relevant experience in the DR's issuance and with a good international credibility (1); this depositary bank will contract a local bank for accepting in their vaults a package of shares (equities/bonds) issued by the biotech company (2); the biotech will issue negotiated number of shares and will deposit them to the indicated local custodian bank (3); the custodian bank will notify the depositary bank about the existing package of shares (4); the depositary bank will sell through an IPO its own depositary receipts (DR) to individual or institutional investors (local on internationally located) (5); the depositary bank will collect the capital against DRs (6) and will transfer it to the biotech company in order to be invested in the growth and development of the business (7); the company will produce and sell products to the market and the profits are converted into dividends that will be deposited to the custodian bank (8); the custodian bank will exchange these dividends into foreign currency and will transfer it to the depositary bank (9); the depositary bank will remunerate the investors in the DRs issued to them (including its own profit) (10).

This long-term financing scheme can have clear advantages for a mature biotech coming from emerging markets: the company gains the access to the international investors and to the important capital resources without being known

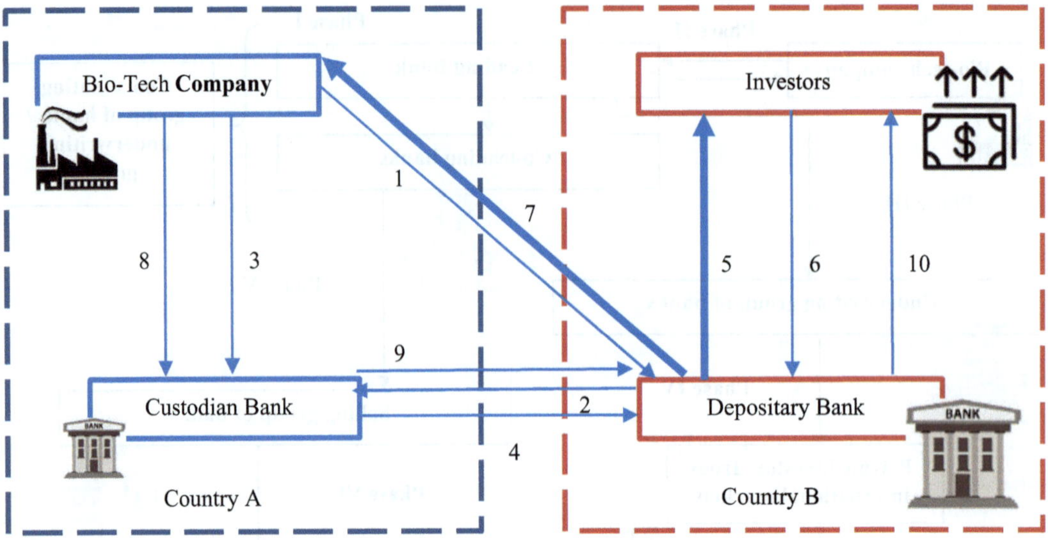

Fig. 10.8 The mechanism initial public offering (IPO) of bonds/equities on local/international capital markets

by them, the systemic risk of the biotech company's country is shared with depositary bank, the biotech company benefits from the international credibility, soundness, and notoriety of the depositary bank, the cost of capital can be significantly improved by depositary bank, etc. (Buljevich and Park 1999; Stowell 2017).

10.4.6 Public Scheme to Finance the Biotech Sector

Public schemes and grants are also important to provide financing for biotech sector. In the case of European Union, the most important financing scheme is the *Program "Horizon 2020"* that is financing the R&D efforts "to develop competitive, sustainable, safe and innovative industrial products and processes and to improve the innovation in this sector" (European Commission 2019). This financing program proposed grants for new medicines, for developing industrial applications in biopharmaceuticals, food and feed production and biochemicals, food security, quality of water resources, environment protection, sustainable agriculture, reduction of CO_2 emissions at the level of European Union, or "green chemistry." The latest available open grants dedicated to biotechnology by this program are as follows: Grant: *Fast Track to Innovation* (with a total budget of 100 million euros/year); Grant: *Open Challenging Current Thinking* focused on novel ideas for radically new technologies (with a total budget of 165 million euros in 2019 and 364 million euros in 2020); Grant: *Regenerative medicine: from new insights to new applications* focused on better health and care, economic growth, and sustainable health systems (with a total budget over 400 million euros in 2019—various schemes including innovation actions, coordination and support actions, and research actions); and the Grant: *Innovation Procurement: Next generation sequencing (NGS) for routine diagnosis* (a budget over 400 million euros in 2019) (European Commission 2019).

Another important public support is provided by European Investment Bank Group. This institution invested 23.3 billion euros in various sectors including agriculture, food and rural development, health and life sciences, forestry, and water and wastewater management. The bank provided financing (loans, microcredits, guarantees) for 374.000 SMEs (in their early stages) employing 5 million people. According to the data provided by EIB, in 2018 the group improved healthcare services for 27.3 million people, backed safer drinking water for 20 million people, and improved sanitation for 10 million people. The main areas financed, in the case of health and life sciences, were health infrastructure, innovation linked to medical research, or fundamental medical research (European Investment Bank 2019).

Besides these regional public schemes, governments can propose national financial schemes that can add more financing resources available for starting and growing a business in the biotech sector. For example, in Romania, the government agency UEFISCDI currently proposed: *Program: complex research frontier* (including life sciences) with a budget of 35 million euros and *Program: Exploratory research* with a budget of 38 million euros that financed 33 research projects in the field of biotechnology (Romanian Government Agency for R&D Financing 2019). In France, the governmental agency CNRS is currently proposing financing for innovative sectors (including biotech) in various schemes such as the *Program: Joint Research Projects (PRC)* or the *Program: International Programs for Scientific Cooperation (PICS)*. In fact, in France the governmental programs are insisting on the international cooperation and on joining R&D scientific laboratories and institutions (CNRS France 2019).

10.5 Concluding Remarks

The capital is always allocated by taking into consideration the risks involved and the cost of capital. Any business is creating economic value when the returns of invested capital are higher than the cost of capital. When a business is started, the access to finance is very limited. Family, friends, and closed partners provide the seed capital for any business idea. Business accelerators, business incubators, research laboratories, R&D networks, and academic partnerships ameliorate the opportunities to find necessary capital to be able to develop a successful innovative product or service for the market. Biotechnology sector is included among the most innovative sectors using advanced technologies. The uncertainty of a start-up in this field is higher than regular businesses. The presence of innovation attracts specific institutional and individual investors in the early stages: business angels, venture capital funds, investment funds, etc. The potential and growth rate of the businesses in such innovative sectors explain this particular interest of such investors that cannot be found in the early stages of common businesses. The financing mechanisms in the last phases of early stages and during the maturity of the businesses are different than the schemes that provided the seed and growth capital. When the business becomes mature, the investors from capital markets replace venture capitalists and business angels that are exiting to finance other innovative companies. Commercial banks and non-banking financial institutions can be also involved in these stages. Due to the specific features of the biotech sector, the public support is vital in all business stages. In many countries, the governmental agencies are providing various financing schemes for SMEs, for innovative sectors, and for exporting of high-tech products and services.

References

Atche J-M (2017) Strategy for SME financing. Éditions universitaires européennes

Baker K, Martin G (2011) Capital structure and corporate financing decisions: theory, evidence, and practice. Wiley

Bierman H (2002) The capital structure decision. Springer Nature

Blackstaff M (2001) Financing and leasing. Springer, London. https://doi.org/10.1007/978-1-4471-0689-0_8

Bryant T (2014) The entrepreneur's guide to raising capital from angel investors. Wiley

Buljevich EC, Park YS (1999) Project financing and the international financial markets. Springer, Boston. https://doi.org/10.1007/b102491

Cardullo M (1999) Technological entrepreneurism: enterprise formation, financing and growth (Engineering management series, 5), Research Studies Press

Chesini G, Giaretta E, Paltrinieri A (2018) Financial markets, SME financing and emerging economies. Palgrave Macmillan

Cifci D, Ispir MS, Yetkiner H (2017) Financial development and economic growth: some theory and more evidence. J Policy Model 39(2):290–306

Corelli A (2018) Analytical corporate finance. Springer Nature, Cham. https://doi.org/10.1007/978-3-319-95762-3

Cumming D, Hornuf L (2018) The economics of crowdfunding: startups, portals and investor behavior. Palgrave Macmillan

Daxhammer RJ, Resch A (2017) Initial public offerings – an inside view. UVK Verlagsgesellschaft mbH
Deltuvaitė V, Sinevičienė L (2014) Investigation of relationship between financial and economic development in the EU countries. Proc Econ Financ 14: 173–180
Freeman I (2012) Seeking synergy in SME financing: an examination of the dichotomy affecting banks and the real needs of small-to-medium business enterprises. LAP LAMBERT Academic Publishing
Kirzner I (1978) Competition and entreprenurship. The University of Chicago Press, Chicago
Minnesota Department of Employment and Economic Development (2005) A guide to biotechnolgy finance. Minnesota Department of Employment and Economic Development
Mishkin F, Eakins S (2017) Financial markets and institutions, 9th edn. Pearson
Munro D (2013) A guide to SME financing. Palgrave Macmillan, New York, NY. https://doi.org/10.1057/9781137373786
Rossi S (2018) Access to bank credit and SME financing. Palgrave Macmillan
Rothbard M (2007) Man, economy and state: a treatise on economic principles. Ludwig von Mises Institute, Auburn, AL
Saunders A, Cornett MM (2018) Financial markets and institutions, 7th edn. McGraw Hill
Stowell DP (2017) Investment banks, hedge funds, and private equity, 3rd edn. Academic
von Mises L (2010) Human action: a treatise on economics. Liberty Fund Inc, Carmel
Willsher R (1995) Export finance: risks, structures, and documentation. Palgrave Macmillan

Internet Sources (Websites, Pages, and Portals)

CNRS France (2019, February) CNRS financing portal. Preluat de pe CNRS. http://www.cnrs.fr/en/research
European Commission (2019, February) Horizon 2020 framework programme (H2020). Preluat de pe. https://ec.europa.eu/info/funding-tenders/opportunities/portal/screen/opportunities/
European Investment Bank (2019, February) Financing projects by sectors. Preluat de pe European Investment Bank. https://www.eib.org/en/projects/sectors/
OECD – Organisation for Economic Co-Operation and Development (2004) Financing innovative SMEs in a global economy. OECD Publishing
OECD – Organisation for Economic Co-Operation and Development (2015) New approaches to SME and entrepreneurship financing: broadening the range of instruments. OECD Publishing.
UEFISCDI Romania (2019, February) UEFISCDI financing portal. Preluat de pe UEFISCDI Romania. https://uefiscdi.ro/p4-cercetare-fundamentala-si-de-frontiera

Key Elements of the Biotech Market 11

Alexandra Perju-Mitran

Abstract

Addressing the uncontrollable factors influencing the European biotech market is critical to the commercial success of emerging biotech companies, as the accelerating pace of biotechnological research allows for the development and implementation of new products and services. Modern biotechnology demands rigorous oversight of macroenvironmental factors, as each of these uncontrollable macroenvironmental factors affect new ventures in biotech; change occurs when companies manage to favorably respond to the challenges imposed by each factor, constructing a different representation for the public, and thus shaping the development of biotechnology itself.

With biotechnology used in a wide variety of sectors, it is imperative to analyze market factors in several fields, with focus on one hand on specific European factors and on the other hand on global factors that either pose threats or opportunities for new ventures.

As in all cases, factors of the macroenvironment are beyond the influence capacities of individual companies and will always present themselves as either opportunities or threats. Given the specificity of biotechnology, there will always be needs that are not catered to.

Keywords

European biotech · External factors · Biotech start-up · Social factors · Economic factors · Political factors · Legal factors · Technological factors · Biotech marketing

11.1 Introduction

Many companies view macroenvironmental forces as uncontrollable elements to which they must react and make efforts to adapt. They design strategies to avoid the threats and seize the opportunities that the market provides, while others take a proactive stance toward these changes and develop strategies to create and shape new industries and business models.

By analyzing the needs and trends of the market, researchers generally reach the conclusion that successful companies are those with the ability to recognize unsatisfied needs and trends and profitably respond to them. Biotech companies stand to earn fabulous amounts of money should they find solutions to issues such as medical treatment for psychological disorders, cures for terminal diseases, food products that are tasty and fat free, and alternately powered transportation.

In formulating a market strategy, decision makers must analyze the forces internal to the

A. Perju-Mitran (✉)
Faculty of Management-Marketing, Romanian-American University, Bucharest, Romania

F. Matei, D. Zirra (eds.), *Introduction to Biotech Entrepreneurship: From Idea to Business*,
https://doi.org/10.1007/978-3-030-22141-6_11

organization, the micro factors that the company should be able to influence, as well as conditions in the external environment which are beyond the company's influence but act as broader forces that influence the micro factors. In the following sections, macro forces shaping the European biotechnology industries' activities will be examined.

11.2 The Social and Cultural Environment Influencing European Biotech Companies

Because advancement in biotechnology has shown fast growth in recent years, it has been intensely debated and it has raised a lot of controversy in advanced countries. Studies on public attitude toward biotechnology generally show multiple similarities to risk perception studies. The good news is that, although the European population is generally inattentive to biotechnology advances, they are globally positive about biotechnology (Pardo and Calvo 2006). Previous research in risk perception also reveals several differences between experts and laypeople in their perception of risks, depending on the type of hazard, as laypeople tend to rely on aspects such as catastrophic potential of the effects, while the experts are more prone to rely on observed or expected fatalities (Savadori et al. 2004). In practical terms, this implies that individuals will process scientific information based on their own values and beliefs, so when starting a company in biotech, one must be familiar with and research the cultural and ethical effects of the technology on the public. To do so, several aspects of the social and cultural environment for each country in company's operation area should be investigated.

The first aspect to investigate is the demographic environment, comprising the population targeted by the company or situated in their operation area. It is characterized by population number, population density, geographical location and shifts, how age groups are structured, gender, ethnicity, family size, occupation, income distribution, and wealth concentration. Analysis of models and trends in population structure makes it possible to anticipate consumer behavior on a market, as generational cohorts tend to express similar needs and wants.

There is a tendency toward differences in opinions between men and women on technology and science-related topics (Amin et al. 2007). Biotech acceptance also implies connections with knowledge and learning; therefore, the level of education is also an important factor. For instance, it was found that in Europe, declared supporters of biotechnology are more likely to be younger than 25, male, and well educated, while declared rejecters are more likely female and over 35 (Gaskell et al. 2000).

General public perceptions, their understanding, and acceptance of biotechnology may both promote and restrict or disrupt ventures in various applications of biotechnology. An important factor in the emergence of controversial subjects concerning the market introduction and adoption of biotech products has been the desuetude of the needs, interests, and concerns of biotech's primary stakeholders, the laypeople.

The cultural environment is made up of areas that concern the value system, customs, traditions, shared beliefs, and regulations shaping people's status in society. Changes in tastes and fashion trends are also included in the social and cultural factors category. If we take into account earlier research perspectives, attitudes toward biotechnology would tend to follow the same trajectory as attitudes toward science and technology in general, views and opinions regarding the natural environment, technological progress, religious beliefs and moral persuasions, and confidence in the specific industry and its regulations (Amin et al. 2007).

Kotler and Armstrong (2015, p. 114) define the main cultural factors influencing decision making in marketing. These factors are as follows: the continuity of cultural values, changes occurring in secondary cultural values, self-assessment, opinions concerning others, views on organizations, society, nature, and the universe. Marketing operators must be aware of these factors, as well as of differences in their manifestations at the level of communities making up the company's market. The natural environment consists of the natural resources required for market operators to perform their activities or that influence company activities. As far as the influence of this environment on biotech companies is concerned, the following trends characterizing the current situation of the natural environment are the most common: a raw material shortage (resource depletion), environmentalism movements, increasing power costs, waste management, the loss of natural recreational areas, declining health caused by pollution and low-quality food, and state interventions in the management of natural resources.

Derived in part from culture, ethical beliefs about how biotech companies should operate influence the ways in which people respond to new ventures. This is where the issue of social responsibility and ethics arises for a start-up biotech company, where the management approach should be fueled by a vision of sustainability. In clearer terms, managers should develop a strategic framework for making a profit while also sustaining the environment. Companies emphasizing pollution prevention by practicing product stewardship, for instance, also use their environmental strategies in their marketing strategy, to create and emphasize sustainable value and establish customer relationships.

Clean-tech, a philosophy used by investors seeking to profit from environmentally friendly companies, is a term stemming from the venture capital (VC) investment community, different from green businesses, which are focused on sustainability more than profitability. It characterizes companies striving to increase productivity, efficiency, and performance by minimizing their impact on the environment. Clean-tech has grown to include industries like biofuel, water purification, and solar and wind energy.

Although the European Commission and many European leaders have been supportive of the further growth of biotechnology, green biotechnology has become a much more controversial issue in Europe than initially expected, the main reason for which European agriculture remains largely free of genetically modified crops (Aerni et al. 2015) is due to the force of publics, when a coalition of environmental NGOs, consumer groups, and small traditional or organic farmers, whose beliefs also found expression in public opinion surveys, started an anti-biotech movement (Kurzer and Cooper 2007).

11.3 The Technological Environment of Biotech Companies

The technological environment influences organizations in multiple ways. An innovation or development can have dramatic effects on the industry and on the demand for an organization's products or services.

Technological changes can affect company operations such as processing methods, service delivery, and use of materials and end up changing the balance between competitors and even countries, by shifting demand from one company's product to another.

Technological advancement speed varies considerably across industries. For example, in fields ranging from neuroscience to computing, change is rapid and constant, but in medicine change is slower and more gradual. However, technological change can offer major opportunities for development or threaten the existence of the company. Therefore, building the company's ability to react to changes in the technological environment involves:

- Forecasting and identifying technological developments related to the company's activity
- Assessing the impact of technological developments on the company's present operations
- Identifying or defining the opportunities presented

In the field of biotechnology, such abilities should lead to the creation of a technological strategy or some form of technological forecasting, encompassing product design and development, technology required to up the production capacity, focus on flexibility to potentially introduce new products, resource and development management, and funding.

Technological forecasting in the field of biotechnology is a constantly evolving task which can help protect and improve the profitability of companies where improvement and innovation takes place. For example, pharmaceutical companies now pursue research in medicine and diagnosis by implementing biotechnology (Tyagi et al. 2018). Entrepreneurial potential in all areas of biotechnology is exceedingly influenced by factors such as the development of technology as the single solution for current problems or the more desirable solution for presently available technologies. Testing and regulatory authority approval are prerequisites to commercial production and marketing. For this reason, the field of biotechnology poses great opportunity although market uncertainty and new venture failure is also high.

Another aspect to be taken into account is the explosive growth in digital technology in the face of today's marketing landscape. Companies are now compelled to use digital marketing tools in order to take advantage of global opportunities provided by the Internet (Kotler and Armstrong 2015, p. 54).

Within the 53 European Countries there is a reported 85.2% Internet penetration rate in December 2017 and a 41.2% Facebook penetration rate (Internet World Stats 2019). This phenomenon, alongside the global improvements in communication technology, has generated major shifts in digital marketing strategies (Tiago and Veríssimo 2014).

Digital marketing technologies that can be used by a company consist of digital profiling tools, segmentation tools, websites, search engine marketing, campaign management tools, content management, social media tools, mobile applications, digital collaboration, and analytics tools.

To conclude, technological factors are important to biotech companies for the following reasons:

- They come up with new sources for customer satisfaction.
- They help identify latent needs.
- They can assist in discovering new markets.
- They alter demand patterns.
- They can shift competition in an industry.
- They can increase the efficiency of marketing activities.

11.4 The Economic Environment Influencing European Biotech Companies

Biotechnology tries to solve some of the major challenges faced by society, and by doing so, it can act as a driving force for economic growth and job creation.

The economic environment comprises the elements that make up economic life where the company operates. In its characterization, the branch structure of the economy, the overall development level, labor force occupancy, currency values, and the financial situation per country are all to be taken into account.

The state of the economy influences the level of income available to consumers and companies in order to make purchases. The disposable income consumers set aside for personal expenditures and the income generated by a company shapes demand for products and services and the willingness to make transactions. The most important economic variables that biotech companies must observe in order to be able to make timely strategic decisions:

1. Income distribution: As income increases, the amount spent on luxury items increases, whereas food expenditures tend to remain stable.
2. Recession: When the total amount of income and expenditure of a country decreases, buyers reduce their expenses and limit their purchases to essential or substitute goods.
3. Taxation: Direct taxes determine how much money is available for goods and services, while indirect taxes, such as the value-added tax, influence the price of goods that the consumer must pay.
4. Interest rates can have a major impact on consumer willingness to make non-essential purchases. A high interest rate will not only decrease current expenditures but also prevent further loans, which also implies less expenditures for perceived non-essential items.
5. Inflation endangers foreign markets. Once a growing demand allows companies to increase their prices while maintaining sales volume in one country, the same company will register a

decrease in sales where countries do not face inflation pressures, as they turn to cheaper alternatives.

6. The exchange rate: another economic variable affecting international exchange. As the company's home-country currency grows in comparison to currencies of other markets where the company sells, products become more expensive for the respective markets, yet if the currency depreciates, export prices decrease and demand will likely increase.
7. The "feel-good" factor: referring to consumer trust in the state and future of an economy, how prosperous consumers feel, which may or may not be reflected by the economic indicators.

11.5 The Political and Legal Environment of European Biotech Companies

The political environment comprises society structures, social classes and their parts in society, political forces and their relations, as well as the degree of stability of the domestic political element. The institutional (legal) environment is made up of laws, government agencies, and pressure groups influencing and limiting the freedom of action for companies and individuals in society. Factors shaping the present institutional environment are trade legislation, the high number of public interest groups, and a higher importance placed on ethics and social responsibility (Kotler and Armstrong 2015, p. 111).

Political and legal forces can also be stretched to include election outcomes, court decisions, and decisions rendered by commissions and agencies. At European level, there are major differences between local, national, and international sectors of the political environment. Among the more influential government actions, we find government spending, regulation, taxation, policies concerning the use of natural resources, and takeovers. Public policy is developed by governments to regulate commercial activity for the good of society. Legislation affecting biotechnology businesses has increased steadily over the years, with regulations constantly changing. However, biotechnology is faced with strong opposition, especially concerning its applications in agriculture, possibly the main reason why European agriculture makes very little use of genetically modified crops (Aerni et al. 2015).

The primary objective in regulation and government intervention in the biotechnology industry should be to ensure that society is protected from products that may prove unsafe or inefficient, while not suppressing scientific innovation (Kearney 2018). Political activity in many countries may also include the monitoring of government policy toward income tax and monitoring the influence of unions.

While controls are inevitable for biotechnology, control implementation will play a major role in the development of the biotechnology industry. Regulation is often written to combat public perceptions. Industry and the technical community have been involved in setting regulatory policy, and as a result, incentives are incorporated into the regulations in order to foster innovation on the European market.

As a threat to starting a business on the European biotech market, very few intergovernmental documents related to the "green economy" or "climate-smart agriculture" express the green potential of biotechnology and rarely do we find clean-tech investment portfolios or climate change funds that list biotechnology-based products (Aerni et al. 2015; Knight 2010).

Thus, regulation of the biotechnology sector targets what scientific risk assessments identify to present risks to consumers and the environment (Bernauer and Meins 2003). Further business legislation is enacted toward fair competition practices.

Without placing focus on the specificity of a certain biotechnology sector, the laws any company in biotechnology should be aware of concern:

1. The protection of intellectual rights (e.g., Directive 98/44/EC, the so-called Biotech Directive)
2. The Consumer Protection Act (European Community Directive 85/374/EEC)

3. The Companies Act (with its various new provisions for private and public companies)
4. The existing regulatory commission or regulatory agency
5. The environmental protection laws
6. Regulations on takeovers and mergers (such as Directive 2005/56/EC on cross-border mergers and the Economic Concentration Regulation 139/2004)
7. Laws concerning personal rights, media freedom, and advertising
8. Regulations and laws on exchange control

In the field of biotechnology, there are three government functions of significant importance which influence the market on which companies operate:

1. The supplier function. This implies decisions concerning creation and accessibility of private companies to state-owned natural resources and agricultural products. This affects competition and the viability of company strategies.
2. The customer function. Market opportunities and threats arise out of government demand for biotechnology products and services.
3. Government as a barrier. Government agencies can impose licenses and permits, thus limiting or even refusing entry. National governments regularly make use of fees and trade restrictions to raise the entry barriers for new foreign companies. Knowledge of government strategies can help a company avoid unfavorable confrontation.

Ecological issues relevant to European biotechnology companies are resource depletion, where the use of materials to develop products can lead to the depletion of natural resources and pollution concerns (noise, environmental, and visual pollution).

Concerning ethics, the most basic issues have been formalized through laws and regulations to conform to the standards of society. Companies are expected to at least obey the laws and regulations that have been set in place. However, it is important to realize that business ethics go beyond the legal minimum, as ethical business decisions lead to mutual trust.

References

Aerni P, Gagalac F, Scholderer J (2015) The role of biotechnology in combating climate change: a question of politics? Sci Public Policy 43(1):13–28

Amin L, Jahi JM, Nor ARM, Osman M, Mahadi NM (2007) Public acceptance of modern biotechnology. Asia Pac J Mol Biol Biotechnol 15(2):39–51

Bernauer T, Meins E (2003) Technological revolution meets policy and the market: explaining cross-national differences in agricultural biotechnology regulation. Eur J Polit Res 42(5):643–683

Gaskell G, Alum N, Baouer M, Durant J, Allansdottir A, Bonfadelli H, Boy D, Cheveigne DS, Fjaestad B, Gutteling JM, Hampel J, Jelsoe E, Jesuino JG, Kohring M, Kronberger N, Midden C, Nielsen TH, Przestalski A, Rusanen T, Sakellaris G, Torgersen H, Twardowski T, Wagner W (2000) Biotechnology and the European public. Nat Biotechnol (18):935–938

Internet World Stats (2019). https://www.internetworldstats.com/europa2.htm. Accessed 10 Mar 2019

Kearney C (2018) Regulation and market influences on innovation in biotechnology. J Commer Biotechnol 24 (3):57–62

Knight ERW (2010) The economic geography of clean tech venture capital. Oxford University Working Paper Series in Employment, Work and Finance. Available at SSRN: https://ssrn.com/abstract=1588806 or doi: https://doi.org/10.2139/ssrn.1588806

Kotler P, Armstrong G (2015) Principles of marketing-global edition. Pearson

Kurzer P, Cooper A (2007) Consumer activism, EU institutions and global markets: the struggle over biotech foods. J Public Policy 27(2):103–128

Pardo R, Calvo F (2006) Are Europeans really antagonistic to biotech? Nat Biotechnol 24(4):393–396

Savadori L, Savio S, Nicotra E, Rumiati R, Finucane M, Slovic P (2004) Expert and public perception of risk from biotechnology. Risk Anal 24(5):1289–1299

Tiago MTPMB, Veríssimo JMC (2014) Digital marketing and social media: Why bother? Bus Horiz 57 (6):703–708

Tyagi P, Kumar A, Sengar RS (2018) The opportunities to develop a successful entrepreneurship and business model in biotechnology: an overview. J Commer Biotechnol 24(3):63–68

Intellectual Property and Transfer of Innovation in Biotechnology

12

Cristina Popa, Narcisa Elena Babeanu, and Ovidiu Popa

Abstract

Intellectual property biotechnology, with special emphasis on biotech patents, is considered a strong tool in the development of biotechnological R&D and biotech industry. In the European patent system, more than 50% of the granted biotech patents cover pharmaceutical products, around 40% refer to industrial processes in biotechnology, and the rest are patents granted for application of biotechnology in agriculture and environmental protection.

The authors propose to reveal specific insights into patenting biotechnological inventions, according to the European Patent System, emphasizing what can be protected or not protected by patents in the area of biotechnology, considering the legislation in force in this field, and ethical and moral disputes on this subject.

The authors give special attention to aspects regarding the transfer of innovation in biotechnology, referring to its very specific problems and how intellectual property could interfere with facilitating this transfer of knowledge.

Keywords

Intellectual property · Biotech patents · Transfer of innovation

List of Abbreviations

BoA	Board of appeal
EBoA	Enlarged board of appeal
EC	European Commission
EPC	European Patent Convention
EPO	European Patent Office
EU	European Union
G	Decision of the EBoA
GL	Guidelines for examination under the European Patent Convention
IP	Intellectual property
IPR	Intellectual property right
OECD	Organization for Economic Cooperation and Development
SPC	Supplementary protection certificate
T	Decision of the technical board of appeal
UPOV	Union pour la Protection des Obtentions Vegetales
WIPO	World Intellectual Property Organization

12.1 Short Overview of Biotechnology

Biotechnology can be defined as the science using living systems and organisms to develop or make products intended to improve the quality of human life (OECD 2005). However it is defined, biotechnology has experienced impressive

C. Popa (✉)
Rominvent SA, Bucharest, Romania

N. E. Babeanu · O. Popa
Faculty of Biotechnologies, USAMVB, Bucharest, Romania

F. Matei, D. Zirra (eds.), *Introduction to Biotech Entrepreneurship: From Idea to Business*,
https://doi.org/10.1007/978-3-030-22141-6_12

growth throughout the world during the last few decades.

Despite the fast development of biotechnology over recent years, humans have been practicing biotechnology for thousands of years to make food such as bread and cheese, to produce wine, or to derive medical products from plants.

It could also be considered that the first biotechnologists in the history of mankind were the farmers who improved species of plants and animals by cross-breeding. The inventors in this area filed patent applications even before 1800; thus, the first known patent in the field was granted in the year 1787 to Blunt, by the British Patent Office, under the number GB 178701625, for "A new-invented composition to be used as yeast" (Gelinas 2010).

The most famous scientist from the beginning of the classical biotechnology era is Louis Pasteur, but his first patents obtained in France were granted for processes for producing alcohol, wine, beer, or vinegar, and not for a specific microorganism.

Our century is already referred to as the "Century of Biology" (Venter and Cohen 2004), the amazing development of the modern biotechnology being the main reason for this nomination. What is understood by "modern biotechnology" is better revealed by the list of definitions of biotechnology techniques, as is shown in Table 12.1.

The business of biotechnology started with small entities or university teams. A biotech firm could be a firm acting in one of the biotechnological sectors and using one of the biotechnological techniques, as presented above. It could be a firm in the area of production that produces goods or services using specific biotechnology techniques. Or it could be a R&D biotechnology firm, in which most R&D activities are dedicated to biotechnology.

The classification of biotechnology sectors, as presented in Table 12.2 is generally accepted (DaSilva 2004).

Not only the European economy relies on biotechnology as an important industrial sector, providing growth and employment opportunities, along with useful products for its inhabitants, including medical care and environmental protection.

On the other hand, progress in this cutting-edge field incurs considerable costs; thus, the expected rewards must be tempting for investors and investment security.

One way to secure the investment is to protect your business through intellectual property rights. Patent protection grants an inventor/applicant exclusive rights over his invention for a specific time period.

12.2 Intellectual Property Rights in Biotechnology

Intellectual property (IP) plays an important role in promoting innovation by providing a basis for return on investment in research and development. This is particularly the case where that certain technology advances rapidly but where recovery of the investments could be slow, as in

Table 12.1 Main areas of biotechnology and specific techniques applied

No.	Subject of biotech technique	Specific techniques applied
1	DNA/RNA	Genomics, pharmacogenomics, gene probes, genetic engineering, DNA/RNA sequencing/amplification, profiling of gene expression, use of antisense technology
2	Proteins/peptides/ hormones	Proteins/peptides sequencing/synthesis/engineering; improved methods for large molecule drugs transportation, proteomics, signaling/identification of cell receptors
3	Cell/tissue culture and engineering	Cell/tissue culture, tissue engineering, cellular fusion, vaccine/immune stimulants, embryo manipulation
4	Biotechnological processes	Fermentation techniques in bioreactors, bioprocessing, bioleaching, biodesulphurization, biopulping, bioremediation, biofiltration, phytoremediation
5	Gene and RNA vectors	Gene therapy, viral vectors
6	Bioinformatics	Databases for genomes, protein sequences, modeling complex biological processes

Table 12.2 Classification of biotechnology sectors

Sectors of biotechnology	Exemplary components requiring intellectual protection
Red biotechnology (pharmaceuticals, diagnostics, health)	Biomedicaments, new compounds based on genetical engineering, targeting rare diseases (orphan drugs), substances used in diagnostic methods
Yellow biotechnology (food, nutrition science)	Enzymes, other active substances used in the food industry
Blue biotechnology (aquaculture, marine biotech)	Engineered organisms from marine environment (algae, protozoa)
Green biotechnology (agriculture, environmental biotechnology)	Improvement of plant characteristics, such as resistance to disease or hard environmental conditions, tolerance for herbicides, higher production yields, biofuels, biofertilizers
Brown biotechnology (desert and arid zone biotechnology)	Engineered plants with specific characteristics for arid zones
Black biotechnology (bioterrorism, biowarfare, anticrop warfare)	No intellectual property protection
White biotechnology	Gene-based bioproducts
Mustard biotechnology (nanobiotechnology)	Nanomaterials, processes for obtaining them
Grey biotechnology (classical fermentation and bioprocess technology)	Bioethanol, single-cell proteins, antibiotics

the biotechnology field. This industry has grown exponentially in the last several years; consequently, the patents granted and patent applications also increased by 15% every year in USA between 1990 and 2000 (OECD 2004). The Annual 2017 Report published by the European Patent Office (EPO 2018a) on 22.01.2018 reveals a growth of biotech patent applications filed with EPO in 2017 versus 2016 of 14.5%, considering herewith patent applications related to peptides, microbiology, and genetic engineering, for example. The increase in pharma patent applications filed at the EPO in 2017 vs 2016 was of 8.1%, whereas the total number of patent applications filed with the EPO, including all the technical fields, has increased by 6.2%.

The definition of Intellectual Property Rights (IPRs) is undergoing continuous transformation. First, IPRs are property rights, protecting human innovation and creation and conferring to the owners moral and material rewards. In exchange for their creative work, the creators receive an exclusive right to their creations, limited in time and territory.

Initially, there were two distinct terms in use: "industrial property" and "intellectual property"; the former was used to refer technology-based subject areas such as patents, designs, and trademarks, whereas the latter used to refer copyright or other artistic creations. But, over time, this distinction has diminished and the modern convention is to use "intellectual property" to refer to both industrial and intellectual property (Strenc et al. 2005).

The first international regulations of IPRs were adopted by the Paris Convention in 1883, and since then have been repeatedly revised. The most important revision was in Stockholm on 14 July 1967, when the World Intellectual Property Organization (WIPO) was founded by way of the Convention. In Article 2(viii) the main areas that can be protected by intellectual property are provided:

- Literary, artistic or scientific
- Artistic performances, phonograms, and broadcasts
- Inventions in all fields of human knowledge
- Scientific discoveries
- Industrial designs
- Trademarks, service marks, and commercial names and designations
- Protection against unfair competition
- All other rights resulting from intellectual activity in the industrial, scientific, literary or artistic fields (WIPO Convention 1967)

Specific IP laws provide protection to the creators of the mentioned work by granting them the legal right to control the use of their creative work for a certain period of time. These exclusive rights are not granted for the physical creation of certain goods, but for the intellectual work incorporated in that product.

Intellectual property takes a number of different forms, each with its own specific manner of protection. The same subject matter may attract more than one form of protection.

Some examples of the forms that IPRs embrace in the biotechnology field are presented in Table 12.3.

12.3 Short Presentation on the Main IPRs

Patents A patent is a right granted by the official national or international authority to the patent owner for an invention, meaning a new product or process, that provides a new technical solution to a problem. This title allows the patent owner to exclude others from commercially exploiting an invention within a certain territory, i.e., producing, using, selling, offering for sale or importing a product, a process or a product made by a process protected by that patent within its lifetime. This exclusive right can be enforced only after the granting of the patent. Over

Table 12.3 Main intellectual property rights (IPRs)

IPRs	Subject matter protected by the specific IPR in the biotechnology field	Time of protection	International conventions
Patents, supplementary protection certificates	Isolated polynucleic acids, peptides and polypeptides, microorganisms, viruses, vectors, genes, antibodies, vaccines, compositions, expression systems, cell lines, methods for preparation or the use thereof, medical devices, plants (in certain legal conditions)	20 years from the international filing date; a maximum of an additional 5 years protection through a SPC (supplementary protection certificates) for pharmaceuticals or phytosanitary products	Paris Convention; Patent Cooperation Treaty; European Patent Convention; Budapest Treaty; EU Reg. 1610/96 and 469/2009
Industrial designs	Medical devices, biochemical/ biophysical apparatus	Usually up to 25 years, renewal fees for each 5 years	Hague Agreement; Locarno Agreement, Geneva Act of the Hague Agreement
Plant breeder's rights	Plant varieties, propagating/ harvested material derived from them	20 years from the date of granting; for trees and wines 25 years from the date of granting of the breeder's right	Union pour la Protection des Obtentions Vegetales
Trademarks	Names (words), graphical signs, multimedia elements associated with biotechnological products/ processes/uses	Usually 10 years, renewed indefinitely on payment of additional fees	Madrid Agreement Nice Agreement Protocol adopted at Madrid Vienna Agreement Trademark Law Treaty (EU)
Confidential information	Knowhow notes, customer information, internal documentation of processes, exclusivity data of clinical tests provided in the market approval process for new drugs		EC Directive 83/2001 (for medicines market in Europe)
Copyright	Computer programs, graphical designs, databases, manuals, marketing and promotional materials		Berne Convention

time, the granting criteria became well established for technical disciplines and evolved along with technical progress.

Industrial Designs An industrial design is an IPR granted to a creative person to protect the visual design of an object, which has no technical function. It refers to the aspect and the aesthetical appearance of a device, to the artistic effect this object produces on a certain person.

Plant Breeder's Rights Also known as plant variety protection rights, they are granted to a breeder for a new plant variety that is distinct, uniform, and stable. They confer the breeder an exclusive right over the propagating material, including seed, parts of the plant, such as cuttings, tissue cells, and harvested materials (flowers, fruits, foliage) of a new variety for a certain period.

Trademarks The trademark is a distinctive sign capable of identifying and differentiating certain goods or services as those produced or provided by a specific person or enterprise. It could be represented by letters, words, or graphical signs, even specific sounds or scents, or a combination of these. They can be renewed for a lifetime or longer if certain conditions are fulfilled.

Confidential Information Although considered within the family of IPRs, no rights are granted by the government in this specific of confidential information. It comprises secret business information, such as know-how and trade secrets, which are valuable as they are kept secret.

Copyright Copyright is granted for creative works in areas such as literature, music, movies, and software, but does not cover ideas, methods of operation or mathematical concepts as such. They provide the creator with the right to exclude others from copying their own creations.

12.4 Biotech Patents

Although biotechnology and IPRs have existed separately for many years, even centuries, they came together only recently, in the last decades of the twentieth century. Modern biotechnology started developing during the second part of the last century in universities and the academic world. At that time, the first start-up firms and small and medium-sized enterprises (SMEs) came into existence, which are active in different sectors of modern biotechnology, some of them transformed today in major global companies worldwide. The scientific world and the business community understood that the development of DNA technology and other related biotechnological techniques would lead to significant financial rewards if this last frontier of science could be converted into goods and services on the market.

In biotechnology, the most valuable form of IPRs are patents, although other forms are also used in practice. They offer economic incentive for the scientists, legal protection for the owners, and at the same time, they encourage public disclosure of the invention so that the whole of society can benefit.

According to the European Patent Convention (EPC 2016a), "biotechnological inventions" are inventions that concern a product consisting of or containing biological material or a process by means of which biological material is produced, processed or used (Rule 26(2) EPC). The same legal text provides the definition for "biological material" as being any material containing genetic information and capable of reproducing itself or being reproduced in a biological system (Rule 26 (3) EPC). This definition generally covers living organisms and DNA.

Like any other inventions, a biotechnological invention must meet some criteria to be granted: to be new, to involve an inventive step, and to be susceptible of industrial application. These criteria are the same in all the jurisdictions, even the wording may be different, and, in most countries, they are examined to be fulfilled by the competent authorities before a patent is granted.

There are still some differences regarding the subject matter allowed to be patented: although in USA, Australia, and a few other jurisdictions, a patent for a method of treatment or diagnosis of the human body is permitted. In Europe, most of the Asian countries and other worldwide countries, claims drafted for methods of treatment of the human body per se are not permitted, but a device to be used in a method for the treatment of the human body or a compound used in the manufacture of a medicine for the aforementioned method, could be allowed (Blatmann et al. 2008).

In Europe, a debate on biotechnology patents started in the late 1980s with the aim of clarifying the distinction between what is patentable and what is not, and harmonizing the laws of EU member states in this area. It was clear at that time that the whole community needs legal provisions to solve ethical considerations relating to the granting of patents for biological inventions/living matter.

This resulted in the adoption on 6 July 1998 of EU Directive 98/44/EC on the legal protection of biotechnological inventions. Originally proposed in 1988, it was eventually accepted 10 years later after consistent ethical debate related to the patentability of living matter (McNamara and Booth 2018). Even so, the Directive was still contested by several ecological, religious, anti-genetic engineering non-governmental organizations, and even by the governments of some European states. In October 1998, an action was opened by the Dutch Government against the European Parliament and the Council of the European Union before the Court of Justice of the EC, an action later also supported by Italy and Norway. The grounds for this request to cancel the EU Biotech Directive were that it breaches obligations under international law, such as the Convention on Biological Diversity, for example. However, the Court of Justice of the EC dismissed the action in June 2001 (Dutfield 2009).

The Directive provides specific regulations regarding the patentability of many different categories of biological materials, ranging from certain elements isolated from the human body, to plants and animals, and to plant breeding (including the patentability of genetically modified organisms) (EU Commission 2016).

The Directive contains provisions regarding exceptions from patentability for moral reasons, such as methods for cloning human beings, and the use of human embryos for commercial and industrial proposes. It is also mentioned that only inventions concerning an isolated element from the human body/nature or produced by a technical process are allowed, including gene sequences or fragments of these, with the proviso that their function (use) is indicated (Directive 98/44/EC 1998).

The directive has now been implemented by all EU member states. In 1999, the contracting states to the EPC decided to incorporate the directive into the Implementing Regulations to the EPC. This is aimed at clarifying which inventions are patentable and which are not on ethical grounds, giving legal certainty to the interested entities in the field that are required to attract the considerable investment needed for innovation in this particular sector (EU Commission 2019). The EPC, the national laws of the individual European states together with these specific rules now provide the basis for deciding on the patentability of biotechnology patent applications in Europe (Zekos 2006).

The biotech patent applications must fulfil the same legal requirements provided by the EPCs as all the other patent applications in any technical field. After performing the substantive examination, the EPO grants or rejects a patent. Once granted, the European patent must be nationalized in any European country in which the owner wants to protect his invention, and which had previously been designated by the applicant. The enforcement of said European patent in a country's part of the EPC will be governed by the national laws, as with any other further procedure after the nationalization of the European patent (infringement, licensing, and assignment of the rights conferred by the patent) (EPC 2016b). The EPO provides an opposing procedure, by which any person can contest before the EPO the granting of the European patent, by filing a reasoned appeal during the first 9 months starting from the publication date by the EPO of the mention of the granting (EPC 2016b). The EPO has an appeal system, consisting of an Enlarged Board of Appeal (EBoA), a Legal Board of Appeal, and 26 Technical Boards of Appeal. Their decisions are considered to be legal provisions to be followed by the EPO and even the national courts tend to voluntarily consider their judgments.

Patentable Subject Matter The general provision regarding subject matters allowed to be granted under the provisions of the EPC are provided in Article 52, which mentions "European patents shall be granted for any inventions which are susceptible of industrial application, which are new, and which involve an inventive step" (EPC 2016b). The general patentability conditions, as set forth in the EPC in Articles 54 (Novelty), 56 (Inventive Step), and 57 (Industrial Application), are quite similar to the requirements of novelty, non-obviousness, and utility provided by US patent legislation (Singh 2015).

A specific legal definition of novelty was developed over the years, teaching that "new" means "made available to the public." This interpretation has a specific understanding in the field of biotechnological inventions: for example, a new gene is considered that previously existed in an organism, but it is "hidden" to the public as its existence was not known. It could be patentable if the gene were isolated from that organism or if it could be produced by means of a technical process, and all other requirements, mentioned above, would be fulfilled.

According to the EPC, any invention is eligible as patentable subject matter unless it falls within the list of excluded inventions provided in it. Articles 52 (2), (3) and 53 of the EPC say what can and cannot be patentable (EPC 2016c).

Basically, biotech inventions are patentable, but the following are considered not patentable:

- "Discoveries (e.g., the discovery of natural substances, such as the sequence or partial sequence of a gene) are not patentable because, without a description of the technical problem they are intended to solve and a technical teaching, they are not regarded as inventions" (Art. 52 (2)(a) EPC).
- "Any invention whose commercial exploitation would be contrary to 'ordre public' or morality" (Art. 53 (a) EPC).
- "Plant and animal varieties" (Art. 53(b) EPC).
- " Essentially, biological processes to produce plants and animals" (Art. 53(b) EPC, e.g., classical breeding, crossing, and selection).
- "Methods for the treatment of the human or animal body by surgery or therapy, and diagnostic methods practiced on the human or animal body" (Art. 53(c) EPC).

All these provisions gave rise to controversial discussions, opposition to the EPO or referrals addressed to the Court of Justice of the EU, resulting in decisions of the Board of Appeal of the EPO, or Court of Justice of the EU, now guiding the legal interpretation of the EPC. These decisions were partly included in the Implementing Rules of the EPC. The EPO also provides the Guidelines for Examination (GL) to give substantial help in granting inventions, which is updated every year with the decisions of the boards of EPO.

The GL, GII-35 (G-II, 3.1, 3.2, 3.3) (Guidelines for Examination 2018) specifies that the subject matter provided in Article 52 (2) (a), meaning that discoveries, scientific theories, and mathematical methods are not excluded from patentability if the invention discloses a technical application of these for solving a technical problem. For example, the discovery of a protein that specifically binds to a given receptor will not be patentable, but a medicine comprising the mentioned protein binding to the specific ligand and used for treating a certain disease, could be patentable. Similarly, a gene that is discovered to exist in nature may be patentable if a technical effect of this is revealed, such as its use in making a certain polypeptide or in gene therapy.

Rule 28 EPC, as amended by the Decision of the Administrative Council of the EPO on 01.07.2017 (EPO 2017b), describes in more detail the exceptions to patentability, as set out in Article 53(a) EPC:

1. Under Article 53(a), European patents shall not be granted in respect of biotechnological inventions which, in particular, concern the following:
   (a) Processes for cloning human beings
   (b) Processes for modifying the germline genetic identity of human beings
   (c) Uses of human embryos for industrial or commercial purposes

(d) Processes for modifying the genetic identity of animals that are likely to cause them suffering without any substantial medical benefit to man or animal, and also animals resulting from such processes.'

Many conflicting discussions have come about and divergent viewpoints have emerged around Rule 28 (1) (c), (d). In 2002, the much discussed "Edinburgh patent" (EP0695351) was maintained in an amended form as result of the opposition proceedings before the EPO. The patent in the name of the University of Edinburgh was granted for the invention with the title "Isolation, selection, and propagation of animal transgenic stem cells." At that time, the patent caused a major public debate on the patentability of stem cells. The amended patent no longer contains human or animal embryonic stem cells, but still covers modified human and animal stem cells other than embryonic ones (EPO 2018b). The dispute regarding whether human embryonic stem cells are patentable was finally resolved by a Decision of the EBoA G 0002/06/Use of embryos/WARF (Wisconsin Alumni Research Foundation) (EPO 2009). By the answer provided in Question 2, it was concluded that the EPC forbade the patenting of inventions covering products that could be prepared exclusively by a method that supposedly required the destruction of human embryos from which the said products were obtained, even if after the date of filing said patent application, the same products could be manufactured without having to imply a method necessarily comprising the destruction of human embryos. Accordingly, the exclusion from patentability of any use of human embryos according to R.28(1) (c) is effective for all patent applications under examination before the EPO. However, the EBoA pointed out that its decision did not concern the general question of human stem cell patentability.

Rule 29 EPC (EPC 2016d) states that the human body and its elements, defined at point 1 the general principle "The human body, at the various stages of its formation and development, and the simple discovery of one of its elements, including the sequence or partial sequence of a gene, cannot constitute patentable inventions." However, at point 2 of the same Rule 29, the patentable exceptions are defined, such as an element being isolated from the human body, e.g., "a sequence or a partial sequence of a gene, isolated or produced by a technical process, may be patentable, even if said element is identical to one found in nature."

Many anticancer drugs are based on patented human gene sequences, e.g., Herceptin for breast cancer or Avastin for the treatment of colon cancer. Humira, a medication covered by a patent protecting human gene sequences, is used to treat auto-immune disease (e.g., arthritis), and was the world's best-selling medicine in 2014. Most of the top 10 best-selling medications in the world are from a biological source and covered by patents (EPO 2017a).

Another subject that had provoked many disputes and finally determined some legally binding decisions is the patentability of plant or animal varieties and of the essentially biological processes/products obtained by such processes. The subject is also extensively addressed in the EU Biotech Directive. Article 4 of the Directive directly refers to the patentability of plants and animals, specifically excluding plant and animal varieties from the scope of patentable subject matter. It also establishes that "essentially biological processes for the production of plants and animals" are not patentable. Article 2 of the Directive defines an *essentially biological process* as "consisting entirely of natural phenomena such as crossing and selection" (Directive 98/44/EC 1998). However, the Directive does not specifically stipulate whether plants or plant material (fruits, seeds, etc.), or animals/animal material obtained through essentially biological processes, can be patented. The main provisions of the Directive are now part of the EPC 2000.

Considering the legal provision in force at that time, December 2010, the Enlarged Board of the EPO stated in two of its important Decisions that essentially biological processes making use of gene markers for selection were not patentable subject matter, yet these decisions did not pronounce on products obtained from these processes (EPO 2012). In March 2015, the EBoA of the EPO decided in the cases "Tomatoes" (G2/12) and "Broccoli II" (G2/13) that "a patent may be granted for plants/plant materials obtained from essentially biological processes if the basic requirements of patentability are fulfilled" (EPO 2016).

However, the story was not yet at an end. In December 2016, the EPO decided to rule a stay of all proceedings in examination and opposition cases in which the subject matter was a plant or animal obtained by an essentially biological process (Saez 2017). The reasoning for this decision was revealed by the EPO notice dated 24.11.2016: "In its practice the EPO applies the Biopatent Directive which was introduced into the European Patent Convention by the decision of the EPO member states in 1999 and which has no explicit provision in relation with plants or animals .... Should the EPO member states follow the interpretation offered by the European Commission Notice, the EPO will implement their decision."

This communication was followed in June 2017 by the Decision of the Administrative Council of the EPO amending Rules 27 and 28 of the Implementing Regulations to the EPC (EPO 2017b). Article 2 states on point 2: "The following new paragraph 2 shall be added: 'Under Article 53(b), European patents shall not be granted in respect of plants or animals exclusively obtained by means of an essential biological process.'"

The GL provide an illustrative explanation of the meaning of the new R28(2) (Guidelines for Examination 2018): if a technical feature of a claimed plant or animal, e.g., a single nucleotide exchange in a genome, might be the result of either a technical human action (direct mutagenesis) or an essentially biological process (natural allele), a disclaimer is necessary to delimit the claimed subject matter to the technically produced product. The Rule 28(2) as amended in 2017 conflicts with the decisions G2/12/and G2/13, which stated the possibility of obtaining protection by patent for plant/plant material obtained by essentially biological processes, even if said essentially biological processes cannot be protected per se.

Eventually, in December 2018, during the oral proceedings held in the appeal case T1063/18, the Technical Board of Appeal of the EPO, consisting of three legally qualified members and two technically qualified members, reached the decision that "Rule 28(2) adopted by the Administrative Council of the EPO in June 2017 was in conflict with Article 53(b) of the EPC, and thus was invalid." In its reasoned written decision in the mentioned case T1063/18 issued on 05.02.2019 (EPO 2019), the Technical Board stated that "Rule 28(2) EPC could not be interpreted in such a way that it was not in conflict with Article 53(b) EPC as interpreted by the Enlarged Board of Appeal." The Board concluded that it was no reason to deviate from the previous interpretation of the Enlarged Board, as was stated in the decisions G2/12 and G2/13. The Board also stated that the provisions of Article 164(2) EPC have to be enforced, meaning that the provisions of the Convention prevailed over its Implementing Regulation. At least for now, plants/plant materials obtained even by essentially biological processes can be protected by patent in Europe.

To conclude, we present in Table 12.4 some examples of patentable and non-patentable subject matter from the biotechnology field.

It is understood that all the inventions, including those from the biotechnological field, to be granted, must be new, inventive and to have an industrial application, also including here agriculture. Some specific mentions related to the novelty condition (EPC 2016e):

- The so-called "purpose-limited compound protection" – a known substance could be still patentable if no medical use of this is known so far – "1st medical use" (Article 54 (4) EPC)
- "2nd medical use" – a known substance could be patentable if the medical use is known, but not yet stated for specific medical use.

As already mentioned, the greatest number of patents granted in the biotechnological area are in the pharmaceutical and health care field (55%). Patents are crucial for promoting medical progress. One of the first biomedicines was insulin and since then, many new improved forms of insulin

Table 12.4 Some examples of patentable/non-patentable subject matter in biotech patent applications

Patentable biotech subject matters		Non-patentable biotech subject matters	
General subject	Examples	General subject	Examples
Genes, nucleic acid molecules	Genes related to a specific disease, used in diagnosis; siRNA molecules in therapy	Sequences without a known function	ESTs resulting from automated sequencing
Proteins	Insulin, erythropoietin in therapy	Genetically modified animals suffering without any benefit for them or for humans	Genetically modified animals used in cosmetic testing
Enzymes	Proteases in the detergent industry or food processes	Plant varieties	Protected under UPOV
Antibodies	Medicines in cancer treatment	Animal varieties	Certain cattle species
Viruses/virus sequences	HCV and HIV in diagnostic tests, development of new vaccines	Human embryos	They and processes involving their destruction
Cells	Hematopoietic stem cells in leukemia treatment	Human germ cells	Sperm, oocytes
Microorganisms	Bacteria in bioremediation, yeast in the food industry	Human/animal chimera	
Plants	Plants resistant to certain herbicide, to drought		
Animals	Disease models in research, dairy animals for producing milk enriched in certain drugs		

siRNA small interfering ribonucleic acid, *HCV* hepatitis C virus, *HIV* human immunodeficiency virus, *EST* expressed sequence tag, *UPOV* Union pour la Protection des Obtentions Vegetales

have been developed, and are covered by patents. A new direction for medical treatment is currently arising: the so-called personalized medicine using targeted treatments. This type of medical treatment is based on a person's individual genetic code to identify his risk for developing certain diseases (EPO 2017c). It is a promising new world powered by genetics, meaning genetic testing and DNA sequencing. The future of personalized medicine relates mainly to two factors:

- Increased access to DNA testing through low costs
- Improved, deep understanding of DNA markers associated with specific diseases

Developing a new innovative drug and putting it on the market are highly expensive activities, performed over a long period of time. The necessary funds are mostly provided by powerful investors. To support the highly costly innovative research and clinical trials involved in launching a new drug, even the big drug companies must recoup their investments by claiming their exclusive rights provided by a patent covering the new drug. Patents are, at the same time, the most appropriate means of preventing illicit copying of the original medicines and the health risks associated with unauthorized fake medicines.

Patents are an exclusive right for only a limited time period: 20 years from the date of filing the patent application, after which the invention can be used by anyone without paying royalties. After the protection period conferred by the patent has expired, the generic drug companies have the right to produce cheaper versions of the original medicines.

12.5 Transfer of Innovation in Biotechnology

In an economy based on the latest contest of science and technology, biotechnology is proving to be the new "wonder child," just after information technology, because of its growing contribution to development. A report by the European Commission (2002) estimated that the global biotechnology market could reach over 2000 billion Euros in the second decade of this century. Biotechnology has proved to be the source of remarkable developments in pharmaceuticals, agrochemicals, energy and environmental industries. This important trend is also confirmed by the growth rate of patent applications in this field filed with the EPO over the last few years. The latest available statistics (EPO 2018c) revealed that biotechnology covers part of the top ten fields, representing 53% of the total patent applications filed at the EPO. Moreover, the number of biotech patent applications filed with the EPO shows the greatest growth rate of 14.5%, versus 8.1% in pharmaceuticals, 6.2% in medical technology, and 5.7% in digital communications. The same trend is also shown at the United States Patent and Trademark Office, where the number of patents granted in the biotechnology field showed 15% growth between 1990 and 2000, versus an overall increase of 5% in patents from all fields of technology, according to an Organization for Economic Cooperation and Development (OECD) report (2002). Without doubt, the patents have proved to be crucial in the development of the biotechnological sector and their importance in the global policy of a company owning the patent is much greater than in other scientific or economic areas. This is because biotechnology is one of the most research-based industries. The biotech companies often invest more than 40% or even 50% of their revenues in R&D, compared with 5% in the chemical industry or around 15% in pharmaceutical companies (Burrone 2006).

Obtaining patent protection for their inventions is a key element in the strategy of any company or research entity involved in biotechnology. The main reasons for this are:

- The high costs necessary for the development of new innovative products, which are cumulative with the increased risks involved in each research project, and very powerful competition, whereas the results are relatively easy to copy. An appropriate IP protection is a must to ensure their survival and further development.
- The close connection existing in this field between fundamental and applied research. Great biotech companies are often based on the fundamental research performed at universities and public R&D institutions; meanwhile, in some cases, basic research teams give rise to SMEs in the biotechnology field that are based on one or a few patents previously developed.
- In some cases, the final product of small biotech R&D companies are actually the patents covering their innovative products or processes and which they plan to license to powerful companies able to put the product on the market. The revenues obtained are further invested in R&D to create more new innovative products.

A study published in 2003 under the umbrella of the OECD (Arundel 2003) identifies means of transferring innovation in the biotechnology area, mainly by the contribution of public policies:

- "Diffusing biotechnology knowledge and expertise" with the help of public policies supporting collaboration between private firms and R&D public institutions and universities, by granting funds to increase the number of lucrative contacts between different actors implicated in the biotech sector, collaborative networks.
- "Commercializing biotechnology research results": several EU countries provide grants to increase the number of start-up companies and to support SMEs in the biotech field as a means of bringing to the public the latest results of biotech R&D in the form of marketed products and services.
- "Encouraging the application in production and end uses of biotechnology," including providing the public and private business sector with information and/or demonstration projects, facilitating access to the appropriate regulatory approval systems for interested SMEs, especially in pharmaceutical, food, plant or agricultural areas.

Besides these policies established at a national or EU level, there are some specific means that facilitate the transfer of innovation by commercializing IP rights, meaning the most valuable assets in patents in the biotechnology field.

The words of the CEO of Novo Nordisk seem very relevant to the challenge this sector faces (Ovlisen 2002): "Anyone, I would imagine, who has tried to create a biotech company knows just how important patents are. You learn when you're studying, and again at your first job, and if you haven't done so before, you realize it the first time you meet potential investors."

The greatest challenge that anyone in this business faces is how to turn the patented invention into a profitable asset, thus ensuring the future development of R&D activity.

Generally, IP commercialization means the exploitation of IP in the market with the aim of generating income.

The successful commercialization of a patent must consider some important aspects, especially in such a cost-intensive field of technology as biotechnology. Some of the issues refer to the nature of the IP (patent) such as:

- The validity of the patent at the said moment, meaning is it still in force in the territory concerned, and what is the remaining life of the patent
- How strong and how broad is the protection conferred by the patent
- Whether the said IP right requires further development
- How long is the life of the product covered by the patent on the market estimated to be

Other aspects concern potential competitors or infringers, for example:

- Who are the potential competitors and how they are known to respect IP rights
- Which costs will imply the defense or prosecution of IP infringement actions

The main forms known for commercialization of IP are the following:

- Direct exploitation by the patent owner
- Licensing
- Assignment
- Newly created subsidiaries of the company owning the patent (spin-off companies)

12.5.1 Direct Commercialization by the Patent Owner Company

In the biotechnology sector, this form of commercialization of IP is the most frequent in the huge global companies, which have the financial resources to support possible additional research and development projects, product development, clinical or other mandatory trials, approvals, scaling-up the production, and conducting marketing research to put the product on the market, especially in the biotech field.

12.5.2 Licensing

An IP license entitles another entity to use said IP for a certain time period, in a certain territory, this being the most common and at the same time most flexible form of commercializing an IP. It has the advantage that the transfer of IP ownership is not permanent. A license may be exclusive or non-exclusive, being granted for a specific activity such as researching, developing, manufacturing or selling products or services covered by the said IP. A license agreement has specific requirements and is legally bound by the legislation of the state where the principal office of the licensor or licensee is located, as agreed by the parties.

Licensing agreements should generally encourage new biotech inventions related to human health care or the development of innovative agriculture; they should be a better guarantee that new medicines, or diagnostic products, or other products or services incorporating genetic inventions, are made available to the public on a more reasonable basis. At the same time, licensing will ensure a more rapid dissemination of scientific information and meanwhile, both parties could obtain returns from their investment, an essential step that will allow further development of innovation in this high-cost R&D field.

12.5.3 Assignment

An assignment in the IP field is a permanent transfer of the ownership of the IP asset (patent) to another entity. This form of innovation transfer could be preferred by small research innovative teams, which, in exchange for the assignment agreement, could have more options: to receive in return financial compensation in the form of a lump sum payment, or in return for royalties or a combination of the two. In any of these situations, the previous owner of the patent no longer has the right to use said IP asset, unless a license-back clause was agreed.

In most jurisdictions, the licenses and assignments must be recorded with the national competent authority, such as the National IP Office.

12.5.4 Spin-Off Companies

The original company owning the IP can decide to delegate the mission of commercializing the IP asset to one of its own subsidiaries, which will become a spin-off company, with a separate legal identity.

This measure presents the advantage of transferring the risks and responsibilities related to bringing the new innovative product to the new spin-off entity. The original company that owns the IP asset will license or assign said patent to the new spin-off company to enable it to commercialize the IP. This type of option will be chosen depending the global policy of the mother company. This is probably the most complex and expensive form of IP commercialization, but it could be the right option, especially when the IP requires extensive additional development.

Any of these forms of commercialization of IP assets may be embraced by an R&D entity in the biotechnology field, or a combination of these.

An illustrative example of a successful biotech company, founded in 2006, may be found on the EPO website (EPO 2017d). This SME is a biopharmaceutical spin-off from the Veterinary University of Vienna. The company developed health care products based on a natural polymer isolated from red algae, which is active against respiratory viruses. The activity of the company is covered by three patents, enforced in almost 100 countries, and a trademark, also registered in over 50 countries. Acting in the biopharmaceutical field, the company challenged the usual risks involved in the drug industry: major R&D investment to discover new active compounds, followed by another expensive and time-consuming phase for their approval.

To achieve financial success, the company was based on its patent portfolio and developed a business strategy focused on the exploitation of its IP. The business strategy comprises two distinct forms of business. The first one is a common license agreement, by which the licensee receives the right to produce, to put onto the market, and to distribute the product in certain countries, in exchange for upfront payments and royalties for the company owning the IP.

The second business form agreed by the SME and its partner is a distribution partnership. According to this latter agreement, the partner has the right to purchase the product direct from said SME and can distribute it within a certain geographical area. The products are, in this case, adapted to complete satisfy the partner's needs, bearing their name and logo.

This example proves that a flexible licensing agreement will result in win-win situations for all the parties involved.

For such an SME involved in the biotechnology R&D field, patents were essential to attract investors from different sectors, such as public funding institutions, venture capitalists or bond investments.

12.6 Some Final Remarks

With respect to IP, the two main ethical issues that are often confronted in the field of biotechnology: (1) the morality of patenting life and (2) whether it is morally acceptable to restrict access to scientific advances that could potentially benefit a large number of people if freely available. Ethical concerns of the leaders of the

Human Genome Project that gene patenting could raise barriers to research and the clinical use of genomics, made the scientists determined to place sequence data in public databases, thus rendering it impossible for them to be patented.

Patenting of life forms is however different from that of classical chemical or mechanical inventions and poses unique problems such as establishing the "inventive" nature of a living subject matter, difficulties in preventing infringements owing to the ability of the organism to self-replicate. On the other hand, the biotech firms often rely on IPRs for their success in business, in particular patents, as these are one of the most valuable assets they own in this business sector that is extremely research-intensive but in which the costs of copying are low.

This chapter was intended to find some answers to tricky questions arising from patenting in biotechnology.

References

Arundel A (2003) Biotechnology indicators and public policy. STI Working Papers. OECD, Paris

Blatmann A, Heuzenroeder P, O'Brien M (2008) Biotechnology intellectual property management manual. Spruson & Ferguson, pp 8–14

Burrone E (2006) Patents at the core: the biotech business. https://www.wipo.int/

DaSilva JE (2004) The colours of biotechnology: science, development and humankind. Electron J Biotechnol 7(3)

Directive 98/44/EC (1998) Directive of the European Parliament and of the Council of 6 July 1998 on the legal protection of biotechnological inventions, published in Official Journal of the European Communities, 30 July 1998, extracted from https://eur-lex.europa.eu

Dutfield G (2009) Intellectual property rights and the life science industries – past, present and future, 2nd edn. World Scientific, Singapore, pp 204–205

EC (2002) Life sciences and biotechnology – a strategy for Europe

EPC (2016a) European Patent Convention 2000, 16th edn, Chapter V, Rule 26, Biotechnological inventions, published on June 2016 on https://www.epo.org/law-practice/legal-texts/epc.html

EPC (2016b) European Patent Convention 2000, 16th edn, Art. 64 (1), (3), published on June 2016 on https://www.epo.org/law-practice/legal-texts/epc.html

EPC (2016c) European Patent Convention 2000, 16th edn, Art. 52–53, published on June 2016 on https://www.epo.org/law-practice/legal-texts/epc.html

EPC (2016d) European Patent Convention 2000, 16th edn, Rule 29, published on June 2016 on https://www.epo.org/law-practice/legal-texts/epc.html

EPC (2016e) European Patent Convention 2000, 16th edn, Art. 54(4) and (5), published on June 2016 on https://www.epo.org/law-practice/legal-texts/epc.html

EPO (2009) Decision of the enlarged board of appeal G2/06, dated 25 November 2008, published in the Official Journal of the EPO 5/2009, p 306

EPO (2012) Decisions of the enlarged board of appeal G2/07 and G1/08, dated 09.12.2010, published in the Official Journal of the EPO 3/2012, p 130, 206

EPO (2016) Decisions of the enlarged board of appeal G2/12 and G2/13, dated 25 March 2015, published in the Official Journal of the EPO 2016, A27, A28

EPO (2017a) Biotechnology patents at the EPO, extracted from http://www.epo.org/news/issues/biotechnology-patents, last updated on 16 January 2017

EPO (2017b) Decision of the Administrative Council of 29 June 2017 amending rules 27 and 28 of the implementing regulations to the European Patent Convention (*CA/D 6/17*), Official Journal of the EPO, A56

EPO (2017c) Is it in my genes? The future of personalised medicine is focused on individual patient DNA, last updated on 24 May 2017, from https://www.epo.org/news-issues/technology/medical-technologies/personalised-medicine

EPO (2017d) Using red algae to fight the flu, *EPO SME case studies/MARINOMED*, published on www.epo.org/sme

EPO (2018a) Annual report 2017 – European patent applications, published on 22 January 2018 on https://www.epo.org/

EPO (2018b) Decision of the opposition division of the EPO published in the European Patent Bulletin 8/44 of 29 October 2018

EPO (2018c) Annual report 2017 – At a glance, published on 22 January 2018, www.epo.org/statics-indicators2017

EPO (2019) Decisions of the technical board of appeal T1063/18, issued on 05 February 2019, extracted from http://www.epo.org/law-practice/case-law-appeals/communications

EU Commission (2016) Notice on certain articles of Directive 98/44/EC of the European Parliament and of the Council on the legal protection of biotechnological inventions, (2016/C 411/03). Off J Eur Union, C411/3, 08 November 2016

EU Commission (2019) Protection of biotechnological inventions, extracted from http://ec.europa.eu/growt/industry/intelectual-property/patents. Accessed January 2019

Gelinas P (2010) Mapping early patents on baker's yeast manufacture. Compr Rev Food Sci Food Saf 9:483–497

Guidelines for Examination (2018) GII-35. https://www.epo.org/law-practice/legal-texts

McNamara K, Booth M (2018) The birth of biotech, extracted from www.vennershipley.co.uk

OECD (2002) Genetic inventions, IPRs and Licensing Practices
OECD (2004) Patents and innovation: trends and policy challenges
OECD (2005) A framework for biotechnology statistics. OECD, Paris
Ovlisen M (2002) In Medicon valley patent guide. http://www.mva.org/media
Saez C (2017) Patentability of some biotech uncertain in Europe as examination, opposition suspended. Intellectual Property Watch. http://www.ip-watch.org
Singh KK (2015) Biotechnology and intellectual property rights – legal and social implications. Springer India, pp 66–67
Strenc AC, Ionescu B, Gheorghiu G (2005) Dreptul Brevetului. Editura Lumina Lex, Bucuresti, pp 55–56
Venter C, Cohen D (2004) The century of biology. N Perspect Q 21(4):73–77
WIPO Convention (1967) Convention establishing the world intellectual property organization, Art 2(vii)
Zekos GI (2006) Nanotechnology and biotechnology patents. Int J Law Inform Technol 14:310

European Biotech Entrepreneur Profile: Case Studies

13

Florentina Matei, Flavia Anghel, Mioara Varga,
Mihaela Draghici, Cristina Coculescu, and Oscar Vicente

Abstract

To be entrepreneur it takes a certain type of personality, but there is also a range of skills needed for success. The authors have conducted a survey among European biotech entrepreneurs, trying to profile their skills, knowledge, and competencies adapted to different socioeconomical contexts; an analysis of potential differences between the Western and Eastern Europe sides has also been targeted. A high similarity in answers has been generally noticed, in relation to respondents' competences levels, as well as on the importance of common competences/skills/abilities for being an entrepreneur, like creativity and innovation, professionalism, communication, leadership, and teamwork. An important conclusion is that in biotech companies there is a high demand for higher education graduates and these graduates have to continuously improve their technical and managerial knowledge and skills following training courses. The societal and economic context in the targeted countries has different influence on biotech new ventures. Several European biotech entrepreneurs have provided testimonials on their own experience and useful recommendations for future graduates.

Keywords

Europe · Biotechnology · Entrepreneurship · Challenges · Testimonials

13.1 Introduction

According to STATISTA (www.statista.com), 2259 biotech companies have been registered in Europe in 2016, of which 89.6% were private companies. The same source indicates that the share of biotech companies by focus areas is 9.3% in the health sector, 10.7% in industrial biotech, 9.3% in services, and less than 1% in agrobiotechnology.

Being entrepreneur is not easy to everyone. According to Shimasaki (2009), the biotech entrepreneur is unique from all other entrepreneurs; a too cautious, too analytical, and too practical person will never start a business in biotech. Apart of being independent, being confident, and having willingness to take risks, passion for his work,

F. Matei (✉) · M. Varga · M. Draghici
Faculty of Biotechnologies, University of Agronomic Sciences and Veterinary Medicine of Bucharest, Bucharest, Romania
e-mail: florentina.matei@biotehnologii.usamv.ro

F. Anghel · C. Coculescu
Romanian-American University, București, Romania
e-mail: coculescu.cristina@profesor.rau.ro

O. Vicente
Institute for Plant Molecular and Cell Biology (IBMCP, UPV-CSIC), Universitat Politècnica de València, Valencia, Spain
e-mail: ovicente@upvnet.upv.es

F. Matei, D. Zirra (eds.), *Introduction to Biotech Entrepreneurship: From Idea to Business*,
https://doi.org/10.1007/978-3-030-22141-6_13

and the ability to work long hours, the biotech entrepreneur is usually an accomplished scientist, bioengineer, physician, or businessperson capable of identifying problems but not focusing too long on finding lots of solutions to any situation. Concluding, it takes a certain type of personality to work out independently, but there is also a range of skills needed to find success. In this context, during the implementation of an Erasmus+ project (2017-1-RO01-KA203-037304; www.supbioent.usamv.ro), the authors have conducted a survey among biotech business persons, trying to profile the biotech entrepreneur's skills, knowledge, and competencies adapted to different socioeconomical contexts; in addition an analysis of potential differences between the Western and Eastern Europe sides has also been targeted. Some entrepreneurs have been asked to provide testimonials and their answers were compared with the survey results.

13.2 Profiling European Biotech Entrepreneurs

During the year 2018, a mixed team of biotech and business high education teachers from a European consortium (www.supbioent.usamv.ro) including UASMV Romania, Romanian American University, Universitat Politecnica de Valencia, Universita degli Studi di Perugia, and KU Leuven have conducted a survey to profile the biotech entrepreneur in Romania, Italy, Spain, and Belgium. The survey targeted biotech companies' owners or shareholders and, secondarily, employees with managing responsibilities in biotech companies.

In the first step the **general profile** of the interviewed persons was established. All the respondents declared that they are university graduated, but only 27% were owning a doctoral degree; however, 78.4% of them have followed advanced training courses. Most of those advanced courses were in the topics of management and marketing, followed by biotechnology and entrepreneurship. Our results are in line with results reported by other authors (Jimenez et al. 2015) which emphasize the demand in biotech companies for staff with high-level training (PhD and postdoctoral graduates), instead of high-school or bachelor graduates. In terms of age, 61% were over 45 years old and a relative gender balance has been noticed (52% men to 48% women), while generally men are recognized as having much more disposition to start a business; this can be correlated with the last decade policy efforts to support women in entrepreneurship (Braidford et al. 2013).

Coming to the respondents' **companies profile**, they are active in different field of biotech (Fig. 13.1). The first place was taken equally by the pharmaceutical industry and agriculture (18% each), followed by the chemical industry (15%) and food industry (13%). If taken together, medicine (11%) and veterinary medicine (9%) have actually the first place (20%). Considering the distribution by nation, pharmaceutical and food industries were predominant in Romania, chemical industry and agriculture in Italy, and pharmaceutical industry and veterinary medicine in Spain. The majority of the respondents from Belgium belong to human medicine and pharmacy.

More than half (51%) of the companies have been running over 10 years and only 3% are young companies. Regarding the distribution by nation, it has been noticed that only in Belgium young companies are actually running, while in Romania and Spain more than half of the companies are over 10 years. When asked about the **operational level**, only 20% of the companies

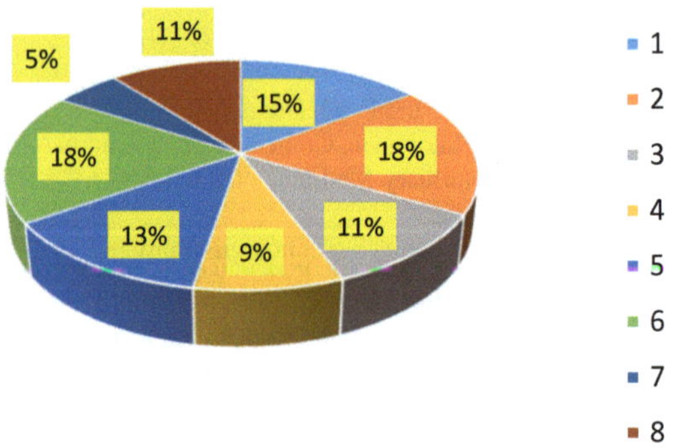

Fig. 13.1 Biotech field of the owned/employing company: (1) chemical industry; (2) pharmaceutical industry; (3) medicine; (4) veterinary medicine; (5) food industry; (6) agriculture; (7) technical Services; (8) other

are operating at the local or regional level, while over half of them (55%) have an international market.

Next, the respondents provided information on their **personal entrepreneurial profile**. The respondents have been asked about the "**idea's sources**" when they started the business. Overall, the main ideas' sources are linked to a previous job or research project (30%) or from the university studies (24%). Media or discussions with job colleagues are not reported as important inspiration source (5% each). The answers were similar in all targeted countries (Fig. 13.2).

When asked what **made them to set up a biotech company**, almost half of the respondents (48%) said that passion was the driving engine; other important reasons were on financial level (moneywise opportunities) or the need for independence ("to be your own boss"). When analyzed by country, slight differences in the answers have been noted. For Romania, the favorites were passion for biotechnology and the desire for autonomy. For the Spaniards, the favorite responses were passion for biotechnology and confidence in this area of the future. Belgian respondents are also passionate about biotechnology, believe it is an expanding field, and want to be their own bosses.

In accordance with Shimasaki's description (2014), over 60% of the Northern-American respondents declared that they are not afraid to take risks and caution is not in their personality; the answers were similar for the European biotech entrepreneurs interviewed in the study. The similarities go further, and over 80% of the respondents are allocating over 10 h a day to run their business. The most appreciated value as being a biotech entrepreneur was considered creativity, followed by money income and independence.

When **setting up their companies**, almost half of the European respondents (43%) were using their own funds, while the other half (48%) were accessing nonrefundable regional or European funding; only 3% made use of a credit. A special case was noticed for Romanian respondents; only 6.7% are accessing nonrefundable funding from national or European programs (Matei et al. 2019). This may be correlated with studies of the European Commission (2016) which reported that a survey conducted by the Romanian Council of SMEs shows that over 81% of SMEs do not intend to access structural funds in the future. One of the reasons may be linked to the rate of success which is very low; according to the same survey, only 0.18% of the companies that try to access funding obtained an approval of their project.

Regarding the moment of setting up their companies, different **challenges** have been indicated by the respondents (Fig. 13.3): scarce

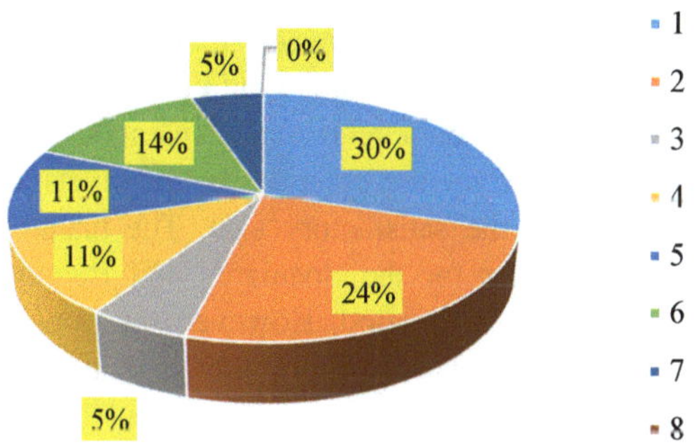

Fig. 13.2 Idea's sources for starting a biotech business: (1) previous job/a research project; (2) university studies; (3) discussion with job colleagues; (4) discussions with university colleagues; (5) discussions with friends; (6) discussions with family members; (7) media (TV, Internet); (8) other

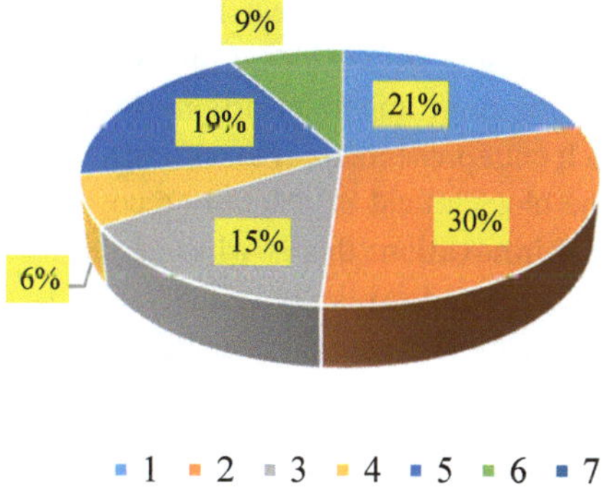

Fig. 13.3 Repartition of challenges when setting up the biotech company: (1) difficulty obtaining authorizations and approvals; (2) scarce capital for financing the business; (3) excessive taxes; (4) lack of managerial skills of planning, coordination, etc.; (5) difficulty in hiring specialists in the field; (6) lack of financial management skills; (7) other

capital for financing the business (30%), difficulty in obtaining authorizations and approvals (21%), difficulty in hiring specialists in the field (19%), excessive taxes (15%), and lack of skills in financial management (9%) or even in general managerial skills, like planning or coordination (6%). Remaining on challenges, different problems are going to be **confronted during business.** The first place is taken by bureaucracy (28%), followed by the difficulty to access financial funding (24%), and the excessive taxes (19%). The picture is completed by the difficulties to employ specialists in the field (15%) and the lack of sale skills (8%). Looking at both situations described above, it looks like that on setting up the company, and running the company, the main challenges are funding, bureaucracy, and high taxation. To solve such issues, networking may be a solution; 72% of the respondents agreed that organizing clubs, hubs, and entrepreneurs' associations have a positive impact on their business's development.

Passing though their own profiles and challenges, the interviewed participants were asked to describe the **ideal profile of a biotech entrepreneur**. What kinds of **skills** are needed for a biotech entrepreneur? Surprisingly, only half of the respondents considered the technical skills being essential, while over 95% agreed that managerial skills are the most important, followed by interpersonal skills (communication, negotiation, teamwork, time management, networking). Actually, leadership is considered all over the world one of the most important driving experience in biotech entrepreneurship (Patzelt et al. 2012).

The biotech field is generally recognized as a field of **innovation**; this is why the respondents have been asked explicitly if innovation is a key element in biotech entrepreneurship. As expected, the answer had a strong orientation toward the positive answer (Fig. 13.4), meaning that 91% of the respondents consider that innovation is one of the most important elements when running a biotech start-up/company.

The study also tried to establish if the **entrepreneurial skills are to be acquired through education**. An important majority of the respondents (77%) were in agreement with the statement, while 14% have been neutral to the statement.

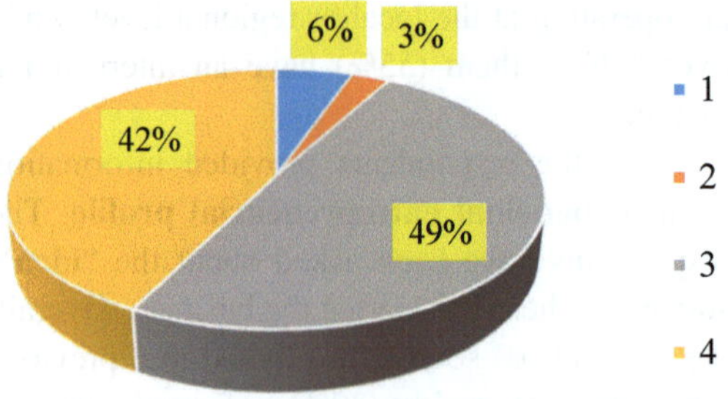

Fig. 13.4 Answers' distribution among European biotech entrepreneur for the statement "Innovation is a key quality of the biotechnology entrepreneur". (1) Disagree; (2) neither agree, nor disagree; (3) agree; (4) strongly agree

After asking the respondents what do they consider as key skills for a biotech entrepreneur, they have been asked how do they fit in this "absolute" profile. On the side of **technical skills**, while 31% considered themselves having an average level of technical skills, an important majority (61%) declare to have good (12%) to very good (49%) knowledge in biotechnology, despite the fact that only 50% of the respondents have declared that technical skills are important for an entrepreneur.

Regarding the **level of expertise in different areas** (management planning, strategic planning, human resources management, financial management, IT), on a scale from 1 to 5 (from very low level to very good level), here are the results. The Romanian entrepreneurs consider that they have an average level for all the itemized skills, the answers being placed between 3.03 and 3.77 on the 1 to 5 scale. The Italian entrepreneurs have self-evaluated their level in management planning as very good; for the other skills, the levels are average, on the satisfactory side. The management planning for the Spaniards and Belgians was considered as good. However, the Spanish respondents declare to have only average-level skills in human resources management and IT and low-level skills in financial management and strategic planning). Meanwhile, the Belgians consider their skills level in human resources management and IT as low and very low in the case of strategic planning.

In addition, the entrepreneurs have been asked to evaluate their **level in transversal competences/skills/aptitudes** (communication, teamwork, creativity, negotiation, problem-solving, networking, time management, leadership). Regarding respondents' nationality, there is a less significant variation among the answers (3.00–4.57), which indicate average to good and very good level of such skills. The Romanians and Belgians indicate to have a good level in communication, team working, creativity, and problem-solving and an average level in negotiation, conflict solving, networking, team management, and leadership. The Spanish respondents indicate an average level of skills about networking, and a good level for all the other skills, while the Italians consider having a good level on leadership and average for all the other skills.

Further, it was important to have an image of **biotech entrepreneurship in the societal context**. Some questions related to the **society's perception about the entrepreneurs** have been addressed to the respondents. Only 35% of the respondents have agreed that "*the entrepreneur has a positive image in the society, enjoying media*," more than half (55%) being neutral about the statement and some even disagree (10%). An important majority of the respondents (65%) agreed with the idea that entrepreneurship is a valuable profession for society, but still, 32% have a neutral position on the statement. The distribution of the answers regarding the statement "*Entrepreneurship is considered a remedy type of professional choice, is a form of self-employment adopted when individuals don't have better working options*" has provided some surprises; a relatively important part of the respondents (35%) did not agree with the statement, and 40% were neutral. Only a quarter of the respondents did agree that entrepreneurship is a remedial choice of professional life.

Another approached issue was the **causes of business failure**. When asked if the lack of technical skills can be a cause of business failure, half of the respondents were neutral (53%) and 31% agreed with the statement. Talking about the lack of managerial skills, the neutral respondents decrease in favor of positive respondents (44% agree with the idea). When asked if business failure can be a valuable feedback for personal and professional growth, few respondents disagree (9%), while more than half (61%) agreed with the idea; still, 30% of the respondents are in a neutral position. Further, when asked if such failure occurs and can this be a barrier for new entrepreneurial initiatives, only 30% agreed with the statement; an important majority was neutral on this issue (43%). Meanwhile, it was important to find out if the way business failure is perceived in society will influence other potential entrepreneurial initiatives. The answers were relatively balanced; while 40% believe that the social perception has little influence, 47% are on the opposite side, and a few (13%) have a strong belief that this will influence future entrepreneurial initiatives.

National culture has been considered as an important issue and the respondents have been asked if such a culture encourages the entrepreneurial initiative (Fig. 13.5). Overall, more than half respondents (60%) agree with the statement, while 22% are neutral and only 18% are in disagreement. In the case of Romania, it has been noticed that the average distribution is somewhat

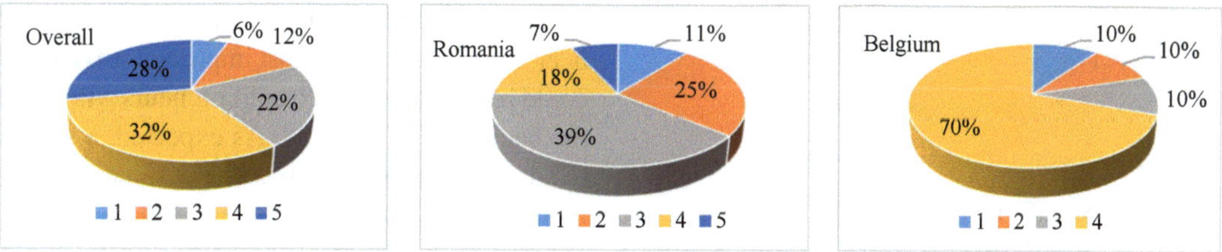

Fig. 13.5 Answers' distribution on how national culture encourage entrepreneurial initiatives. (1) Strongly disagree; (2) disagree; (3) neither agree, nor disagree; (4) agree; (5) strongly agree

different, probably due to the interferences of the economic context in which people lived in the last few decades. For example, only 25% agreed that the Romanian national culture encourages entrepreneurial development and some measures should be taken to solve the situation. On the other extreme is Belgium, in which 70% of the respondents have a strong belief that the national culture encourages entrepreneurial initiatives.

Talking about specific measures to develop entrepreneurial initiatives, (Fig. 13.6), when asked which are the most common ways to access information or to acquire skills and knowledge for business development, the answers' repartition is quite equilibrated. Overall, the same importance is given to attending courses, coaching, and networking (21–23%), followed by reading specialized books and mentoring (14–16%). Regarding nations, it can be noticed that in Spain the most important source is networking (46%), while in Italy (37%) it is coaching.

The respondents have also been asked to respond to some "free answer questions." One question was about the **most important principle/rule that guide the respondents' personal life.** A large variety of answers have been received which are in relation to family and moral values/ethics, as well as courage, optimism, and happiness. Some answers were expressed more reflexively, like "*Life is volatile like a shadow*," "*If it is too easy, it is not worth to do it*," or "*I try to see only the good things around me*."

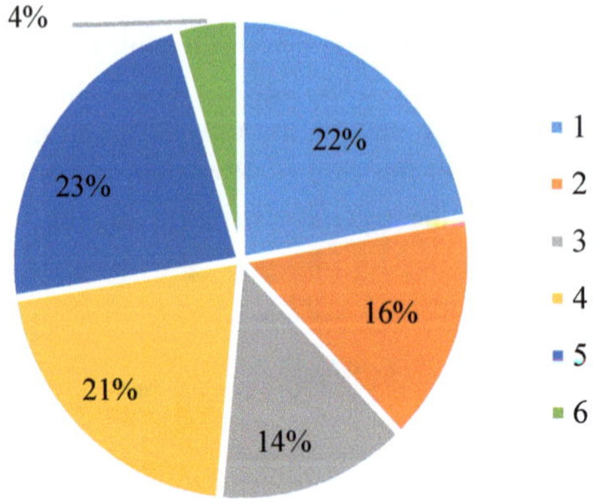

Fig. 13.6 Specific measures to develop entrepreneurial initiatives—a European perspective: (1) attending courses; (2) reading of specialized books; (3) mentoring; (4) coaching; (5) networking—discussions with other entrepreneurs; (6) Other

Another question was about the **most important principle/rule that guide their business**. Also here, a wide variety of answers has been registered, relating to action and dynamism, vision, initiative and assuming risks, invest and develop, competences and professionalism, ethics and fair play, commitment and perseverance, innovation and creativity.

A challenging question was "**If you would have the power to make a change in your country, what would you change to encourage entrepreneurship**?" Same as in the precedent free questions, the area of the answers was very wide. In **Romania,** the intervention areas would be the following: measures to support young entrepreneurs, less bureaucracy and taxes, changes in work legislation, professionalism in funding agencies, more training opportunities and written guides, and changes of the negative society's perception about the entrepreneurship. In **Italy**, the respondents identified issues which support changes related to high taxes, more opportunities for the young generation, and much more need for research and rewards for the innovative companies. The **Spanish** respondents will make changes in curricula and education, will encourage public funding for companies, and will ask for national pacts between research-development and politic power. In **Belgium,** some specific issues have been detected, considering that entrepreneurship is well developed in the country; they are asking for governmental recognition to avoid potential failures, governmental support for all companies, not only for start-ups, more efforts on high education improvements, and tax reduction.

To sum up their opinions, the respondents have been asked to give the most important **advice** that could be given to biotech entrepreneurs who are at the beginning of the road. As expected, a variety of advices have been proposed, but some are common and rise often, like "to be" active, patient, optimistic, responsible, hard worker, professional, anchored in reality, "to have" courage, "to don't" despair, to believe in their products, to access the existent know-how and ask the experts, to do a deep market analysis and develop a good business plan before, to keep learning, and to take care of potential coworkers.

The study was completed by interviewing some of the European biotech entrepreneurs about their **personal experience in entrepreneurship**. They have provided both written (http://supbioent.usamv.ro/wp-content/uploads/2018/07/O1_Success-stories-in-European-Biotech-Entrepreneurship-Final.pdf) and video testimonials on YouTube channel of the project Erasmus+ **2017-1-RO01-KA203-037304** "*Supporting biotechnology students oriented towards an entrepreneurial path* - SupBioEnt" (https://www.youtube.com/channel/UCb149acJwLHj4Spi59tWsug). They were representing each country involved in the study (Romania, Italy, Belgium, and Spain), and they are active in different biotech fields: health and pharmacy industry, veterinary biotech, plant biotech, and biotech services (molecular biology analysis and equipment providers). They have provided several recommendations to the young biotech students, potential entrepreneurs, like to believe in their own ideas, to be patient and to try to make a balance between risks and benefits, *play a team game to succeed*, learn and work in an international environment, keep contact with your mother university, look always for innovation and improvements, and, of course, "*never give up!*"

As **general conclusion**, even though this study intended initially to target a higher number of respondents, because of the questionnaire's high complexity only a relatively small number of respondents have answered to all the questions. However, taking into account that the study targeted mainly biotech entrepreneurs from countries involved as partners (Romania, Italy, Belgium, and Spain) in the project Erasmus+ **2017-1-RO01-KA203-037304** "*Supporting biotechnology students oriented towards an entrepreneurial path*—SupBioEnt," the number of respondents may be considered as significant for the niche of European biotech entrepreneurs.

During the study, a high similarity in answers has been generally noticed, in relation to respondents' competences levels, as well as on the importance of common competences/skills/abilities for being an entrepreneur, like creativity and innovation, professionalism, communication, leadership, and teamwork. An important conclusion is that in biotech companies there is a high demand for tertiary level graduates (master's, PhD), and these graduates have to continuously improve their technical and managerial knowledge and skills following training courses. This is in line with data reported by Muscio and Ramaciotti (2019) for Italy, in which evidence shows that both university- and course-level factors have a fundamental impact on graduates' decisions to start new ventures. Among the respondents' biotech companies, the main profile was health/pharma, followed by food industry and agricultural goods and services, on both national and international levels. When asked about the national societal context, most of the Romanian biotech entrepreneurs agreed that the national culture doesn't encourage the entrepreneurial initiatives, while the opposite answer came from the Western European respondents (Italy, Spain, Belgium).

Instead of ending with our conclusions, we choose to put here conclusions from the biotech entrepreneurs' testimonials. One of the most important said the biotechnology business is an extremely beautiful business, even if it is challenging; there are plenty of opportunities; what is important is to take the chance and never give up.

References

Braidford P, Stone I, Tesfaye B (2013) Gender, disadvantage and enterprise support – lessons from women's business centres in North America and Europe. J Small Bus Enterp Dev 20(1):143–164

European Commission (2016) The Romanian entrepreneurial ecosystem background report

https://www.statista.com/topics/3399/biotechnology-in-europe/. Accessed 11 Mar 2019

Jimenez A, Palmero-Camara C, Gonzales-Santos MJ, Gonzalez Bernal J, Jimenez Eguizabal JA (2015) The impact of educational levels on formal and informal entrepreneurship. BRQ Bus Res Q 18:204–212

Matei F, Anghel F, Ghidurus Panagopoulu M, Varga M, Draghici M, Coculescu C, Matei B, Garais GE, Toma RC (2019) A study on the Romanian biotech entrepreneurs profile and challenges. Sci Pap Ser MEEARD 19(2)

Muscio A, Ramaciotti L (2019) How does academia influence Ph.D. entrepreneurship? New insights on the entrepreneurial university. Technovation 82–83:16–24
Patzelt H, Schweizer L, Behrens J (2012) Biotechnology entrepreneurship. Found Trends® Entrep 8(2):63–140
Shimasaki CD (2009) The business of bioscience: what goes into making a biotechnology product. Springer Science+Business Media. doi:https://doi.org/10.1007/978-1-4419-0064-7_2
Shimaski CD (2014) Biotechnology entrepreneurship. Starting, managing, and leading biotech companies. Elsevier, Academic
SupBioEnt, Supporting biotechnology students oriented towards entrepreneurial path www.supbioent.usamv.ro. Accessed 1 Feb 2019

MIX
Papier aus verantwortungsvollen Quellen
Paper from responsible sources
FSC® C105338
FSC
www.fsc.org

If you have any concerns about our products,
you can contact us on
ProductSafety@springernature.com

In case Publisher is established outside the EU,
the EU authorized representative is:
Springer Nature Customer Service Center GmbH
Europaplatz 3, 69115 Heidelberg, Germany

Printed by Libri Plureos GmbH
in Hamburg, Germany